# Fundamentals of Statistical Quality Control

ROBERT J. POND, C.Q.E.
CENTRAL OHIO TECHNICAL COLLEGE

Merrill, an imprint of
Macmillan College Publishing Company
NEW YORK
Maxwell Macmillan Canada
TORONTO
Maxwell Macmillan International
NEW YORK OXFORD SINGAPORE SYDNEY

*To Harriett and Bill*
*for bringing me up with quality*

Editor: Steven Helba
Production Supervisor: Betsy Keefer
Production Manager: Francesca Drago
Text Designer: Robert Freese
Illustrations: Precision Graphics

This book was set in 10 pt. Palatino by Bi-Comp, Inc.
and was printed and bound by R.R. Donnelley & Sons, Inc.
The cover was printed by Phoenix Color Corporation.



Printed in the United States of America



Macmillan College Publishing Company
866 Third Avenue, New York, New York 10022

Macmillan College Publishing Company is part of
the Maxwell Communication Group of Companies.

Maxwell Macmillan Canada, Inc.
1200 Eglinton Avenue East
Suite 200
Don Mills, Ontario M3C 3N1

**Library of Congress Cataloging-in-Publication Data**
Pond, Robert J.
Fundamentals of statistical quality control / Robert J. Pond.
p. cm.
Includes index.
ISBN 0-02-396034-5
1. Quality control—Statistical methods. I. Title.
TS156.P657 1994
658.5′62′015195—dc20 93-33889
CIP

Printing: 1 2 3 4 5 6 7 Year: 4 5 6 7 8 9 0

# MERRILL'S INTERNATIONAL SERIES IN ENGINEERING TECHNOLOGY

## INTRODUCTION TO ENGINEERING TECHNOLOGY

Pond, *Introduction to Engineering Technology, 2nd Edition*, 0-02-396031-0

## ELECTRONICS TECHNOLOGY

### Electronics Reference

Adamson, *The Electronics Dictionary for Technicians*, 0-02-300820-2
Berlin, *The Illustrated Electronics Dictionary*, 0-675-20451-8
Reis, *Becoming an Electronics Technician: Securing Your High-Tech Future*, 0-02-399231-X

### DC/AC Circuits

Boylestad, *DC/AC: The Basics*, 0-675-20918-8
Boylestad, *Introductory Circuit Analysis, 7th Edition*, 0-02-313161-6
Ciccarelli, *Circuit Modeling: Exercises and Software, 2nd Edition*, 0-02-322455-X
Floyd, *Electric Circuits Fundamentals, 2nd Edition*, 0-675-21408-4
Floyd, *Electronics Fundamentals: Circuits, Devices, and Applications, 2nd Edition*, 0-675-21310-X
Floyd, *Principles of Electric Circuits, 4th Edition*, 0-02-338531-6
Floyd, *Principles of Electric Circuits: Electron Flow Version, 3rd Edition*, 0-02-338501-4
Herniter, *Schematic Capture with PSpice*, 0-02-353771-X
Keown, *PSpice and Circuit Analysis, 2nd Edition*, 0-02-363526-6
Monssen, *PSpice with Circuit Analysis*, 0-675-21376-2
Murphy, *DC Tutor: A Software Tutorial Using Animated Hypertext*, 0-02-385141-4
Murphy, *AC Tutor: A Software Tutorial Using Animated Hypertext*, 0-02-385144-9
Tocci, *Introduction to Electric Circuit Analysis, 2nd Edition*, 0-675-20002-4

### Devices and Linear Circuits

Berlin & Getz, *Fundamentals of Operational Amplifiers and Linear Integrated Circuits*, 0-675-21002-X
Berube, *Electronic Devices and Circuits Using MICRO-CAP II*, 0-02-309160-6
Berube, *Electronic Devices and Circuits Using MICRO-CAP III*, 0-02-309151-7
Bogart, *Electronic Devices and Circuits, 3rd Edition*, 0-02-311701-X
Bogart, *Linear Electronics*, 0-02-311601-3
Floyd, *Basic Operational Amplifiers and Linear Integrated Circuits*, 0-02-338641-X
Floyd, *Electronic Devices, 3rd Edition*, 0-675-22170-6
Floyd, *Electronic Devices: Electron Flow Version*, 0-02-338540-5
Floyd, *Fundamentals of Linear Circuits*, 0-02-338481-6
Schwartz, *Survey of Electronics, 3rd Edition*, 0-675-20162-4
Stanley, *Operational Amplifiers with Linear Integrated Circuits, 3rd Edition*, 0-02-415556-X
Tocci, *Electronic Devices: Conventional Flow Version, 3rd Edition*, 0-675-21150-6
Tocci & Oliver, *Fundamentals of Electronic Devices, 4th Edition*, 0-675-21259-6

### Digital Electronics

Floyd, *Digital Fundamentals, 5th Edition*, 0-02-338502-2
Foster, *Digital Logic Tutor I: An Introduction to Combinational Logic*, 0-02-338735-1
Foster, *Digital Logic Tutor II: An Introduction to Sequential Logic*, 0-02-338731-9
McCalla, *Digital Logic and Computer Design*, 0-675-21170-0
Reis, *Digital Electronics through Project Analysis*, 0-675-21141-7
Tocci, *Fundamentals of Pulse and Digital Circuits, 3rd Edition*, 0-675-20033-4

### Microprocessor Technology

Antonakos, *The 68000 Microprocessor: Hardware and Software Principles and Applications, 2nd Edition*, 0-02-303603-6
Antonakos, *An Introduction to the Intel Family of Microprocessors: A Hands-On Approach Utilizing the 8088 Microprocessor*, 0-675-22173-0
Brey, *8086/8066, 80286, 80386, and 80486 Assembly Language Programming*, 0-02-314247-2
Brey, *The Advanced Intel Microprocessor*, 0-02-314245-6
Brey, *The Intel Microprocessors: 8086/8088, 80186, 80286, 80386, and 80486: Architecture, Programming, and Interfacing, 3rd Edition*, 0-02-314250-2
Brey, *Microprocessors and Peripherals: Hardware, Software, Interfacing, and Applications, 2nd Edition*, 0-675-20884-X
Driscoll, Coughlin, & Villanucci, *Data Acquisition and Process Control with the M68HC11 Microcontroller*, 0-02-330555-X
Gaonkar, *Microprocessor Architecture, Programming, and Applications with the 8085/8080A, 2nd Edition*, 0-675-20675-6
Gaonkar, *The Z80 Microprocessor: Architecture, Interfacing, Programming, and Design, 2nd Edition*, 0-02-340484-1
Goody, *Programming and Interfacing the 8086/8088 Microprocessor: A Product-Development Laboratory Process*, 0-675-21312-6
MacKenzie, *The 8051 Microcontroller*, 0-02-373650-X
Miller, *The 68000 Family of Microprocessors: Architecture, Programming, and Applications, 2nd Edition*, 0-02-381560-4
Quinn, *The 6800 Microprocessor*, 0-675-20515-8
Subbarao, *16/32 Bit Microprocessors: 68000/68010/68020 Software, Hardware, and Design Applications*, 0-675-21119-0

### Electronic Communications

Monaco, *Introduction to Microwave Technology*, 0-675-21030-5
Monaco, *Preparing for the FCC Radio-Telephone Operator's License Examination*, 0-675-21313-4
Schoenbeck, *Electronic Communications: Modulation and Transmission, 2nd Edition*, 0-675-21311-8
Young, *Electronic Communication Techniques, 3rd Edition*, 0-02-431201-0
Zanger & Zanger, *Fiber Optics: Communication and Other Applications*, 0-675-20944-7

### Microcomputer Servicing

Adamson, *Microcomputer Repair*, 0-02-300825-3
Asser, Stigliano, & Bahrenburg, *Microcomputer Servicing: Practical Systems and Troubleshooting, 2nd Edition*, 0-02-304241-9
Asser, Stigliano, & Bahrenburg, *Microcomputer Theory and Servicing, 2nd Edition*, 0-02-304231-1

### Programming

Adamson, *Applied Pascal for Technology*, 0-675-20771-1
Adamson, *Structured BASIC Applied to Technology, 2nd Edition*, 0-02-300827-X
Adamson, *Structured C for Technology*, 0-675-20993-5
Adamson, *Structured C for Technology (with disk)*, 0-675-21289-8
Nashelsky & Boylestad, *BASIC Applied to Circuit Analysis*, 0-675-20161-6

### Instrumentation and Measurement

Berlin & Getz, *Principles of Electronic Instrumentation and Measurement*, 0-675-20449-6
Buchla & McLachlan, *Applied Electronic Instrumentation and Measurement*, 0-675-21162-X
Gillies, *Instrumentation and Measurements for Electronic Technicians, 2nd Edition*, 0-02-343051-6

### Transform Analysis

Kulathinal, *Transform Analysis and Electronic Networks with Applications*, 0-675-20765-7

### Biomedical Equipment Technology

Aston, *Principles of Biomedical Instrumentation and Measurement*, 0-675-20943-9

### Mathematics

Monaco, *Essential Mathematics for Electronics Technicians*, 0-675-21172-7
Davis, *Technical Mathematics*, 0-675-20338-4
Davis, *Technical Mathematics with Calculus*, 0-675-20965-X

## INDUSTRIAL ELECTRONICS/INDUSTRIAL TECHNOLOGY

Bateson, *Introduction to Control System Technology, 4th Edition*, 0-02-306463-3
Fuller, *Robotics: Introduction, Programming, and Projects*, 0-675-21078-X
Goetsch, *Industrial Safety and Health: In the Age of High Technology*, 0-02-344207-7
Goetsch, *Industrial Supervision: In the Age of High Technology*, 0-675-22137-4
Goetsch & Davis, *Introduction to Total Quality: Quality, Productivity, and Competitiveness*, 0-02-344221-2
Horath, *Computer Numerical Control Programming of Machines*, 0-02-357201-9
Hubert, *Electric Machines: Theory, Operation, Applications, Adjustment, and Control*, 0-675-20765-7
Humphries, *Motors and Controls*, 0-675-20235-3
Hutchins, *Introduction to Quality: Management, Assurance, and Control*, 0-675-20896-3
Laviana, *Basic Computer Numerical Control Programming*, 0-675-21298-7
Pond, *Fundamentals of Statistical Quality Control*, 0-02-396034-5
Reis, *Electronic Project Design and Fabrication, 2nd Edition*, 0-02-399230-1
Rosenblatt & Friedman, *Direct and Alternating Current Machinery, 2nd Edition*, 0-675-20160-8
Smith, *Statistical Process Control and Quality Improvement*, 0-675-21160-3
Webb, *Programmable Logic Controllers: Principles and Applications, 2nd Edition*, 0-02-424970-X
Webb & Greshock, *Industrial Control Electronics, 2nd Edition*, 0-02-424864-9

## MECHANICAL/CIVIL TECHNOLOGY

Dalton, *The Technology of Metallurgy*, 0-02-326900-6
Keyser, *Materials Science in Engineering, 4th Edition*, 0-675-20401-1
Kokernak, *Fluid Power Technology*, 0-02-305705-X
Kraut, *Fluid Mechanics for Technicians*, 0-675-21330-4
Mott, *Applied Fluid Mechanics, 4th Edition*, 0-02-384231-8
Mott, *Machine Elements in Mechanical Design, 2nd Edition*, 0-675-22289-3
Rolle, *Thermodynamics and Heat Power, 4th Edition*, 0-02-403201-8
Spiegel & Limbrunner, *Applied Statics and Strength of Materials, 2nd Edition*, 0-02-414961-6
Spiegel & Limbrunner, *Applied Strength of Materials*, 0-02-414970-5
Wolansky & Akers, *Modern Hydraulics: The Basics at Work*, 0-675-20987-0
Wolf, *Statics and Strength of Materials: A Parallel Approach to Understanding Structures*, 0-675-20622-7

## DRAFTING TECHNOLOGY

Cooper, *Introduction to VersaCAD*, 0-675-21164-6
Ethier, *AutoCAD in 3 Dimensions*, 0-02-334232-3
Goetsch & Rickman, *Computer-Aided Drafting with AutoCAD*, 0-675-20915-3
Kirkpatrick & Kirkpatrick, *AutoCAD for Interior Design and Space Planning*, 0-02-364455-9
Kirkpatrick & Kirkpatrick, *AutoCAD for Interior Design and Space Planning, Release 12 Version, 2nd Edition*, 0-02-364471-0
Kirkpatrick, *The AutoCAD Book: Drawing, Modeling, and Applications, 2nd Edition*, 0-675-22288-5
Kirkpatrick, *The AutoCAD Book: Drawing, Modeling, and Applications, Including Release 12, 3rd Edition*, 0-02-364440-0
Lamit & Lloyd, *Drafting for Electronics, 2nd Edition*, 0-02-367342-7
Lamit & Paige, *Computer-Aided Design and Drafting*, 0-675-20475-5
Maruggi, *Technical Graphics: Electronics Worktext, 2nd Edition*, 0-675-21378-9
Maruggi, *The Technology of Drafting*, 0-675-20762-2
Sell, *Basic Technical Drawing*, 0-675-21001-1

## TECHNICAL WRITING

Croft, *Getting a Job: Resume Writing, Job Application Letters, and Interview Strategies*, 0-675-20917-X
Panares, *A Handbook of English for Technical Students*, 0-675-20650-2
Pfeiffer, *Proposal Writing: The Art of Friendly Persuasion*, 0-675-20988-9
Pfeiffer, *Technical Writing: A Practical Approach, 2nd Edition*, 0-02-395111-7
Roze, *Technical Communications: The Practical Craft, 2nd Edition*, 0-02-404171-8
Weisman, *Basic Technical Writing, 6th Edition*, 0-675-21256-1

# Preface

The need for quality improvement in both service and manufacturing industries is no longer questioned. However, the universal implementation of effective quality systems is yet to be realized. This text, *Fundamentals of Statistical Quality Control,* will be used to prepare today's student for quality problem solving that will provide improved implementation in the future.

This text is specifically designed to meet the needs of technology or business management students enrolled in a one-term course. The text may also be used to support training programs serving business or manufacturing organizations. *Fundamentals of Statistical Quality Control* may also be used to prepare managers and technologists for the Certified Quality Technician (CQT) Examination, offered by the American Society for Quality Control.

## To the Student

The many new quality textbooks on the market attest to the need for training, as well as to the abundance of individual approaches to this vast, constantly evolving subject. Many textbooks have been written with the assumption that you have had at least one college-level statistics course prior to enrolling in the quality course. Yet, there is a significant number of quality students in business, industrial, and engineering technologies who do not have a statistics course in their curriculum. *Fundamentals of Statistical Quality Control* assumes that you may not have completed a college-level statistics course.

The quality system presented also includes many of the practical, nonmathematical, human aspects of quality control. Discussed in Chapter 1 and throughout are the commonly accepted "seven tools of quality." These are as follows: graphs of all types, cause-and-effect diagrams, histograms, check sheets, Pareto charts, control charts, and scatter diagrams. Other problem-solving tools are introduced as well (e.g., flowcharts and brainstorming techniques). These tools are often referred to as SQC problem-solving tools, and they will help you to communicate with people, for the implementation of quality depends upon your ability to work with people.

Real-world applications are taken from both manufacturing and service sectors. Examples include such diverse areas as automobile suppliers, health care facilities, office operations, and power companies. The text also includes actual case studies, most of which were performed by the author's students during the one-term offering of the course Statistical Quality Control at Central Ohio Technical College. The case studies may be used by your instructor as items for discussion or as models for actual projects in service or manufacturing industries. Your instructor may also choose to work with a local service or manufacturing industry in solving a process control problem. This real-world experience will supplement the material in the text.

The numerous Section Review Problems are designed to offer you a chance to review the most important ideas developed in each section. Experience solving the simpler Section Review Problems combined with the experience of solving more complex Chapter Review Problems will aid you in retaining vital information and preparing for examinations. All Section Review Problems are answered at the end of the text, and the answers to the odd-numbered Chapter Review Problems can be found in a separate section at the end of the text. Hopefully, *Fundamentals of Statistical Quality Control* will be helpful to you not only now but also in the future, as you participate in solving your company's quality problems.

## To the Instructor

The material for this text, *Fundamentals of Statistical Quality Control,* has been developed and used in the author's classroom and laboratory for the past decade. Many of the process control studies have actually been performed by the students; for instance, the fill-weight example that begins with the Case Study in Chapter 1 and is further explained in Chapter 4 as an example of variable control charting.

The laboratories included at the end of each chapter have been performed by the author over the years during which he has instructed the quality course. The laboratories are only recommended and merely outlined, so that you may exercise personal creativity and judgment as to the actual laboratories the student experiences. Also, a quality project with a local business may replace the suggested laboratories in later chapters. The author has found business and industrial studies to be rewarding, even though they are somewhat difficult to manage while presenting the necessary theory for the course. Many local companies will welcome your requests for projects.

*Fundamentals of Statistical Quality Control* is supported by an instructor's manual containing transparency masters and solutions to all problems.

## Organization

Chapter 1 uses a decade-by-decade approach to place SQC and total quality management (TQM) within the historical context of the "quality movement." Also included is an industrial case study that provides an early introduction to the Shewhart cycle and the quality tools necessary to provide project management for quality. The recommended laboratory is a data collection experience.

Chapter 2 is unique in that it reviews good charting techniques for the student who may have forgotten some of the standards of quality charting. The chapter goes on to discuss scatter diagramming (a method of experimentation and calibration), as well as histogram construction and interpretation. Pareto diagramming is also presented. The students are to construct a histogram of the data collected in the Chapter 1 laboratory.

Variable data calculations—the mean, median, mode, and standard deviation—are discussed in Chapter 3. The normal distribution is presented, and Appendixes A and B are referred to as the student's introduction to probability distributions. Even though students may have had a previous course in statistics, the application approach used in this chapter will supplement their understanding of the importance of variable data distributions and of putting data in a consumable form. The Chapter 3 laboratory involves a short gauge-capability study. Repeatability and reproducibility (R&R) studies are discussed along with the concepts of accuracy and traceability.

Chapter 4 begins with a discussion of assignable and chance causes and the problem of overcontrol. The heart of the chapter is the $\overline{X}$ and $R$ chart example taken from an industrial process study. (The $\overline{X}$ and $s$ chart is discussed in Chapter 9.) Capability ratios ($C_p$ and $C_{pk}$) are introduced in this chapter. Appendix C, "Control Chart Factors," is referred to in this chapter. The laboratory has students perform a long R&R study.

Chapter 5 introduces charts for individuals, median charts, trend charts, and precontrol charts. Advantages and disadvantages of each of these optional control charts are discussed. The laboratory involves the construction of an individual chart from data collected by the students.

The probability of attributes is discussed in Chapter 6. This chapter is as focused as possible to allow the student to grasp the important fundamentals of probability and move on to attribute control charts and acceptance sampling. The difference of the population and the sample are introduced. Appendix D, "The Poisson Distribution," is referred to in this chapter. The laboratory introduces the computer and SQC software to the student.

Chapter 7 covers attribute control charts—$p$, $np$, $u$, and $c$ charts. The use of check sheets is discussed. The students construct a $p$ chart in the laboratory (on a computer using SQC software if available). The control limits for the $u$ and $p$ charts will be constructed by the 25 percent rule presented in the chapter.

Chapter 8 blends acceptance sampling plans with appropriate inspection techniques. Sampling is presented along with the concepts of AQL, LQL, and AOQL. MIL–STD–105 is explained. The laboratory confirms the efficacy of the Poisson distribution as a predictive tool.

Chapter 9 discusses the use of the computer and SQC software in quality problem solving. Automated data collection by means of a gauge talker is introduced. Expert systems and closed-loop process control are presented. The laboratory involves the use of a data collector to enter data directly to the computer.

In closing, the text returns to the evolution of quality by presenting total quality management (TQM). Chapter 10 is written for the technologist or middle manager. The suggested laboratory is a student presentation to managers at a local industry.

The most frequently used statistical quality control equations are found inside the front cover. A glossary of SQC symbols and abbreviations used in the text appear inside the back cover. The author has attempted to use a minimum number of symbols throughout the text.

# *Acknowledgments*

I would like to thank the following persons and their organizations for contributing to this text: Melenda Ball, PQ Systems, Inc.; Roger Bentley, Walker Manufacturing; Jerry Besanceney, Holophane Company, Inc.; David Biros, ABB Process Automation, Inc.; Alan Deeds, Rockwell, International; John Daugherty, Owens-Corning Fiberglass; Don Frerichs, Bailey Controls Company; Mike Jenkins, Holophane Company, Inc.; Michael F. McCalley, DataMyte Division of Allen-Bradley Company, Inc.; Carl Nocera, AT&T; Dennis Osborn, Tamarack Farms Dairy of Newark, Ohio; Robert N. Platt, American Society for Quality Control; John C. Schmidt, Johns Hopkins University; and Bob Zaborski, Tamarack Farms Dairy of Newark, Ohio.

The following reviewers provided helpful comments and suggestions: Lindell Holtzmeier, Hocking College; Tom Lavender, Catawba Valley Community College; Kenneth Rennels, Purdue University—Indianapolis; and Gary Winek, Southwest Texas State University.

Also, I wish to thank Steve Helba, Editor at Macmillan College Publishing in Columbus and Betsy Keefer, Production Supervisor.

Specials thanks goes to the students at Central Ohio Technical College who used the manuscript during the Winter Quarter of 1993—Andy Cox, Gary Davis, Craig Fink, and Tom Whittle.

# Contents

*Preface* *v*

## 1 SQC Project Management 1

1.1 Quality—A Continually Evolving Concept 1
*Prior to 1920—Inspection in Manufacturing* 4
*1920—Quality Control* 6
*1930—Statistical Process Control (SPC)* 7
*1940—Acceptance Sampling* 8
*1950—Total Quality Control (TQC)* 9
*1960—Quality Problem Solving* 10
*1970—The Voice of the Customer* 11
*1980—Total Quality Management (TQM), or TQC Revisited* 11
1.2 Quality Project Management 14
1.3 Quality—A Vital Career Field 19
*The American Society for Quality Control* 20
Summary 22
Case Study 23
Chapter Review Problems 33
Laboratory: Data Collection 35

## 2 Graphical Analysis of Data 37

2.1 Graphing Fundamentals 37
2.2 Experimenting by the Use of Scatter Diagrams 41
2.3 Preparing Histograms 47
2.4 Troubleshooting with Histograms 54
2.5 Pareto's Principle 60
Summary 64
Case Study 65
Chapter Review Problems 68
Laboratory: Putting the Data into Consumable Form 73

## 3 The Normal Distribution 75

3.1 Central Tendency 75
3.2 Variability 81
3.3 Shape of Distributions 84
*Calculating Skewness and Kurtosis* 85
3.4 The Normal Curve 88
3.5 Working to Target 96
Summary 101
Case Study 102
Chapter Review Problems 105
Laboratory: Gauge Capability Study (Short R&R Study) 108

## 4 Variable Control Charts 113

4.1 Preparation for Charting 113
*Actions Prior to Implementation* 115
4.2 Constructing $\overline{X}$ and $R$ Charts 117
4.3 Chart Analysis 130
*Out-of-Control Signals* 130
*Averages and Individuals* 136
4.4 Process Capability 139
*The Capability Index* $C_{pk}$ 142
*Continuing Improvement* 143
Summary 149
Case Study 150
Chapter Review Problems 153

Laboratory: Gauge Repeatability and Reproducibility (R&R) Study—Long Method 158

## 5 Optional Control Charts 165

5.1 Charts for Individuals 165
*Chart Construction* 166
*Chart Interpretation* 168
*Advantages of Individuals Charts* 169
*Disadvantages of Individuals Charts* 169
5.2 Median Charts 170
*Chart Construction* 170
*Chart Interpretation* 173
*Advantages of Median Charts* 175
*Disadvantages of Median Charts* 175
5.3 Trend Charts 176
*Chart Interpretation* 179
*Advantages of Trend Charts* 179
*Disadvantages of Trend Charts* 180
5.4 Precontrol Charts 180
*Precontrol Rules* 181
*Precontrol Chart Analysis* 184
*Advantages of Precontrol* 186
*Disadvantages of Precontrol* 186
Summary 187
Case Study 188
Chapter Review Problems 188
Laboratory: Using the Individuals Chart 191

## 6 Probability of Attributes 193

6.1 Fundamentals of Probability 193
6.2 Probability Flowchart 197
*Mutually Exclusive Events* 198
*Non–Mutually Exclusive Events* 200
*Joint Probability—Dependent Events* 201
*Joint Probability—Independent Events* 202
6.3 Permutations and Combinations 206
*Permutations* 206
*Combinations* 207

6.4 Selected Discrete Probability Distributions 208
*Hypergeometric Probability Distribution* 209
*Binomial Probability Distribution* 211
*Poisson Probability Distribution* 215
*The Normal Distribution* 218
Summary 220
Case Study 222
Chapter Review Problems 224
Laboratory: Statistical Software for the Computer 226

# 7 *Attribute Control Charts* 227

7.1 Preparation for Charting 227
*Selection of Attribute Study* 228
*Type of Attribute Chart* 229
*Defining the Nonconformance* 230
*Data Collection Techniques* 230
7.2 Constructing Charts—Nonconforming Units 234
*Calculating the p Chart Parameters* 236
*Constructing the p Chart* 237
*Interpreting the p Chart* 240
*Constructing the np Chart* 243
*Interpreting the np Chart* 246
*Completing the Shewhart Cycle* 247
7.3 Constructing Charts—Nonconformities 249
*The u Chart* 249
*The c Chart* 253
Summary 256
Case Study 257
Chapter Review Problems 258
Laboratory: Attribute Charting Using the Computer 263

# 8 *Acceptance Sampling and Inspection* 265

8.1 Acceptance Sampling 265
*100 Percent Sampling* 270
*No Inspection* 271
*Constant Percentage Sampling* 271
*Collecting Random Samples* 271

*Operating Characteristic Curves* 274
*Type I and Type II Errors* 278
8.2 MIL–STD–105 283
8.3 Nondestructive Testing (NDT) 290
Summary 293
Case Study 294
Chapter Review Problems 295
Laboratory: Acceptance Sampling 297

**9** *Computer-Aided SQC* ______ *299*

9.1 Quality Control Software 299
*R&R Studies* 302
9.2 Computer Analysis 305
*Electronic Data Collectors* 305
*Computer Analysis* 309
*Purchasing Computer Equipment* 315
9.3 Automated Process Control 316
*Computer-Integrated Manufacturing* 323
*When to Automate* 324
Summary 325
Case Study 327
Chapter Review Problems 327
Laboratory: Automatic Data Collection with Computer Charting and Capability 329

**10** *Total Quality Management (TQM)* ______ *331*

10.1 Elements of the TQM System 331
*The Organization* 332
*Sales and Marketing* 333
*Business Functions* 336
*Research and Development (R&D)* 340
*Manufacturing or Service* 342
*Sales and Marketing* 344
10.2 Recognizing Quality Efforts 346
*ISO 9000 (ANSI/ASQC Q90)* 347
*Quality Awards* 348
10.3 Forming a Personal TQM Philosophy 351

Summary 352
Case Study 353
Chapter Review Problems 354
Laboratory: Project Report to Industry 357

*APPENDIXES*

**A** *The Normal Distribution* 359

**B** *Frequently Occurring Probability Distributions* 365

**C** *Control Chart Factors* 369

**D** *The Poisson Distribution* 371

*Answers to Section Review Problems* 381

*Answers to Odd-Numbered Chapter Review Problems* 389

*Index* 395

# 1

# SQC Project Management

**Quality—A Continually Evolving Concept** 1.1
**Quality Project Management** 1.2
**Quality—A Vital Career Field** 1.3

## 1.1 Quality—A Continually Evolving Concept

All of us at one time or another have discussed quality. Most of us feel we have a good understanding of what it is. However, when we attempt to agree with others on a suitable definition, we find a great deal of disagreement. Some common definitions follow:

- The product works, and it lasts a long time.
- The product is attractive and functional.
- Service is quick and courteous.

These commonsense definitions are not as functional as we might like. Quality practitioners have created the following universal definitions of quality:

*Universal Definitions of Quality*

- Make the product or service right the first time.
- Make the product or service fit for use in society.
- Provide a product or service with zero defects.
- Satisfy the customer.

Each of these definitions may be contemplated at length. But to even begin to understand quality, another important concept must be understood—the concept of the quality characteristic.

The total quality of any product or service can be broken down into simple elements. The **product quality characteristic** is the simplest element used to assure product quality.

Of course, one product may have hundreds or even thousands of product quality characteristics. The space shuttle would serve as an example of a very complex system possessing more than 100,000 quality characteristics.

Here are some examples of typical product quality characteristics:

- time of service at a restaurant;
- the number of squeaks in an automobile's dashboard;
- the accuracy of your bill when departing a motel;
- the weight (fill weight) of a milk product.

Many others may be listed. The list shown here demonstrates how important it is to define a quality characteristic of a product. Note also that all characteristics are specific as well as measurable, for the manufacturer of a product or the business offering a service must somehow be able to translate the needs of the customer into the quality characteristics of the product or service. Simply saying a product is "good" or "bad" is too general an idea for action to be taken. Specific product or service quality characteristics must be identified. (In the case of attribute studies, a product may be classified as good or bad based on more than one quality characteristic. However, concrete action can be taken on only one quality characteristic at a time.)

EXAMPLE 1.1

As a customer shopping for a shirt or blouse, you take one item down from the rack. You find the material feels good, and after trying it on you see it fits well. What other quality characteristics could you look for in the product?

*Solution:* There are many other considerations that careful, experienced shoppers could point out:

- Is the material of high quality (long lasting)?
- Has the garment been prewashed, and is it colorfast?
- Is the blouse or shirt made by an established manufacturer and sold by an established retailer (yielding a better chance of return)?
- Is the stitching correct and the weave tight?

An experienced salesperson could add much more to this list (for example, Are the collars and sleeves bonded to the inner lining?), but the point is made: After close inspection and having the right questions answered about specific quality characteristics, the product originally thought to be of good quality might be of questionable or even poor quality.

The more operational the definition, the easier it is to measure quality. For instance, the quality characteristic of fill weight could be written as, "The average fill weight of the 6-ounce yogurts will be 170 g (6 oz.), and individual containers will not vary more than 8.5 g (0.3 oz.) from the average." Considerations to keep in mind when writing operational definitions are available technology, safety, reliability, attractiveness, and maintainability.

A **process quality characteristic** is any individual feature of the process that contributes to the quality of the product. Examples are as follows:

- the speed of a drill or the speed at which the drilling operation is performed;
- the heat of an oven or of a grill preparing a food item;
- the method of measuring the chemical content of a product.

*It is important to distinguish between process quality characteristics and product quality characteristics as we plan for quality.*

Looking back to our "Universal Definitions of Quality," and considering the definitions with an understanding of what a quality characteristic is, the following definition agreed to by both the American

National Standards Institute (ANSI) and the American Society for Quality Control (ASQC) seems to be most appropriate:

> The totality of features of a product or service that bear on its ability to satisfy stated or implied needs.

Most importantly, *quality must begin with the customer.* Few of today's quality experts would disagree with this most basic understanding of quality. But this view was not always widely accepted. Let us review the history of quality to understand its present form.

Table 1.1 summarizes the history and evolution of a worldwide quality concept. *The student of quality should be able to place the phases of quality evolution in chronological order.* The list is not meant to be exhaustive, but is intentionally short and simple. Some of the citings occurred in some countries or in a particular industry before the decade shown. Yet, listing the watershed quality concepts by the decade of their universal awareness aids the quality student in placing present quality efforts in an appropriate context.

## *Prior to 1920—Inspection in Manufacturing*

Early producers knew their customers well and profited from immediate feedback on the quality of their work. These blacksmiths, millwrights, tailors, and other craftspersons had a strong work ethic and orientation. Their training often involved long apprenticeships working with a master craftsman. Their "pride of workmanship" took the form of simple statements, such as the following:

- When you think you know everything there is to know about a job, it's time to quit.
- You have to love your work to be one of the best.

Most of the early craftspersons' lives were spent earning a living, which depended on pleasing their customers. The quality characteristics of their products, as defined by the customers they saw frequently, were an ingrained part of their thinking.

The industrial era (circa 1900) and the introduction of assembly lines resulted in workers being further removed from their customers.

TABLE 1.1
**A Decade-by-Decade History of the Quality Revolution**

| *Decade** | *Quality Concept Introduced* | *Importance* |
|---|---|---|
| Prior to 1920 | Inspection in manufacturing | Increasing recognition of the need to inspect mass-produced products. |
| 1920 | Quality control | First defined. |
| 1930 | Statistical process control (SPC) | Recognition that process quality characteristics are variable. Statistics must be used to understand variability and provide appropriate feedback to manufacturing. |
| 1940 | Acceptance sampling | Sampling procedures to accept or reject product by lot—used extensively in the United States to assure high-quality munitions for the military. |
| 1950 | Total quality control (TQC) | Design, purchasing (materials), manufacturing, marketing and sales, etc.—all are involved in quality improvement; Japanese involvement. |
| 1960 | Quality problem solving | Team building and involving the worker in the solution of quality problems; extensive training of workforce. |
| 1970 | The voice of the customer | Defining of quality characteristics from the customer's perception of quality. |
| 1980 | Total quality management (TQM) | Worldwide recognition and implementation of the tools needed to successfully rebuild industry. |
| 1990 | Quality continues its evolution | |

* Dates shown are approximate dates of implementation, not dates of discovery or invention.

Also, products like the automobile became more complicated. There was a need for a more formal understanding of quality, and thoughts on quality began to change. These changes led to a need for improved measurement techniques and standardization of measurement systems (metrology). Quality characteristics were operationally defined. Jigs and fixtures were produced so that variability could be reduced, although product variability could never be totally eradicated.

Jigs and fixtures were also designed to aid in inspection. Early inspectors were the foremen who normally kept tabs on the work in their area. Later, specific personnel were trained to inspect products. These inspectors also became responsible for determining conformance to blueprints, disposing of nonconforming output (scrapping or reworking it), and quality reporting. The simplest definition of **inspection** is the activity of checking a product's quality characteristics for conformance to specification. Inspection resulted in improved quality of product, but top management was not involved, and manufacturing processes were not strongly linked to inspection and quality of products. In fact, most inspection occurred outside the manufacturing area, without adequate and timely feedback to the people responsible for making the product. This type of inspection was called **tollgate** inspection. Quality was not systematically controlled, but only measured.

## 1920—Quality Control

The early years of industrial growth led to the need for recognizing quality as a management function. Quality managers had to be independent in their decision-making role. Otherwise, foremen—responsible for both quality and production—might be mainly concerned with production, not with making the product right. The term "quality control" was probably used for the first time in the 1920s, in order to distinguish its importance as a management function.

**Quality control** is defined as all of the functions necessary to obtain quality of product, and the regulatory process by which quality of product is attained. Early on, quality control was thought of as only taking place in the manufacturing or "make" operation, because that was the only place control of the product's parts or assembly could take place. Today, the term is more universally applied across all elements of the organization. In order to more surely control quality, however, the use of statistics was needed.

Joseph M. Juran has been a most influential leader in quality management since 1928. His teachings on management of quality control have benefited many quality practitioners over more than six decades. One of Juran's most important contributions is the *Quality Control Handbook* (first edition 1951), which has served for many years as a reference work for managers, supervisors, and engineers in industry.

## 1930—Statistical Process Control (SPC)

In the 1920s emerging products such as telephones and automobiles demanded a lengthy chain of suppliers. This "supplier chain" required that variability be studied in a more formal manner. The resulting work was known as statistical quality control (SQC), or the use of statistics to assure a reduction in variability.

Statistically quality control began with the concepts of a team of statisticians, engineers, and scientists at Bell Laboratories, a division of the Bell Telephone System. As a result of this effort, one of the team, Walter Shewhart, published an important work in the 1930s that described statistical process control. **Statistical process control (SPC)** is the use of statistics in controlling the quality of a process while it is producing goods. SPC has become the most important element of SQC. Shewhart's book *Economic Control of Quality of Manufactured Product* described for the first time how manufacturers could not only measure a process or product quality characteristic, but also control the quality during the manufacturing operation. This control of quality provided for a statistically determined feedback loop for the manufacturing personnel directly involved that could be used to reduce product variability.

The result of Shewhart's work is the process control chart. The **control chart** is a graphical method for evaluating whether or not a process is in statistical control. In Figure 1.1, an automobile's gas mileage (the quality characteristic) is plotted over time. As can be seen by the up-and-down nature of the points in sequence, the mileage is variable; for instance, in-town driving produces poorer gas mileage than long trips on the interstate. But, by viewing the variability statistically (each plotted point is an average of several gas mileage calculations) a definite downward trend can be seen. As the miles per gallon decrease there is a point, known as a control limit, where the cost of a tune-up is justified. (To spend money on a tune-up at, say, point 3 on the chart would be wasteful. The poor gas mileage in this case could be due to greater-than-normal driving time in the city.)

A commonsense problem-solving tool also recommended by Shewhart will be used in this text. Referred to as the **Shewhart cycle,** it involves only four steps—selection of the project (process to be improved), collecting data while attempting improvement, reporting on the success or failure of the improvement, and deciding whether

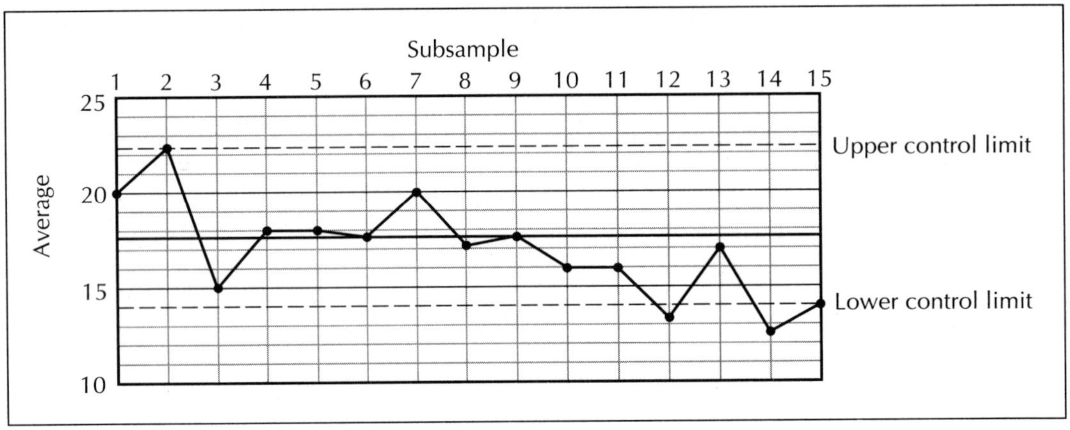


**FIGURE 1.1**
The control chart or run chart tracks miles per gallon and shows a decrease in fuel efficiency.

to continue the same project or to begin another (see Figure 1.3, Section 1.2). The power of this problem-solving model results from the fact that only one quality characteristic is addressed at one time, allowing focus on improving that particular characteristic. (Deming revised this later to be the **plan-do-check-act cycle,** or the **PDCA cycle;** see "1950, TQC.")

Walter Shewhart's work is recognized by today's leaders in the quality field as fundamental to the continuous improvement of the quality of a product. He can truly be called the father of modern-day quality control.

## *1940—Acceptance Sampling*

As Shewhart was implementing SPC techniques, others were working on sampling a few representative products from batches (known as *lots*) of products. The practice known as **acceptance sampling** is to either accept or reject a lot based on a representative sample from the lot. Each piece in the sample is checked for conformance to a particular quality characteristic or to several quality characteristics (Figure 1.2).

During World War II (1940s) both methods—SPC and acceptance sampling—were used successfully in the United States to realize miraculous quality improvements. Acceptance sampling was used most

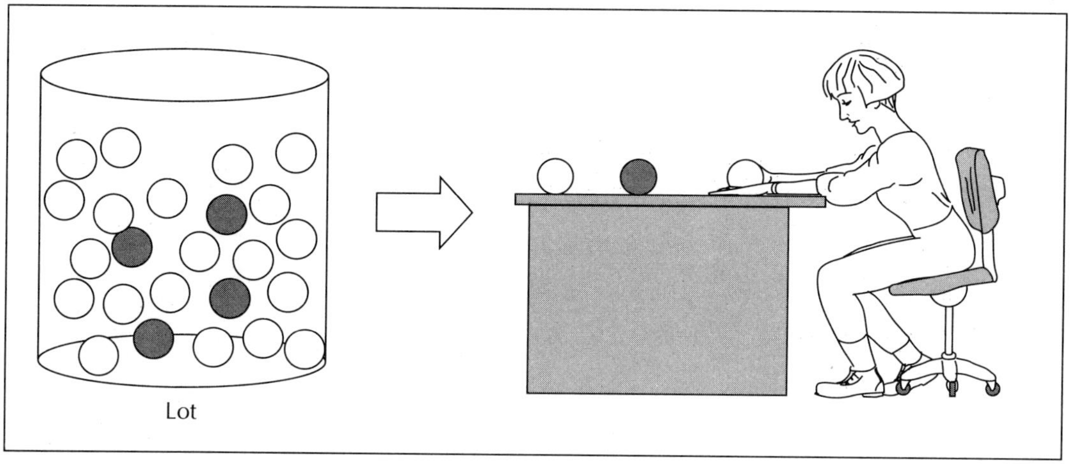


**FIGURE 1.2**
Acceptance sampling involves accepting or rejecting the *lot* based on a small sample.

extensively by miliary suppliers. Ammunition, for instance, went from an average of 15 percent nonconforming to less than 5 percent nonconforming. This situation may not mean much, unless you can imagine yourself in a foxhole and under attack when 15 out of every 100 cartridges didn't fire!

The importance of statistics to quality was underlined by quality control practitioners when they created a professional organization to foster the continuous improvement of quality by the use of statistical methods. The new organization that began in the 1940s is known today as the American Society for Quality Control (ASQC).

After the war and into the 1950s, U.S. industry continued to use SPC and acceptance sampling, with most manufacturers depending mainly on acceptance sampling. The use of control charts was discontinued in many industries until the Japanese demonstrated their effectiveness in continually improving product, and in doing so became one of the world's greatest manufacturers.

## *1950—Total Quality Control (TQC)*

Japanese manufacturers, living in a war-torn country and anxious to rebuild, were willing to listen to quality representatives from the United

States. Among the quality experts invited by General Douglas MacArthur to help rebuild Japanese industry was Armand Feigenbaum, who was working out his ideas for what would become known as **total quality control (TQC)** or companywide quality control (CWQC). Feigenbaum's central message was simply to recognize that *quality was everybody's job*. Up to this time quality was perceived as only having a place in the manufacturing arena. Feigenbaum proposed that everyone from design of product through sales of the product be involved in quality. His ideas evolved into the quality assurance (QA) function or quality systems management (e.g., quality costs, auditing of quality control, inspection and test, and reliability). TQC is the marriage of quality assurance and quality control.

Another of MacArthur's expert guests during the late forties and early fifties was W. Edwards Deming. Deming promoted the use of SPC along with other unique ideas such as his *14 points of management*. Deming's highly regarded 14 points viewed the worker as a primary element in the improvement of quality. *The worker was to be well trained, empowered to make decisions, and supported by the management team.*

## *1960—Quality Problem Solving*

The Japanese listened to the honored visitors (the Deming Prize is the highest industrial award for quality in Japan) and added their own ideas in the 1960s. Their ideas included such innovative concepts as the following: putting quality first and not cost (costs will decrease with better quality and less scrap and rework), unflagging eagerness to satisfy the customer, simple procedures of control of quality, and continuous improvement (e.g., SPC), and quality circles (now known as quality improvement teams) offering a team approach to problem solving.

Kaoru Ishikawa led the way in developing problem-solving techniques that were to be used by quality teams on the job (often referred to as SPC problem solving). He recommended the **seven tools of quality**—*graphs of all types, cause-and-effect-diagrams, histograms, check sheets, Pareto diagrams, control charts, and scatter diagrams.* All of the seven tools have one overriding feature—they are visual. Technologists work best when they can visualize a problem. Ishikawa's seven tools will be explained in detail and used throughout the text.

## *1970—The Voice of the Customer*

A profound respect for the customer led to the most significant Japanese quality innovation. The innovation was that the voice of the customer would be heard throughout the industry. Customers were surveyed to determine what characteristics were most desirable in a particular product. Some of these characteristics are high quality with low cost; a product with little piece-to-piece variability, making it more insensitive to variations in the operating environment; little deterioration with use; and low maintenance. Charts were used within the company to identify and improve the quality characteristics most desired by the customer.

Additionally, the whole idea of customer changed from being simply the consumer at the end of the chain. "Customer" could mean—to a company acting as the supplier of parts—the final assembler of the product. For example, Ford is a customer of an exhaust system manufacturer (the supplier), who, in turn, is a customer of the steel company (the materials supplier). Divisions within corporations may think of their customers as other plants or offices in the same corporation.

The concept of the customer was completed by defining the internal customer. The **internal customer** is all of the people responsible for producing the product or service. If you are an operator on an assembly line, your internal customer is the next operator on the line. If you are a quality technologist or manager, your internal customers are the people that depend on your active participation in aiding them to make a better product or perform an improved service.

## *1980—Total Quality Management (TQM), or TQC Revisited*

Unfortunately, many U.S. manufacturers tried to adopt only part of the Japanese quality approach in the 1970s when it was evident that foreign competition dictated the need for a total quality approach. In the 1980s, U.S. trade imbalances were but one measure of the continued inability of U.S. manufacturers to compete. Fortunately, both service and manufacturing industries are now recognizing the need for a total quality effort that should make all industries better able to meet the needs of society in the 1990s. They have recognized that *the essence of total quality management (TQM) is to achieve less variability of product or service by improving simultaneously the technical system and the people system.* The student should be aware of the following key elements of any viable TQM plan:

1. Work to a customer-defined target (continuous improvement) on all quality characteristics, and not to specification limits. Use statistical methods to work to target.
2. Specify defect-free material from suppliers. (This approach could reduce the use of the acceptance-sampling techniques that were typical of the 1940s while increasing the number of suppliers using SPC techniques.)
3. Make managers—who control the system—responsible for quality improvement, and do not blame workers for poor quality.
4. Promote teamwork across all elements of the organization—from marketing to research and design to manufacturing to sales—and recognize the "people system" as just as important as the "technology system." For instance, the supervisor's role could become more of a "coaching" role and less of a "bossing" role.
5. Promote continuous training, so employees may be most effective in their participation.

This brief history mentions most of the primary elements necessary in building a world that recognizes what it takes to deliver quality product to society. *The most important recent advances in improving quality involve the human element:*

- knowing the customer and respecting customer's needs;
- viewing all of the personnel of an industry—service or manufacturing—as important to the total quality effort.

In quality efforts, the definition of a **customer** is anyone who uses what you have created or brought into reality. If you repair guidance systems in the air force, your customer is the pilot who depends on that system to get to the target. If you work on an assembly line, your customer is the next person down the line. If you are an inspector, your customers are the operators in your department who produce goods and need timely, sensitively communicated, accurate information.

The importance of the second part of that vital human element—the personnel making the product reality—may be illustrated by using the Air Force Logistics Command's (AFLC) "Quality Bill of Rights," designed to remind personnel of the importance of all on the "quality team."

> Each member of the Air Force Logistics Command shall have the following quality RIGHTS:
> - the RIGHT to challenge business as usual
> - the RIGHT to be heard
> - the RIGHT to expect commitment to quality
> - the RIGHT to place quality before production
> - the RIGHT to feel genuine pride in AFLC products and services

## Section 1.1 Review Problems*

1. Checking the straightness of the seams on a dress from a rack in a clothing store is checking a
   a. product quality characteristic
   b. process quality characteristic
   c. either of the above may be true
   d. none of the above are true
2. Maintaining the temperature gradient constant in a distillation column is maintaining a
   a. product quality characteristic
   b. process quality characteristic
   c. either of the above is true
   d. none of the above are true
3. In the history of the quality revolution, total quality management came after
   a. SPC
   b. TQC
   c. quality problem solving
   d. all of the above

* Answers are at end of text.

4. To use statistics to accept or reject "lots" of product is known as
   a. SPC
   b. TQC
   c. quality circles
   d. acceptance sampling
5. The 14 points of management, which provided for a well-trained, empowered worker supported by (not bossed by) the management team, were developed by
   a. Feigenbaum
   b. Ishikawa
   c. Deming
   d. Juran

# 1.2 *Quality Project Management*

Improving quality by using a project-by-project approach, where each project is narrowly defined and easily controlled, is one of the surest ways to improve overall quality in a business or manufacturing operation.

Each project requires a project manager with authority (empowerment) to act on a particular problem. The project manager should be familiar with the functions of all individuals represented on the project team. He or she should also be sensitive to each individual's other responsibilities, for the project will only involve a part of each person's overall responsibilities. The project team is perhaps best known as a **quality improvement team.** The most effective quality improvement teams are comprised of eight to ten employees, all of whom have volunteered to work with the team. *Quality improvement teams have been responsible for millions of dollars in savings since their implementation in the early 1960s.*

Of course, the focus in all projects is effective problem solving. Effective group problem solving involves a distinct series of steps, a systematic plan, that all involved are familiar with. In the 1930s (Section 1.1, "1930s"), Walter Shewhart, the father of SPC, developed just such an effective tool for solving quality problems (Figure 1.3). After the characteristic is selected, the improvement step begins. After improvement actions have been taken, a report analyzes and documents the

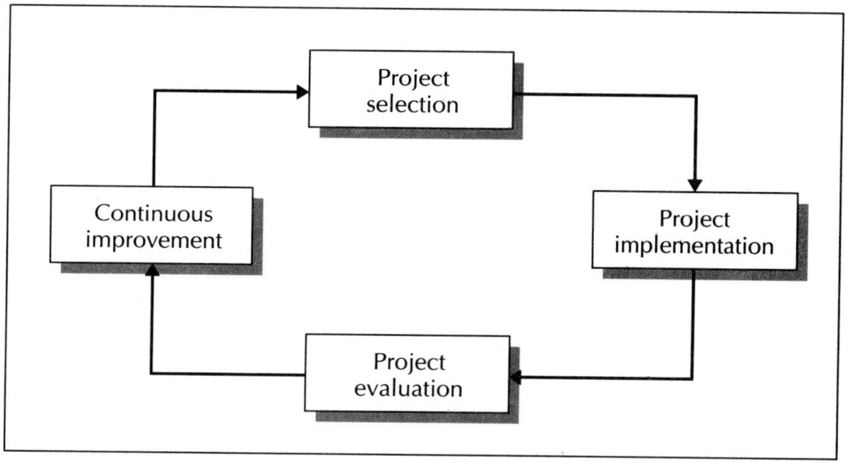


**FIGURE 1.3**
Following the Shewhart cycle assures the project team of closure on any given problem.

revised process. In the final step, that of continuous improvement, recommendations are made either to continue to improve the present product or service characteristic, or to move on to a new project. The Shewhart cycle is a simple but effective problem-solving tool.

This tool was utilized by Deming as he spoke to the Japanese about implementing quality problem solving. The idea, transformed into a **plan-do-check-act cycle** (or **PDCA cycle**) quickly became known as the **Deming cycle.**

The elegance of the Shewhart cycle, which will be used in this text, is its simplicity and its focus on solving one, and only one, problem at a time. *Fire fighting,* which only results in "quick fixes" to problems with no lasting solutions, should no longer be permitted. Problem solving is accomplished on a project-by-project basis, with reports at the end of each project that will ensure that future work will not be hampered by the same problems, repeated endlessly.

The Japanese built upon the idea of a systematic and narrowly defined problem-solving plan by developing the seven tools of quality that would enhance implementation. The seven tools—graphs, cause-and-effect diagrams, histograms, check sheets, Pareto diagrams, control charts, and scatter diagrams—will be linked with the Shewhart cycle as the Case Study illustrates. The **Case Study—Chapter 1** immediately follows the chapter Summary.

| *Basic Tools* | *Chapter* | *Icon* |
|---|---|---|
| Graphs | 2 | 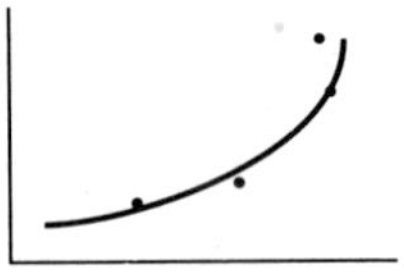 |
| Cause-and-effect diagrams | 1 | 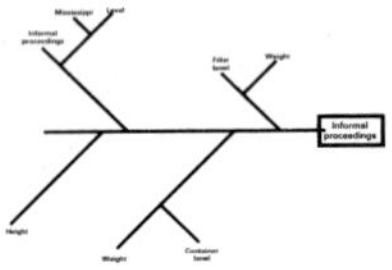 |
| Histograms | 2 | 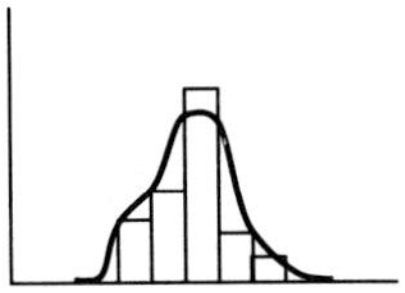 |
| Check sheets | 7 | 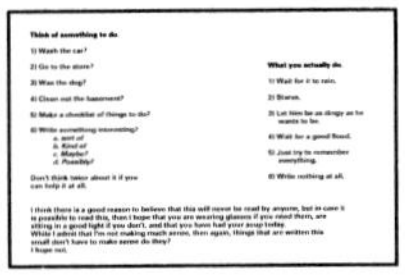 |
| Pareto diagrams | 2 | 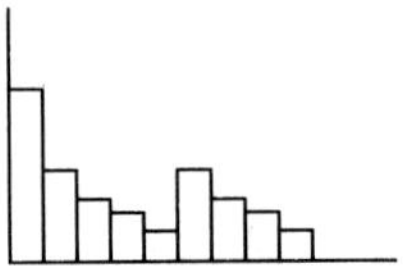 |
| Control charts | 4, 5, and 7 | 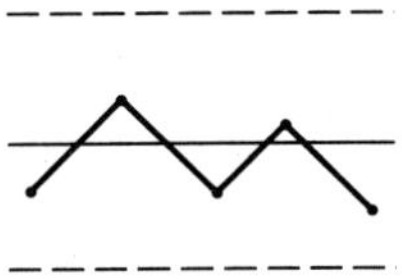 |
| Scatter diagrams | 2 | 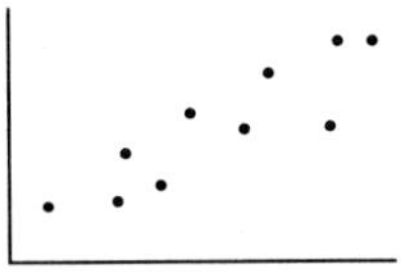 |

**FIGURE 1.4**
The seven tools of quality—chapter reference and icon.

EXAMPLE 1.2

Find the main reference in the text for each of the *seven tools of quality.* List the chapter and draw a representation of the tool (an icon).

*Solution:* The chapter reference and the icon for each of the seven tools are shown in Figure 1.4.

**Data collection** is a key element in successful project management. Data can be categorized as follows:

- Variable data are measurable data (e.g., length, time, pressure).
- Attribute data are either conforming or nonconforming (e.g., rejects, defects, nonconformances, go/no-go).
- Sequences include the 1-to-5 scale, grade points, and such questions as "How does the hamburger taste?"

For effective data collection, the following questions should be answered when preparing a preliminary proposal:

1. What is the purpose of the data collection? Hypothesis?
2. Is the purpose clearly understood by all students and employees involved?
3. Who will coordinate the effort?
4. What are the specific data limits or tolerances?
5. Who is to collect the data?
6. Is any training required? (Training is almost always required.)
7. What is to be used to measure the data? (Is calibrated measurement equipment available?)
8. To what accuracy will the data be measured and reported (Example 1.3)?
9. What forms will be used?
10. Who will analyze the data?
11. Who will prepare the final report?

EXAMPLE 1.3

The accuracy of data collection involves the number of significant figures to which any particular variable measurement should be reported. How do you determine the number of significant figures in a measurement or calculation?

*Solution:* Since the advent of handheld calculators, students have often reported calculations to eight or ten significant figures. In other words, they record all or most of the numbers appearing on their particular calculator's display. Such a practice is very misleading, for no quantity can be more exact or accurate than the numbers used to generate that quantity. Consider the following example of the problems encountered when you ignore accuracy in terms of number of significant figures.

A resistor in a simple one-resistor series circuit has been measured in the laboratory with a relatively inaccurate multimeter as 24 ohms, and the current in the same circuit has been measured with a sensitive galvanometer as 3.6582 amperes. The two numbers must be multiplied to determine the voltage ($E = IR$, Ohm's law). Solving for the voltage, the equation for Ohm's law is used:

$$\begin{aligned} E &= I \times R \\ &= 24 \times 3.6582 = 87.7968 \text{ volts} \end{aligned}$$

But, the 4 in 24 is doubtful. As a result, the true value of the resistor, according to what a doubtful digit implies, could be between 23.6 ohms and 24.4 ohms. What does this mean for our voltage calculation? Let us perform the multiplication again using the two extreme values:

$$\begin{aligned} 23.6 \times 3.6582 &= 86.33 \\ 24.4 \times 3.6582 &= 89.26 \end{aligned}$$

And we see that the true voltage is in *a range of almost 3 volts!*

Clearly, the two results vary so much that to report the data to more than the number of significant figures of the 24 ohms is inappropriate. Such reporting implies a precision measurement greater than exists. To avoid reporting results that are misleading, use the following rules:

1. When multiplying or dividing quantities, record the result to the number of significant figures of the measurement with the *least* number of significant figures.
2. When adding or subtracting quantities, record the result to the decimal place value of the number with the *least* accurate place value.

## *Section 1.2 Review Problems*

Please read the Case Study—Chapter 1, before answering.

1. Problem-solving teams in industry are often known as ____________ ____________ teams.
2. List the seven quality tools used in the Case Study—Chapter 1.

3. Write the descriptors for the main stems of a cause-and-effect diagram (Case Study—Chapter 1).
   $M =$
   $M =$
   $M =$
   $M =$
   $E =$
   $P =$
4. Measuring parts to a ten-thousandth of an inch of accuracy is collecting what kind of data?
   a. sequence data
   b. variable data
   c. attribute data
   d. conforming data
5. Calculate the following to the correct number of significant figures. Use rounding if necessary.
   a. $199 \times 0.3 =$
   b. $0.03402/50 =$
   c. $3100 + 230 + 40 =$

## 1.3 *Quality—A Vital Career Field*

To remain competitive, the world's businesses and industries must continually strive for improved quality. Improved individual processes, illustrated in the Case Study—Chapter 1, can best assure quality improvement. Worker involvement also assures improved productivity. Costs may be reduced with less scrap and rework. Decreasing costs while improving productivity permits organizations to capture more markets. The result is an increase in jobs—more employment for people.

Service industries benefit from improved quality. Florida Power and Light Company won Japan's prestigious Deming Prize in 1989. One central component of the company's plan to compete successfully for the prize was to limit their managers to work on no more than three problems each year. This strong project focus, as illustrated in the Case Study—Chapter 1, is the only way to improve quality continuously.

Businesses can benefit from improved quality. Their processes often directly involve the customer. Therefore, to improve processes, their employees must be well trained and able to deal with problems quickly and effectively. Quality customer service requires intelligent, outgoing, patient people.

For all these reasons, the need for quality managers and technologists is clear. During your quality course you will learn how to define processes objectively by breaking the elements down into their simplest parts. You will use mathematics—especially statistics—to better understand these processes. You will learn the "nuts and bolts" of a system—forms (such as check sheets) and procedures (such as a calibration procedure).

But, no matter how well you learn to break the problem down into its simplest components and analyze it, you must still be able to see the whole system and understand others working within the system. You must be able to break down such barriers as bureaucracy, work with people (fellow workers and customers) who form the major elements of the system, and always keep an eye on the future.

The Shewhart cycle and the problem-solving tools introduced in this chapter will help you to work with people. These tools seem too simple to be effective, but they can solve 95 percent of the quality system problems in an industry. One caution must be given. The problem-solving tools must be used wisely and creatively. Cause-and-effect diagrams may be used in project selection, or in project implementation, or not at all. Some project teams work so well together that special tools are not needed. Using simple tools and simple methods in solving problems will result in improved people systems.

Obviously, quality managers and technologists will be in great demand in order to implement the quality programs of the future. As implementors, technologists will manage the teams that will solve tomorrow's quality problems. They will establish the operational definitions so vital to adequate inspection of manufactured parts and service quality characteristics. Quality technologists will audit processes to ascertain that the processes are capable of delivering the quality necessary for a part or for a service.

## *The American Society for Quality Control*

One of the best resources for those aspiring to work as quality technologists is membership in the American Society for Quality Control (ASQC). Both student and regular memberships are available. Student memberships are quite inexpensive, when compared to regular memberships. Some colleges have student sections where participants attend meetings with others in the quality field, listen to presentations by quality experts, and take tours of local business and industries having strong quality programs. A few of the resources and experiences you can expect from the ASQC include the following:

- sharing ideas with others in the quality field through local section meetings;
- gaining leadership experience by serving your local section as an officer;
- contributing to the development of quality system standards;
- receiving the monthly magazine *Quality Progress,* which covers the latest developments in the quality field;
- an opportunity for membership in specialized divisions, such as aviation, automotive, electronics, and inspection;
- taking part in the Personnel Listing Service and other placement services.

The ASQC offers certification for quality technicians and quality engineers. The **certified quality technician (CQT)** and the **certified quality engineer (CQE)** must satisfy experience and/or education requirements and pass a written exam. The CQT requirement is four years of experience with up to two years of experience waived for those possessing an associate degree in quality or industrial engineering technology. The CQE must possess a bachelor's or advanced degree to sit for the *Engineer-in-Training Examination* and three to four years experience (depending on degree) to sit for the applications portion of the CQE examination.

## *Section 1.3 Review Problems*

1. Improving quality by improving individual industrial processes will
   a. increase worker involvement
   b. decrease costs
   c. improve productivity
   d. accomplish all of the above
2. The only U.S. company to win Japan's coveted Deming Prize is
   a. Cadillac Motor Car

b. Florida Power and Light
c. U.S. Steel
d. the ASQC

3. The abbreviation ASQC stands for the ________ ________ ________ ________ ________.

4. The abbreviation CQT stands for ________ ________ ________.

## Summary

- Quality is determined on the basis of how well the various quality characteristics of a product conform to a given target value (or conformance to specification). The product quality characteristic is an individual feature of a product, well defined and understood by all concerned in producing that product, and should focus on customer satisfaction. Examples of product quality characteristics are length, weight, surface smoothness, time of service, and taste.
- A process quality characteristic is an individual property of a process that produces the product. Examples of process characteristics are speed and feeds of machine tools, cooking time for a hamburger, and fill rate for a fill operation. Process and product characteristics should be carefully differentiated in quality projects.
- Inspection is the activity of checking a product's quality characteristics for conformance to specification (evolved prior to 1920s). Quality is fitness for use in society and customer satisfaction. Quality control (QC) is defined as all of the functions necessary to obtain quality of product, together with the regulatory process by which quality of product is attained (first defined in the 1920s).
- Statistical quality control, the use of statistics in quality control, includes statistical process control (SPC) and acceptance sampling (this combination of statistical techniques was developed in the 1930s). SPC is the use of statistics in controlling the quality of a process while it produces goods. Acceptance sampling is the process

by which statistical sampling plans for "lots" (batches) of finished products are determined to be fit or unfit for use.

- Total quality management (TQM) is a system recognizing that quality is everyone's job. Designer, purchaser, manufacturer, or salesperson—all in the organization are responsible for product quality. There is a strong human factor involved in effective quality efforts. Team building for project-by-project problem solving is necessary for continuing improvement.
- The Shewhart cycle provides a simple problem-solving model for improving quality of product or process. It consists of four steps—project selection, project implementation, project evaluation, and continuing improvement. Project teams benefit from following the steps of the problem-solving model.
- Ishikawa's *seven tools of quality* (Section 1.1, "1960s") are graphs, cause-and-effect diagrams, histograms, check sheets, Pareto diagrams, control charts, and scatter diagrams.
- The basic elements of any process are machinery, materials, measurement, methods, environment, and people. The basic elements, used to form the main stems of the cause-and-effect diagram, can be remembered as the four *M*'s, the *E*, and the *P* (MMMMEP). See the Case Study—Chapter 1.
- Sound data collection techniques must be followed in order to prevent waste and inefficiency.
- The American Society for Quality Control (ASQC) is the professional society for those aspiring to work in—or already working in—the field of quality. Student memberships are inexpensive when compared to regular memberships.

## CASE STUDY · CHAPTER 1

A dairy producer finds that the fill weights (weight of product per container) of several of its products do not meet standards set by the Food and Drug Administration (FDA). If the problem persists, the company could be found out of compliance by the FDA and fined. The problem-solving steps, as outlined by the Shewhart cycle (Figure 1.3), are used to solve the problem.

To begin any problem-solving project, the project team must examine data to enable them to determine what specific problem should be addressed. Knowing what the major problem is, and acting on it, can save needless activity on projects that will not result in measurable savings. (Quality costs are discussed in Section 10.1.) **Project selection** data can be collected from two major sources:

1. from inside the company by analyzing manufacturing or service reports;
2. from the customer by analyzing surveys and field service reports.

**Pareto analysis** is utilized to determine which quality characteristic is the most frequently occurring or the most costly. A bar chart presents a clear picture to the manager or technologist of which problem is causing most of the loss. The product's fill weight can be seen to be the most costly in this case (Figure 1.5). The cost data for the Pareto analysis have been collected from FDA compliance reports. The FDA is the regulator of this industry. Fill weight caused a loss in this reporting period of $432. (See Chapter 2 for details in constructing a Pareto analysis.)

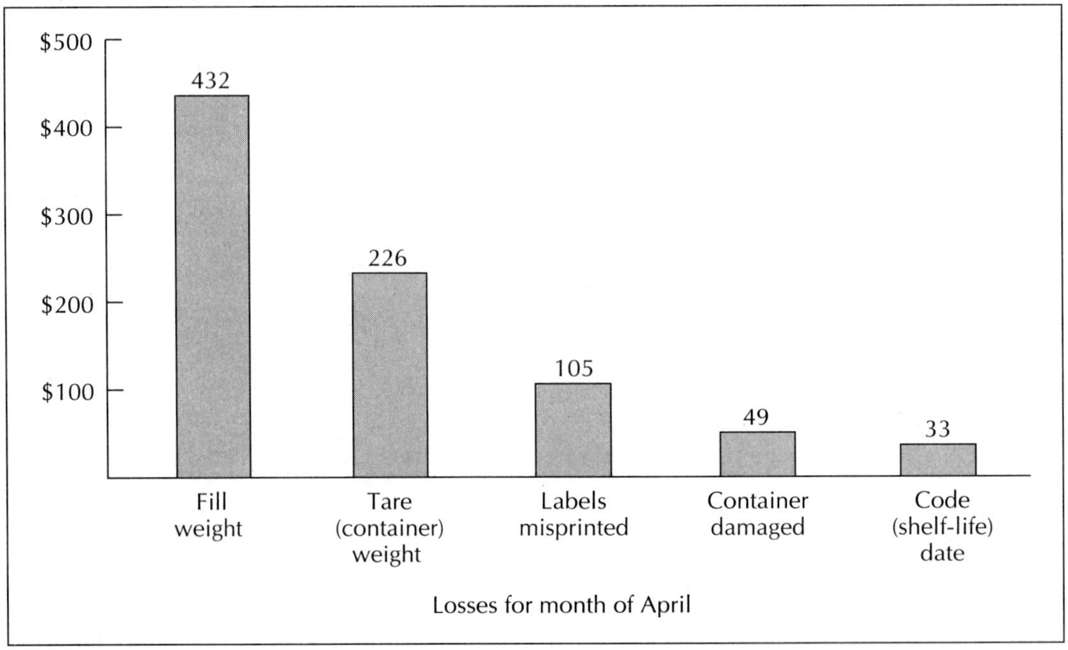


**Figure 1.5**
The Pareto chart depicts the problems that must be addressed first.

Most of the tools utilized in step 1, project selection, are the following:

*Project Selection*

- historical data (reports, surveys, etc.);
- compliance reports from regulators;
- Pareto analysis.

(The cause-and-effect diagram may also aid the project team in project selection. Cause-and-effect diagramming, however, is usually used in project implementation.)

Project selection is as important as, if not more important than, all subsequent steps. The project team must narrowly define the project. Narrowly defined projects are the easiest to solve. Organizations that have not used the project approach before should *start with a project that may be solved easily.* Success on the first project is a great motivator for continuous improvement on later projects.

After determining the nature of the project, project implementation follows. **Project implementation** may involve the following tools:

*Project Implementation*

- flowchart;
- cause-and-effect diagram;
- control chart;
- scatter diagram.

Figure 1.6 shows the actual fill process. The two tanks in Figure 1.6a (right-hand side) are the upper tank and the mix tank. The product is pumped from there to the fill cylinders. (Adjusting screws for the fill cylinders are to the right of the operator's hand in Figure 1.6b.)

The first step in project implementation is to develop a flowchart. The **flowchart** utilizes pictorials or standard symbols to show all of the steps in a process. It allows members of the project team to visualize and agree on the actual process. Managers, technologists, and production personnel are always surprised at how much disagreement and misunderstanding may surround the actual process.

Figure 1.7 is a pictorial of the fill operation. Some problem-solving teams may use pictorial flowcharts, but most use standard flowchart symbols as shown in Figure 1.8.

Examining Figure 1.8, you can see that two tanks are utilized to aid in smooth flow of product during the fill operation, and that the fill cylinder is sometimes adjusted (feedback loop from step 90 to 70) as operators believe adjustments are necessary. In this study, adjustments to the pro-

(a)

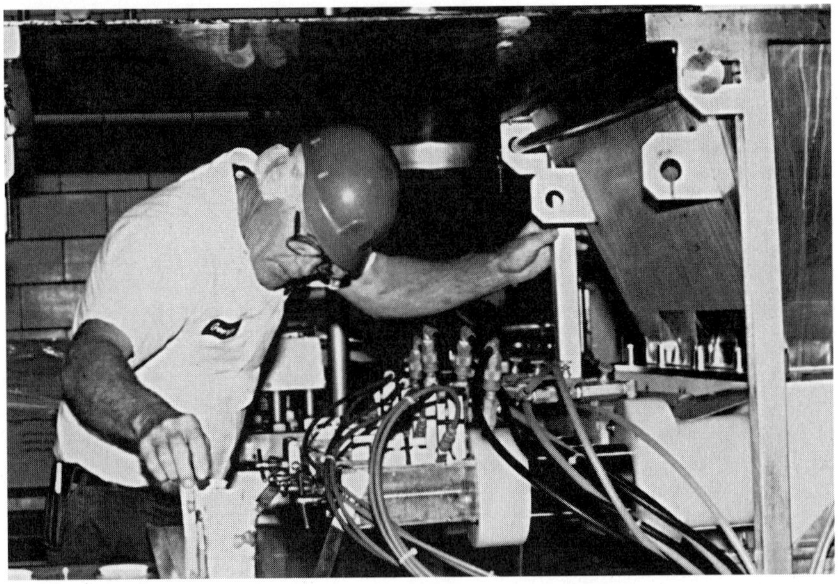

(b)

**FIGURE 1.6**
The vats of yogurt (a) and the adjustment of the fill cylinders (b). (*Source:* Tamarack Farms Dairy, Newark, OH.)

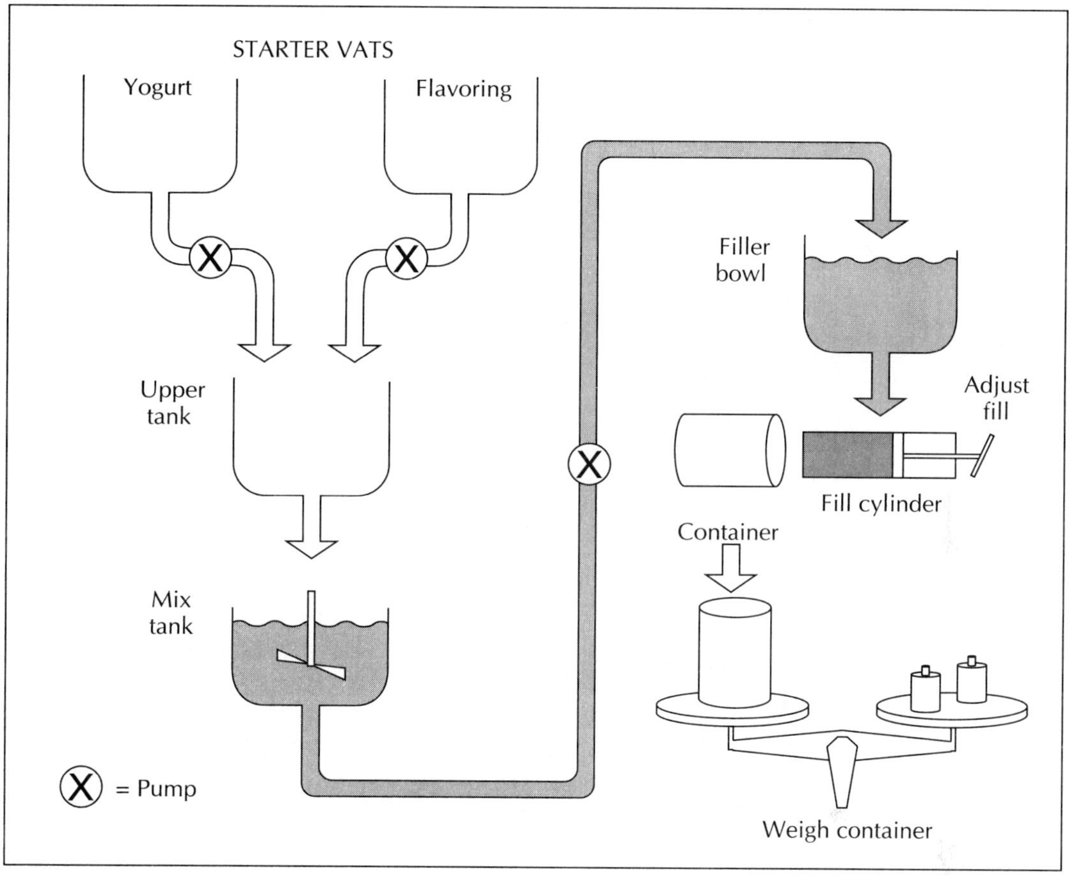


**FIGURE 1.7**
Pictorial flowchart of the fill process.

cess will not be made during the process of filling the containers. Not adjusting the process allows the team to see how the process machinery functions by itself. Overcontrol problems abound in industry, and they will be discussed in Chapter 4.

The **cause-and-effect diagram** may aid the team in selecting parts of the process to change in attempting to improve the process (Figure 1.9). The six main stems of the diagram usually involve the most basic elements of any process—*machinery, materials, measurement, methods, environment, and people.* The basic elements can be remembered as the four *M*'s, the *E*, and the *P* **(MMMMEP).** However you manage it, the six elements should be memorized, for good problem solving requires their use.

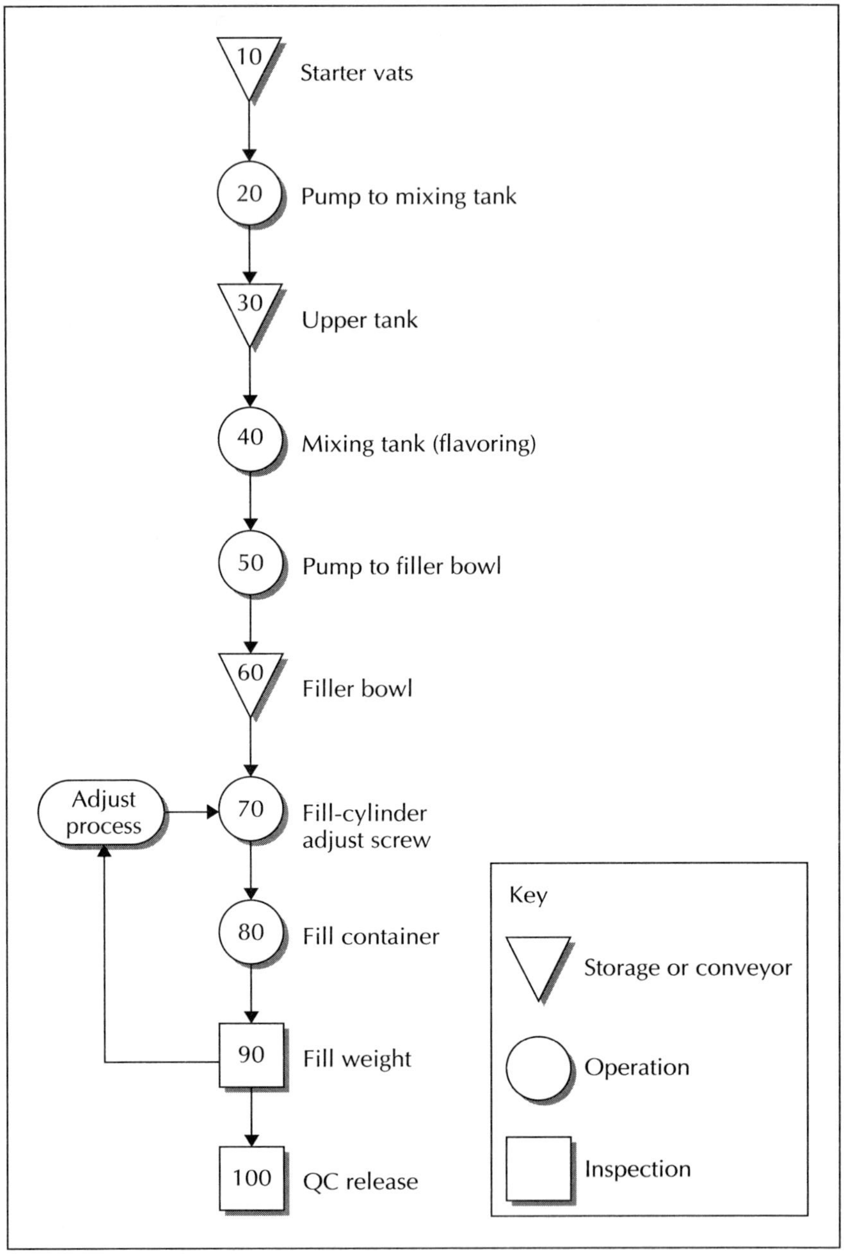


**FIGURE 1.8**
Flowchart of the fill process using standard engineering symbols.

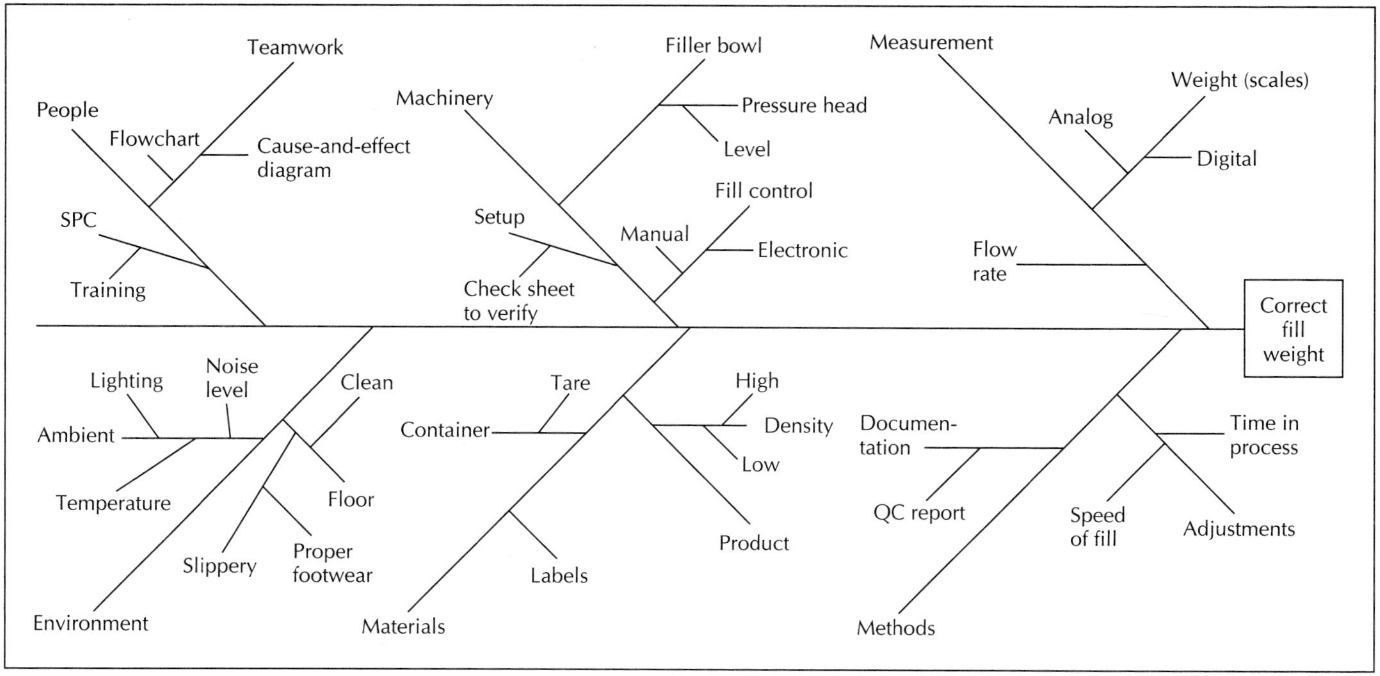


**Figure 1.9**
The cause-and-effect diagram is used to achieve a more consistent fill weight.

In the fill-weight project, the cause-and-effect diagram might be as shown in Figure 1.9. Notice that the central objective, correct fill weight, is connected to the right-hand side of the main stem. A working cause-and-effect diagram is completed by all members of the team working together in concert, each member of the team able to express ideas without fear of criticism by others on the team.

The flowchart and cause-and-effect diagram allow the team to concentrate on the areas in which data should be collected. Process data (as distinguished from project selection data) are meaningless until they are put into a format that makes sense of the process variability. Statistical methods have been developed that separate the **assignable causes** of product variation from the **chance causes** of the variation. Since fill weight is the concern of the present project, the data are the weights of the filled containers. The data are variable (Section 1.2). The **tare,** the weight of the containers, is included. See Figure 1.10 for the control chart illustrating one run. As stated before, the process was not varied by the operators during the run, in order to examine how the equipment functions without operator intervention. In this case, an overfill trend occurs in the later stages of the run. (You may be surprised that overfills are a concern. Would not the customer benefit from overfills? Consider that too much fill would invalidate label amounts [for instance, sodium content], something the FDA strongly disapproves of.)

Based on the overfill trend pushing the weight of containers beyond the upper control limit of 188 grams, the operators could decide to make one adjustment at a critical time during the fill (at point 11 or 12), or they could decide to terminate the fill operation as the filler bowl becomes empty. In this case, the team decides to terminate the operation at the point of the overfill trend because the equipment is not designed to be reliable at low head pressures in the filler bowl.

After several trials (in some projects the time frame could be several months), the process is in control (Figure 1.11). For a detailed explanation of control charts, assignable causes, and in- and out-of-control conditions, refer to Chapter 4.

**Project evaluation** takes the form of a report composed by and agreed upon by all team members. The report is essential for several reasons. Without composing the report the team does not really know how they have done—how well their process is behaving. Recommendations for continuing improvement must flow from a well-organized report that explains what has already been tried and what is yet to be attempted. The report establishes a baseline so that the same team or a different team can continue to improve the process.

Reports should be brief, and certainly not wordy. They will contain enough graphs and other pictorials so the reader may visualize the project. Many businesses and industries have standardized reports that everyone in

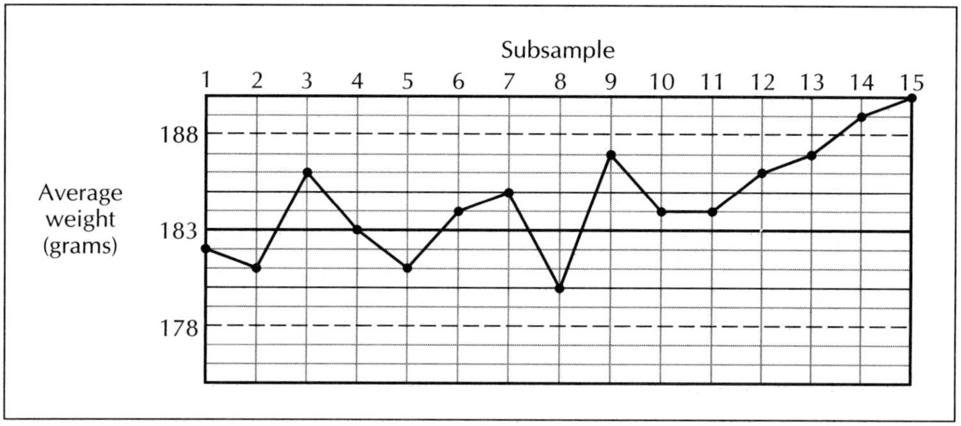


**FIGURE 1.10**
The control chart shows a trend at the end of the fill.

the organization is familiar with, making the reading and interpretation as clear as possible.

A special type of graph is the **histogram.** Figure 1.12 is a histogram of the project explained in this case study. Close examination reveals that the height of the bars displays how many times the quality characteristic (fill weight in this case) occurs.

The histogram shows the average fill weight ($\overline{X}$) to be within the specification limits—*LSL* for lower specification limit and *USL* for upper specification limit—but below the nominal or desired value (*N*). Also, the spread of the individual containers is too large—much larger than the specification limits. Note that the control chart does not use specification limits. Later, the vast difference between control limits and specification limits will be further clarified.

**Process capability** is the ability of the process to be within the variability of the specification limits. From the histogram the capability of the process to make acceptable product may be determined. The process is not capable of maintaining specification limits. See the following elements of project evaluation:

### *Project Evaluation*

- graphs;
- histograms;

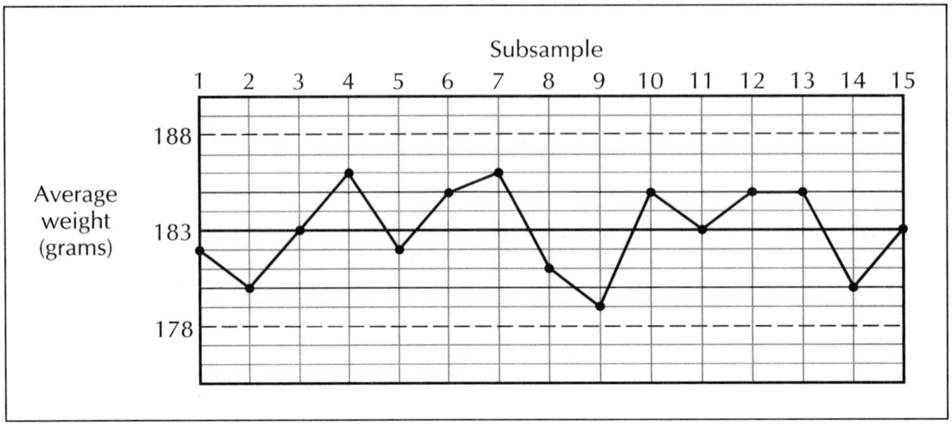


**FIGURE 1.11**
The fill weights are in control.

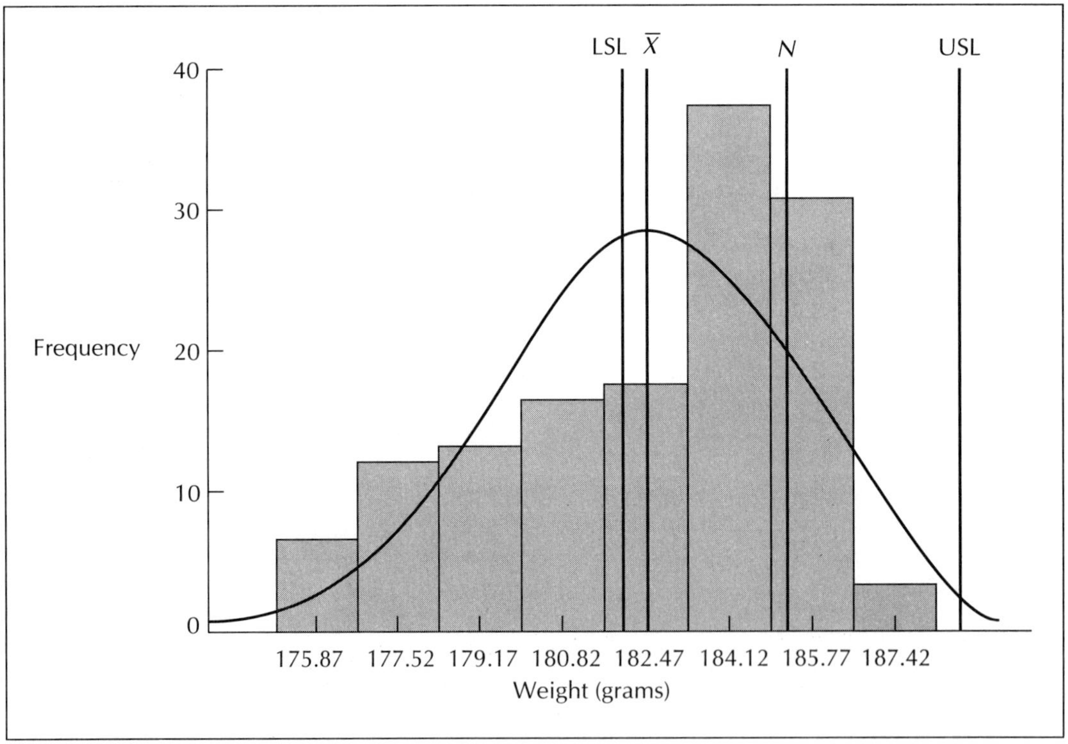


**FIGURE 1.12**
The histogram shows the distribution of the fill weight of the containers.

- determination of process capability;
- written report.

The project team has functioned well together in bringing the process into control with the control chart techniques utilized. **Continuing improvement** is simply a step that requires the team to choose one of the following two alternatives:

1. decide to improve the present process;
2. decide to move on to another project.

Process capability is not as desired in this case, and the team will pursue further measures to reduce the variability of the product. If process capability were acceptable, the team might move on to another project. In any case, the cycle is repeated. *The goal is always the same: never-ending improvement.*

# *Chapter Review Problems*

1. What do you think of when you think of quality? List at least five ideas. Compare your ideas with the commonsense and universal ideas in Section 1.1
2. List five areas of quality improvement in service or product that you consider most important to you, your community, your company, your country, or the world. (Do not forget that governments offer services.)
3. List five product quality characteristics of products or services you are most familiar with (see Section 1.1)—for example, the number of squeaks in an automobile's dashboard.
4. For each of the five product quality characteristics listed in problem 3, write one process quality characteristic. (For example, for automobile dashboards, extrusion temperature and type of thermosetting polymer represent process quality characteristics.)
5. Quality characteristics may be organized under such categories as technological (e.g., hardness, weight), psychological (e.g., taste, beauty), time oriented (e.g., time of service, reliability), contractual (e.g., what you guarantee your customer or what is mandated by government regulators), and ethical (e.g., courtesy of service, honesty in sale of products or service). Group the five quality

characteristics you listed in problem 3 into these categories. For instance, the number of squeaks in a dashboard is a quality characteristic that is technological; it may be remedied by closer fits and materials that expand less with temperature change.

6. Select one of the following processes, and list one process quality characteristic that may be examined to improve the process:

   - welding pipelines;
   - building printed circuit boards;
   - building computer systems;
   - customer service;
   - selling technical equipment to manufacturers (outside sales).

7. Be able to place the phases of quality evolution in chronological order, when examined by the instructor (see Table 1.1).
8. With the class, inspect the classroom or laboratory you are in. List each quality nonconformance you find individually. Then, compare notes with others in your class or with team members (as assigned by the instructor). Attempt to reach agreement on a final list of nonconformances.
9. Statistical quality control may be separated into two separate elements—SPC and acceptance sampling. Describe the two elements in two paragraphs.
10. Be able to draw the Shewhart cycle and list the four elements from memory, at any time when examined by the instructor.
11. Who is the customer of the Internal Revenue Service (IRS)? Who is the customer of the Federal Tax Deposit System (the division that collects withholding statements) of the IRS?
12. Draw a flowchart of the registration process at your college. Compare it with other students' flowcharts. Discuss the value of teamwork in preparing the flowchart.
13. Design a cause-and-effect diagram to achieve the effect of "obtaining good grades." Establish the major categories (or major bones—the cause-and-effect diagram is also known as a fish-bone diagram).
14. Be able to list the major elements of the cause-and-effect diagram (MMMMEP) from memory when examined by the instructor.

15. Research and locate one article from the journal *Quality Progress* that is of interest to you. Write a one-page summary of the article, or copy the article for your instructor (instructor's preference).

## LABORATORY · CHAPTER 1
## Data Collection (Refer to Section 1.2)

Materials needed:

- measurement tools—micrometer or vernier (digital) calipers;
- gage blocks;
- parts to measure (a minimum of 30 to 50 parts for each team);
- data collection form (may be designed by students).

The class will be divided into teams of two. Each team will perform the following:

1. Using an appropriate gage block (one that is close to the part dimension), the partners verify the accuracy of the measurement instruments for measuring the part.
2. Partners practice repeatable readings on five parts, and a decision is made as to which partner is more repeatable.
3. The more repeatable partner will perform the measurements, and the other will record the data.
4. An appropriate form will be completed and submitted to the instructor.

# 2

# Graphical Analysis of Data

**Graphing Fundamentals** 2.1
**Experimenting by the Use of Scatter Diagrams** 2.2
**Preparing Histograms** 2.3
**Troubleshooting with Histograms** 2.4
**Pareto's Principle** 2.5

## 2.1 *Graphing Fundamentals*

One way to present data is the line graph. A **line graph** displays the relationship that exists between two variables. Hopefully, all other possible variables will be eliminated or held constant while data are collected. This process is what is meant by a **controlled experiment.** Figure 2.1 is a multiple-line graph. Some variables are controlled—

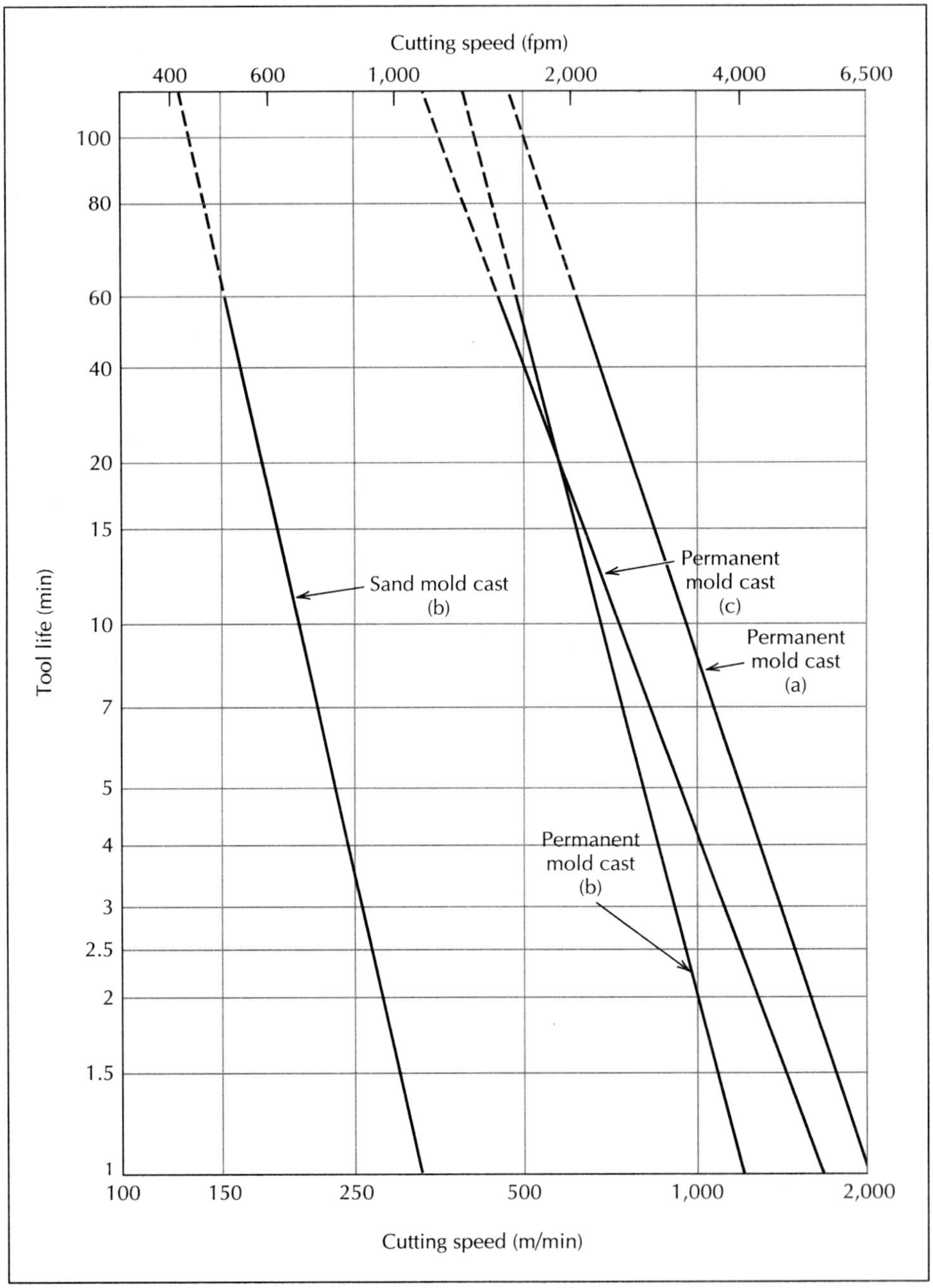


**FIGURE 2.1**
Graph of Si-Al cutting tool—tool life versus cutting speed for various types of castings.

held constant—and are displayed within the graph. The tool is made of a silicon-aluminum (Si-Al) metal alloy.

The variable manipulated during the experiment is called the **independent variable.** This variable will always be placed on the *X* axis or horizontal axis. (The *X* axis is sometimes referred to as the abscissa.) In Figure 2.1 the independent variable is cutting speed in feet per minute (fpm) or meters per minute (m/min).

The other variable, placed on the *Y* axis or ordinate, is the **dependent variable,** for its values *depend* on related values of the independent variable. We usually graph variables that obviously depend on one another. In Figure 2.1 tool life depends upon cutting speed, and tool life is the dependent variable.

Both axes are scaled in a nonlinear format. This practice allows the relationship between the two variables to appear as a straight line. Figure 2.2 shows what one of the "curves" looks like when the axes

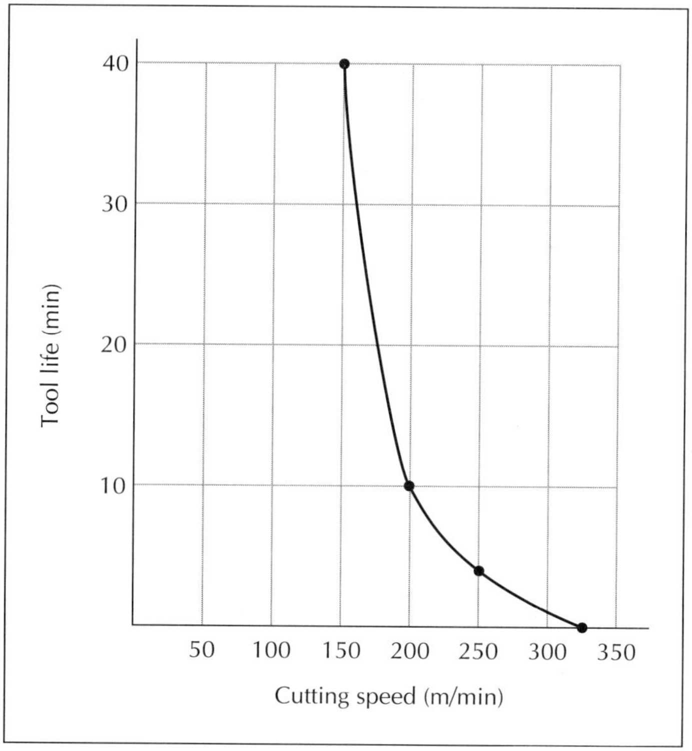


**FIGURE 2.2**
A simpler line graph showing only one of the tools examined in Figure 2.1.

are scaled in a linear fashion. (Recall that a straight line is a curve of infinite radius.)

Interpretation of a graph consists of moving straight up, from a selected point on the *X* axis, to the curve and reflecting from there over to the *Y* axis while maintaining a line parallel to the *X* axis. Figure 2.2, a simpler version of the graph in Figure 2.1, is offered to clarify the interpretation of the graph. Only one sand mold (b) is depicted. From the graph you can predict that tool life will be only 4 minutes at 250 m/min, but it will increase to more than one-half hour at 150 m/min. Of course, the multiple-line graph in Figure 2.1 shows that other molds may be worked (cut or trimmed) at faster feeds.

The two variables—cutting speed and tool life—are negatively correlated. A **correlation** simply describes what one variable does in relation to what the other is doing. If one variable increases as the other increases, we describe the relationship as a positive correlation. If one variable decreases as the other increases, as in the example, it is known as a negative correlation.

## *Section 2.1 Review Problems*

1. The manipulated variable in an experiment is known as the
   a. dependent variable
   b. changed variable
   c. independent variable
   d. none of the above are correct

2. The variable that is dependent on the manipulated variable is placed

   on the ______ axis.
   a. *X*
   b. *Y*
   c. *A*
   d. *Z*

3. If the independent variable increases as the dependent variable decreases, the slope is said to be
   a. tangential
   b. regressive
   c. positively correlated
   d. negatively correlated

## 2.2 Experimenting by the Use of Scatter Diagrams

The best way to improve the quality of a manufacturing process is to attempt to find a relationship between variables. A **scatter diagram** is a plot of many points on the $X$–$Y$ coordinate system. The points represent the relationship between two variables while experimenting with the process. The graphical picture is thereby reinforced.

The advantages of collecting a great deal of data are that the graph is reinforced by more data collection and the data collection can be randomized. With randomized data collection, the independent variable is varied in both directions as data are collected. An example of randomized data collection is to move a machine tool both ways to monitor its actual position compared to its readout's position. One disadvantage occurring in the typical "one-way" academic experiment is that the smooth curve that results changes as the independent variable is decreased, rather than being steadily increased. For one thing, hysteresis errors (caused by a machine's looseness or "play" between its moving parts) are found when varying the independent variable in both directions.

Plotting the points in the scatter-diagram format may or may not reveal a correlation. For instance, in Figure 2.3 we find that the first graph (Figure 2.3a) shows **no (or null) correlation** between the two variables of process temperature and density of product. On the graph in Figure 2.3b we see a **positive correlation** between temperature and viscosity. Finally, the scatter diagram in Figure 2.3c shows a **negative correlation** between the speed of the assembly line and the number of nonconforming units from the line. (Does the negative correlation between these two particular variables make sense?) Note that all three examples have the independent variable, the variable that is adjusted during the experiment, placed on the $X$ axis.

All graphs require that certain conventions be followed so the reader may interpret the graphs. The scatter diagram is no exception. The following checklist will aid you in proofing your final product:

1. Each variable in the study must be carefully determined (which is dependent and which is independent), and the axes must be properly labeled. Both quantity and units must be present (e.g., temperature in Kelvin [K] and density in pounds per

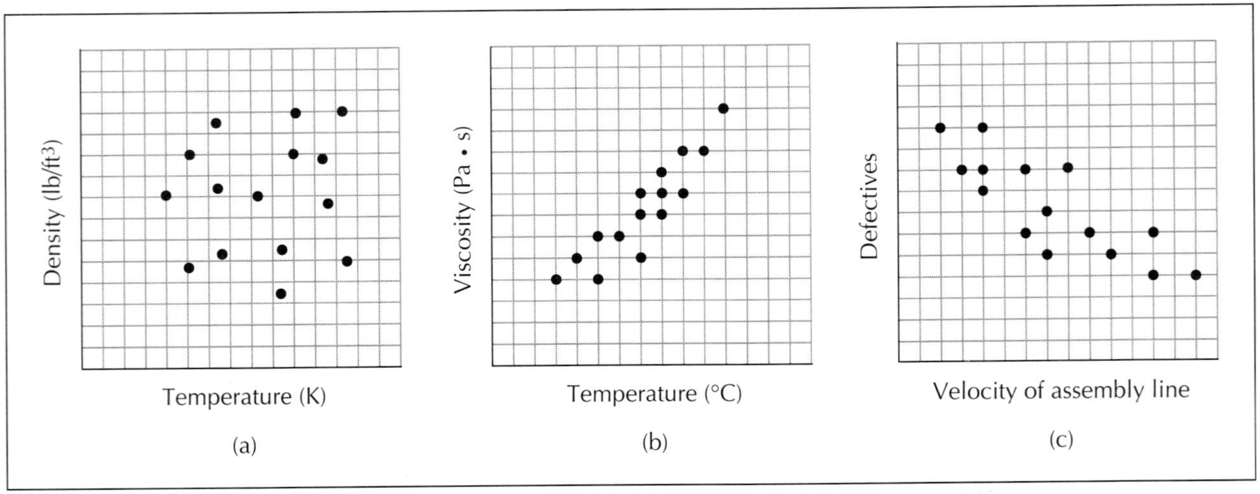


**Figure 2.3**
Graphs showing various correlations: (a) no correlation; (b) a positive correlation; (c) a negative correlation.

cubic foot [lb/ft$^3$] as shown in Figure 2.3a). It is important to be sure all other variables are controlled, or kept constant during the experiment.

2. Each axis must be properly scaled. Data collection practices should be specified. Such items as rounding techniques, range of data, and the accuracy (significant figures) of the measurement equipment (thermometers, viscosimeters, calipers, etc.) must be carefully communicated to those responsible for collecting the data.
3. A line of best fit is drawn through the experimental points. If the data are curvilinear, logarithmic scales may be utilized. (Curvilinear relationships are difficult to analyze with scatter diagrams. The practice is not recommended, unless mathematical analysis is also used.)
4. The graph is titled after completion, when the full objective of the illustration can best be established.
5. The finalized graph is inspected for accuracy, neatness, ease of interpretation, and strength of correlation.

**Correlation analysis** is a statistical method of determining the strength of correlation between the two variables. The calculation returns scattered data to a line of best fit on the $X$–$Y$ coordinate system. The coefficient of correlation, determined by a mathematical formula, is a number between +1 and −1. A value of *+1 means a perfect positive correlation* (all of the points lie on a straight line and the $Y$ value increases as the $X$ value increases), and a value of *−1 means a perfect negative correlation* (the $Y$ value decreases as the $X$ value increases). Many scientific calculators perform correlation analysis. See the statistical section in your calculator manual for further information regarding correlation. **Linear regression** (the least-squares method of calculating slope and intercept information) may also be calculated, as well as curvilinear regression or multiple regression.

EXAMPLE 2.1

The surface finish for an automobile's wheel (brake) cylinder bore (Figure 2.4) is to be studied. The finish of the internal diameter (ID) of the cylinder determines the life of the disk brake system; therefore, a better finish increases customer satisfaction. (The immediate customer is a major automobile manufacturer, and the final customer is the family driving and riding in the car the component is installed in!)

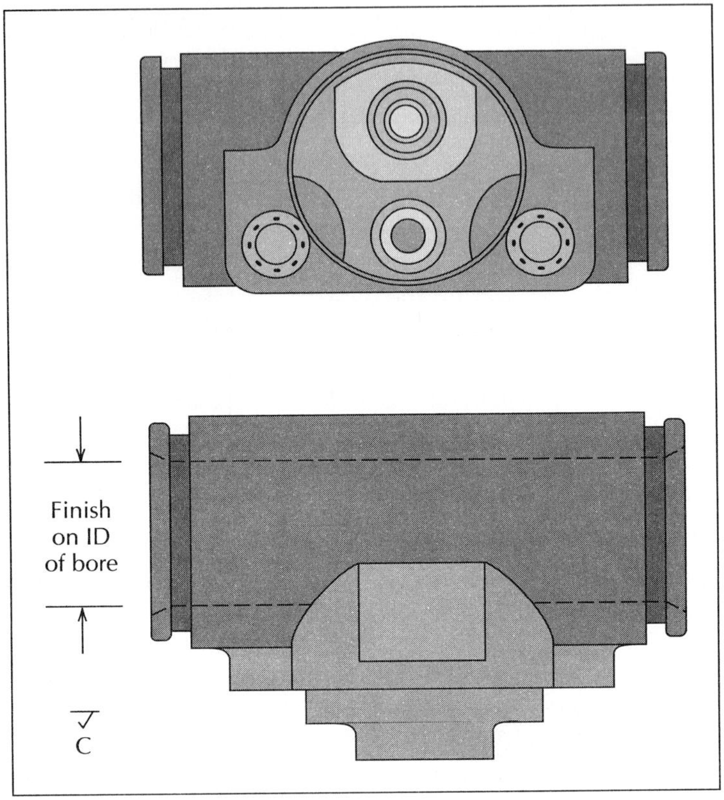


**FIGURE 2.4**
The finish on the internal diameter (ID) of the wheel cylinder bore is a critical quality characteristic, as depicted by the surface finish symbol. "C" designates a circular lay.

*Solution:* In studying the process it was discovered that two machines are involved in producing the bore ID. The first machine roughs out the bore, and the second is used to finish the bore. This procedure is typical in boring operations where the finish must be relatively smooth.

SPC techniques were only being used on the second (finishing) machine. This study involved collecting data on machine 1 (roughing the bore) and determining the relationship of microfinish on machine 1 (independent variable) to microfinish on machine 2 (the finishing machine is the dependent variable).

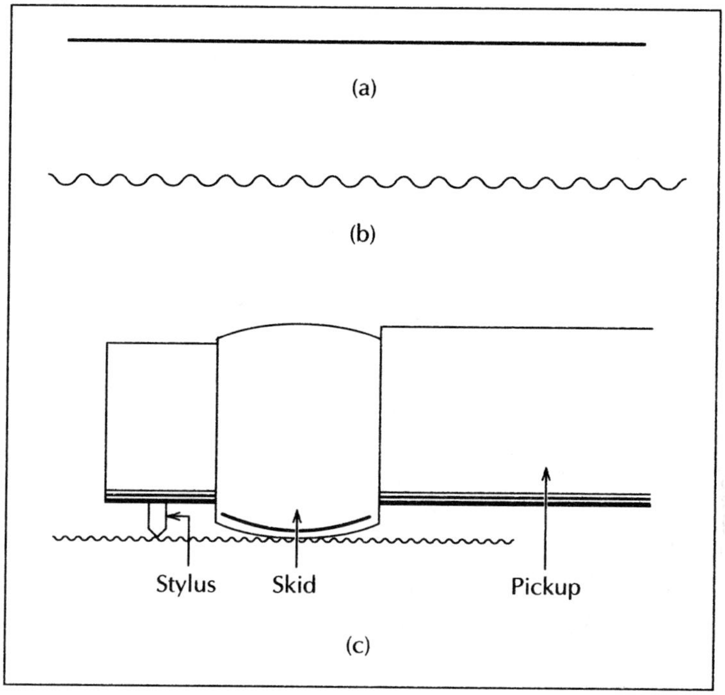


**FIGURE 2.5**
Surface finish is never perfect (a), but exhibits roughness (b). The pickup of the profilometer (c) is supported on the workpiece by a skid.

A profilometer is used to measure the smoothness of the surface. A **profilometer** uses a phonograph-like stylus to measure surface finish in either microinches ($\mu$in.) or micrometers ($\mu$m). An average reading of 0.05 $\mu$m (2$\mu$in.) represents a typical discrimination capability. Refer to Figure 2.5.

The results of the study are depicted in Figure 2.6. The correlation of the data is definitely positive. The coefficient of correlation, determined by the calculator, is +0.9. This number, approaching a maximum of positive one, confirms a strong, positive correlation between surface smoothness on machine 1 and surface smoothness on machine 2. (The 0.9 may be squared—the result is known as the coefficient of determination ($r^2$)—and the resultant 0.81 may be interpreted as "the data show that 81 percent of the variation in surface finish at machine 2 is caused by that from machine 1.")

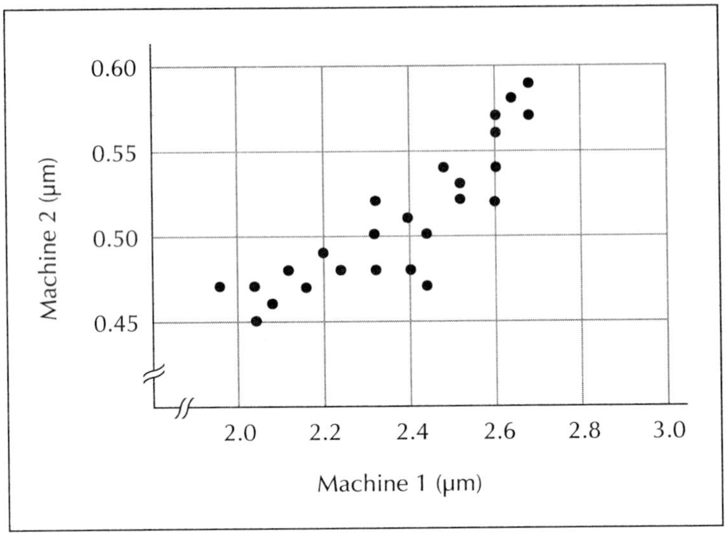


**FIGURE 2.6**
Scatter diagram showing correlation of roughing machine (1) to finish machine (2).

Strong correlations are not conclusive. Sometimes there are other variables operating that we fail to control—or are unable to control—while experimenting. However, given the results of the experiment in the example, it would probably be worthwhile to attempt to improve surface finish on products from machine 1. This objective may be accomplished by studying machine 1 with control charts (Chapter 4). Operators should be trained to use the charts to assure proper centering and to determine when to change tools. (The graph of tool life versus cutting speed in Figure 2.1 was determined by scatter diagramming.)

## *Section 2.2 Review Problems*

1. Hysteresis errors may be detected in an experiment by
    a. varying the independent variable one way
    b. varying the independent variable both ways
    c. varying the dependent variable one way
    d. varying the dependent variable both ways

2. If a scatter diagram's independent variable is increasing and the dependent variable is decreasing, the slope is said to be
   a. tangential
   b. regressive
   c. positively correlated
   d. negatively correlated
3. The profilometer measures
   a. surface finish
   b. part profiles
   c. heat energy
   d. light energy
4. A plot of ambient, outdoor temperatures from 12:00 midnight to 12:00 noon will *always* result in
   a. a positive correlation
   b. a negative correlation
   c. a null correlation
   d. a result cannot be determined with the stated requirements

## 2.3 *Preparing Histograms*

The **histogram** is a special type of bar chart, with the height of the vertical bars representing the number of times a quality measurement occurs. The histogram is usually used with variable data—data with numerical values. The example in Table 2.1 will use the shafts from an electric motor. The critical dimension, length of shaft, in millimeters, is shown for the 30 pieces in the study.

The data may be placed in a histogram format. The picture formed by the histogram, the distribution of the data, leads to conclusions the

TABLE 2.1
**Length of Shaft (mm)**

| | | | | |
|---|---|---|---|---|
| 25.6 | 25.7 | 25.7 | 25.7 | 25.5 |
| 25.3 | 25.3 | 25.8 | 25.7 | 25.6 |
| 25.8 | 25.4 | 25.4 | 25.4 | 25.7 |
| 25.5 | 25.8 | 25.5 | 25.5 | 25.8 |
| 25.7 | 25.8 | 25.6 | 25.5 | 25.8 |
| 25.6 | 25.6 | 25.8 | 25.7 | 25.7 |

raw data would never reveal. A 50-piece study is usually recommended for drawing conclusions beyond the sample data; however, 30-piece studies as in Table 2.1 can be very helpful in early adjustments of a process.

To construct the histogram, use the following steps:

1. Determine the range of the data by identifying the highest and lowest value of data in the distribution. To save on calculating time, first **code the data.** All of the motor shafts in Table 2.1 measure 25 millimeters, so the distribution of the data will be the same whether or not the 25 millimeters is retained in the data set. The result is coded data in tenths of a millimeter (0.1 mm).

| | | | | |
|---|---|---|---|---|
| 6 | 7 | 7 | 7 | 5 |
| 3 | $3^l$ | 8 | 7 | 6 |
| 8 | 4 | 4 | 4 | 7 |
| 5 | 8 | 5 | 5 | 8 |
| 7 | 8 | 6 | 5 | $8^h$ |
| 6 | 6 | 8 | 7 | 7 |

The coded data are then inspected to find the low and high values (see the values with superscripts $l$ and $h$). The **range** is determined by subtracting the lowest value from the highest value. The range of the data on the motor shafts is

$$R = \text{high} - \text{low} = 8 - 3 = 5 \text{ (or 0.5 mm)}$$

2. Next, the number of cells or **classes** that the data can be grouped in will be determined. A good beginning point is to

TABLE 2.2
**Number of Cells in the Histogram**

| *Number of Data (n)* | *Number of Cells* |
|---|---|
| 20–50 | 6 |
| 51–100 | 7 |
| 101–200 | 8 |
| 201–500 | 9 |
| 501–1,000 | 10 |
| Over 1,000 | 11–20 |

approximate the number of cells required by referring to Table 2.2.

For the 30-piece study, six cells are suggested.

3. Divide the range of the data by the number of cells determined from the table. The result is the width of the cell, or the **cell interval.**

$$\text{Cell width (interval)} = \frac{\text{Range}}{\text{Number of cells (table)}} = \frac{5}{6} = 0.833$$

And, the cell interval will be rounded to one (or to 0.1 mm). Rounding may result in a different number of cells in the histogram. In this case the trial value of six cells results in the best histogram. The cell interval will be at the accuracy of the data—0.1 mm.

4. Construct a frequency distribution of the data as shown in Table 2.3.

TABLE 2.3
**Frequency Distribution of Length of Shaft (see Table 2.1)**

| *Measurement (Coded to 0.1 mm)* | *Tally* | *Cell Frequency* |
|---|---|---|
| 3 | II | 2 |
| 4 | III | 3 |
| 5 | 𝍸 | 5 |
| 6 | 𝍸 | 5 |
| 7 | 𝍸 III | 8 |
| 8 | 𝍸 II | 7 |
| | | $n$ = 30 pieces |

5. Scale the $X$ and $Y$ axes using 10 as the maximum value needed on the $Y$ axis, and 3 through 9 (the upper specification limit of 25.9 mm is abbreviated as USL) as the range of values on the $X$ axis. Then, draw in the cells described by the frequency distribution. The result is shown in Figure 2.7. The break in the $X$ axis at 25.3 depicts a nonlinearity of the scaling between 25.3 and the 0 point.

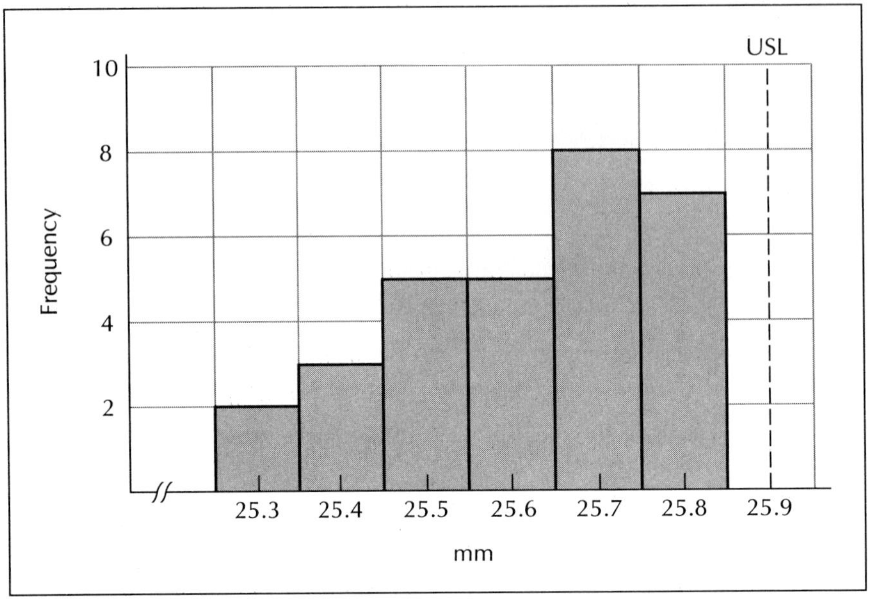


**FIGURE 2.7**
Histogram for length of motor shaft in millimeters.

A summary of the steps used in constructing the histogram follows:

1. Determine the range of the data.
2. Estimate the number of cells by using Table 2.2.
3. Determine the cell interval by dividing the range by the estimated number of cells. (Some rounding and the use of different trial "numbers of cells" are often necessary in this step.)
4. Construct a frequency distribution data table. The data may be ungrouped as in Table 2.3, or grouped as Example 2.2 illustrates.
5. Scale the axes, and draw the bars of the histogram.

EXAMPLE 2.2

A screw manufacturer utilizes a multiple-spindle screw machine to manufacture from 1,000 to 2,000 screws per hour. A random sample of screws from one of the machines results in the data shown in Table 2.4. The maximum value of the quarter-inch screws is ¼ in. or 0.2500 in. (the USL). The minimum value (LSL) is set at 0.2465 in. A process that produces 1,000 to 2,000 screws per hour could generate quite a bit of scrap. Construct a histogram of the process to determine whether or not the process is adjusted correctly, so that no scrap results.

TABLE 2.4

**Diameter of Unthreaded Portion of Shank (¼-in. screws), $n$ = 80**

| | | | | | | | |
|---|---|---|---|---|---|---|---|
| 0.2493$^{x}$ | 0.2492 | 0.2479$^{o}$ | 0.2494$^{x}$ | 0.2479 | 0.2482 | 0.2485 | 0.2480$^{o}$ |
| 0.2492 | 0.2491 | 0.2495$^{x}$ | 0.2485 | 0.2491$^{x}$ | 0.2488 | 0.2492 | 0.2490 |
| 0.2486 | 0.2492 | 0.2484 | 0.2490 | 0.2480 | 0.2486 | 0.2484 | 0.2486 |
| 0.2486 | 0.2490 | 0.2483 | 0.2485 | 0.2490 | 0.2478$^{o}$ | 0.2494 | 0.2485 |
| 0.2492 | 0.2489 | 0.2484 | 0.2476 | 0.2478$^{o}$ | 0.2487 | 0.2492 | 0.2488 |
| 0.2484$^{o}$ | 0.2487 | 0.2489 | 0.2483 | 0.2485 | 0.2496$^{x}$ | 0.2488 | 0.2485 |
| 0.2491 | 0.2489 | 0.2488 | 0.2482 | 0.2481 | 0.2484 | 0.2496$^{x}$ | 0.2487 |
| 0.2487 | 0.2493$^{x}$ | 0.2485 | 0.2489 | 0.2483 | 0.2483 | 0.2484 | 0.2487 |
| 0.2485 | 0.2486$^{o}$ | 0.2494 | 0.2487 | 0.2485 | 0.2489 | 0.2482$^{o}$ | 0.2491$^{x}$ |
| 0.2488 | 0.2480 | 0.2489 | 0.2476$^{o}$ | 0.2486 | 0.2495 | 0.2489 | 0.2488 |

*Solution:* The measurements in Table 2.4 seem almost incomprehensible. To begin to understand the data and the resultant distribution, code the data so the 0.24 is understood. The data are then coded from 0.2400.

*Step 1:* The highest and lowest datum in each column is marked with a superscript $X$ or $O$, respectively, so the range of the data may be determined more easily. From the table, the range of the data is calculated as follows:

$$R = \text{high} - \text{low} = 96 - 76 = 20$$

*Step 2:* Since the number of datum points ($n$) is 80, the number of cells will be approximately seven (refer to Table 2.2).

*Step 3:* The cell interval is found by

$$\text{Cell interval} = \frac{\text{Range}}{\text{Number of cells}} = \frac{20}{7} = 2.86$$

The cell interval is rounded to three ten-thousandths, the accuracy of the data values.

*Step 4:* As an interim step to final preparation of the histogram, a frequency distribution of the "ungrouped" data is shown in Table 2.5.

Obviously, this frequency distribution (the tallies in Table 2.5) is too flat to yield any kind of shape information. The data will next be grouped by every three ten-thousandths of an inch, as was determined by previous calculation. To eliminate any question of where a data point falls, upper and lower **cell boundaries** are established. The cell boundaries, often referred to as **class limits,** are usually recorded as one-half unit greater accuracy than the data. In other words a five is added as a final digit. Also, as may

TABLE 2.5
**Frequency Distribution of ¼-in. Screw, Unthreaded Shank Diameters**

| *Measurement (Coded to 0.2400 in.)* | *Tally* | *Cell Frequency* |
|---|---|---|
| 76 | II | 2 |
| 77 | | 0 |
| 78 | II | 2 |
| 79 | II | 2 |
| 80 | III | 3 |
| 81 | I | 1 |
| 82 | III | 3 |
| 83 | IIII | 4 |
| 84 | ~~IIII~~ I | 6 |
| 85 | ~~IIII~~ IIII | 9 |
| 86 | ~~IIII~~ I | 6 |
| 87 | ~~IIII~~ I | 6 |
| 88 | ~~IIII~~ I | 6 |
| 89 | ~~IIII~~ II | 7 |
| 90 | IIII | 4 |
| 91 | IIII | 4 |
| 92 | ~~IIII~~ I | 6 |
| 93 | II | 2 |
| 94 | III | 3 |
| 95 | II | 2 |
| 96 | II | 2 |
| | | 80 |

TABLE 2.6
**Grouped Frequency Distribution of ¼-in. Screw, Unthreaded Shank Diameters (Coded from 0.2400 in.)**

| *Cell Boundaries* | | | |
|---|---|---|---|
| *Lower* | *Upper* | *Cell Midpoint* | *Cell Frequency* |
| 75.5 | 78.5 | 77 | 4 |
| 78.5 | 81.5 | 80 | 6 |
| 81.5 | 84.5 | 83 | 13 |
| 84.5 | 87.5 | 86 | 21 |
| 87.5 | 90.5 | 89 | 17 |
| 90.5 | 93.5 | 92 | 12 |
| 93.5 | 96.5 | 95 | 7 |
| | | | 80 |

be noted from Table 2.6, the **cell midpoints** (the middle value of each cell) are accurate to the accuracy of the data—one ten-thousandth. *This configuration occurs only when cell intervals are defined in odd-numbered units of the data*—three ten-thousandths in this case.

*Step 5:* The axes of the histogram may now be scaled and the vertical bars drawn up to the frequency of that particular cell. In Figure 2.8 the cell midpoints (often referred to as class marks) are used to describe the value of the cells. (Computer software allows the user to select midpoints or to select upper or lower cell boundaries to use in scaling the *X* axis.)

Now that we have the histogram, it can be easily established that the process is well centered. The clifflike appearance on the left side of the distribution indicates that an assignable cause might be operating. Also, the process variability is too wide. A quality team could study the process and recommend improvement in machinery, materials, measurement, methods, environment, and/or people (MMMMEP).

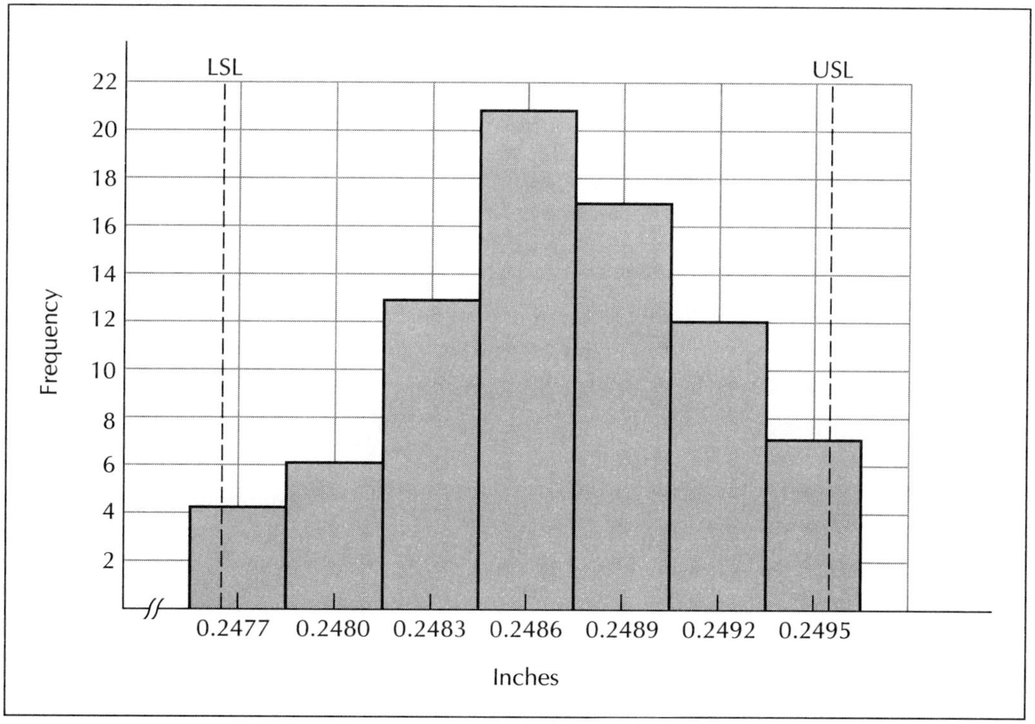


**Figure 2.8**
A histogram of shank diameters (unthreaded portion) of ¼-in. screws.

Care must be taken in graphing a quality histogram; the practice is as much an art as a science. You will benefit most by practice and by in-class discussion with your instructor. However, once the histogram is completed, the results are powerful. Interpretations of distributions can lead to important decisions for your company, especially if the company purchases parts from a variety of vendors.

## *Section 2.3 Review Problems*

1. A bar chart showing the number of times a given quality measurement occurs is called a
   a. nonlinear function
   b. histogram
   c. normal curve
   d. Pareto diagram
2. To subtract the lowest data point in a distribution from the highest data point is to determine the ______________________ of the data.
3. The width of an individual cell in a histogram is known as the cell ______________________.
4. The middle value of a cell is known as the cell ____________ or class ____________.

# 2.4 *Troubleshooting with Histograms*

Histogram shapes may be used by the quality manager or technologist to solve many problems. A few of the commonly occurring shapes of these distributions appear in Figure 2.9. In the figure, LSL stands for lower specification limit, and USL stands for upper specification limit.

A discussion of, and troubleshooting suggestions for, each of the figures follows:

*Figure 2.9a*—This figure depicts a normally distributed process. The **normal distribution** is bell shaped. The bell curve is usually found when a product can be measured along a continuous scale of measure-

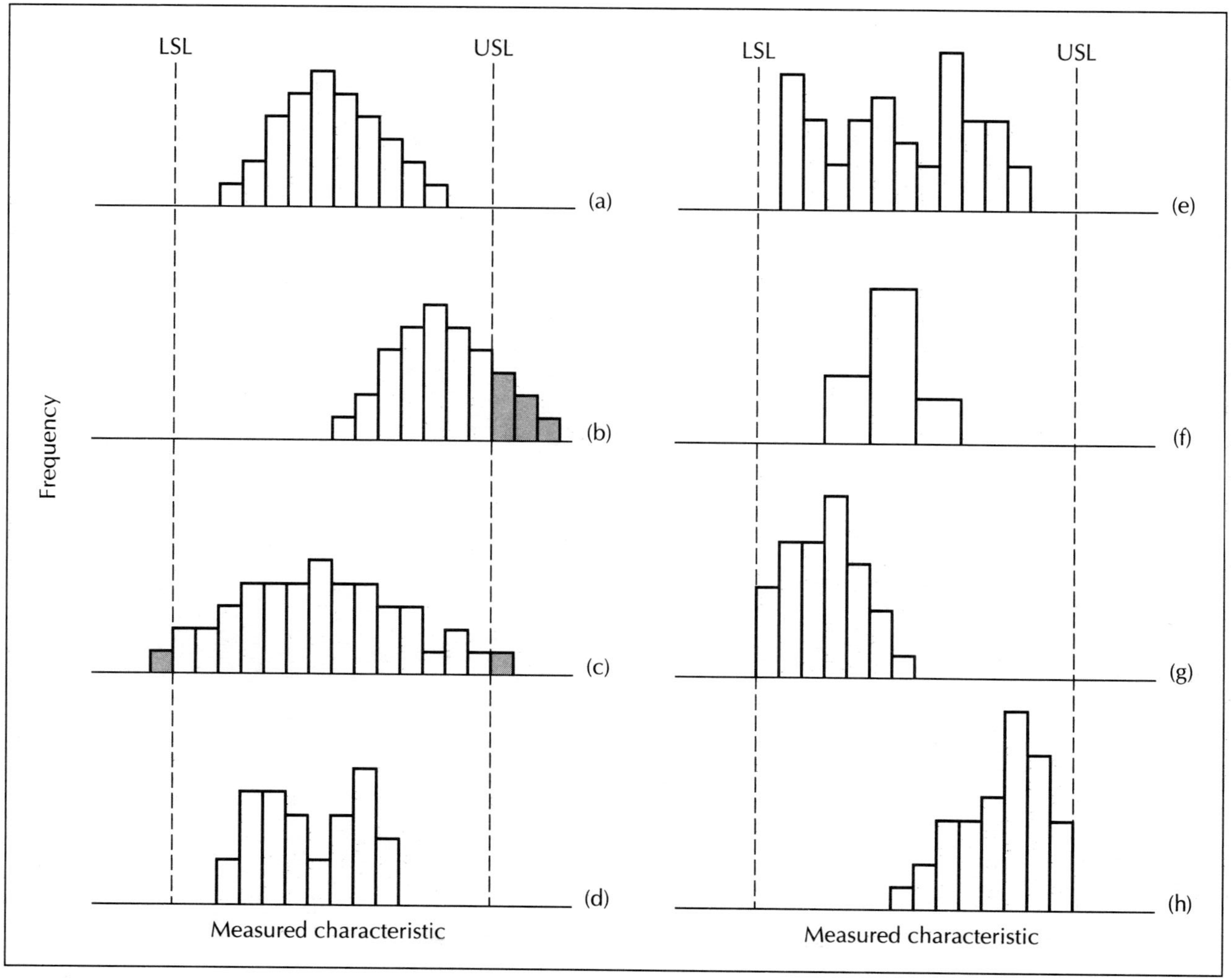


**Figure 2.9**
Commonly occurring process distributions.

ment (definition of variable data). Experience has revealed that many naturally occurring populations—for example, the heights of men, weights of women, IQ scores, or the lengths of screws from an industrial process—take the shape of a normal or bell curve. The normal curve is familiar and desirable to the quality control technologist, for it means a process is changing in a random manner (no assignable causes operating—see Chapter 4). Random variation is due to **chance causes,** factors—generally numerous and individually of relatively small importance—that are not easily identifiable. An **assignable cause** leads to relatively large variation, a cause that should be easily identified. When no assignable causes are operating, the process is predictable over time, and it can be depended on to produce similar parts over time. The desirability of a normal and repeatable process will be discussed further when Figure 2.10 is described.

*Figure 2.9b*—Again, the process is normally distributed and predictable. Notice that the variability of the process is the same as the distribution in Figure 2.9a. Now, however, the process is not centered. If this is a screw machine and the quality characteristic is diameter of shank, some of the screw diameters would be out of specification. The process, when centered, will yield parts within specification.

*Figure 2.9c*—This distribution has no concentration of part characteristics around the center. The cause of such a process could be that many things are changing as parts are produced—many assignable causes are operating. Flatness and peakedness are known as **kurtosis,** and a distribution displaying kurtosis is not as desirable as a normal curve in most processes.

*Figure 2.9d*—This distribution is well centered, but it has two peaks—or modes. *The mode of a distribution is defined as a quality characteristic that occurs most often in a distribution.* The normal curve has only one mode, which indicates that only chance causes are operating. This bimodal distribution indicates that two different causes are operating as the goods are produced. Examples of differing causes are as follows: different work shifts have produced the parts; different methods over time were used to produce the parts; parts are produced on two machines; or the parts have been made of two different materials. For instance, if weight is the quality characteristic, two suppliers delivering materials of different density to the plant results in two modes of weight of product on the histogram in Figure 2.9d.

*Figure 2.9e*—As you have already figured out from the preceding case, the trimodal distribution indicates that three different causes are operating. The multimodal distribution—more than two peaks—offers the quality technologist an opportunity to solve a problem that will increase production of acceptable product. All the technologist has to do is to find which of the nested distributions is the most acceptable one. Then, if the difference is machines, find which machines produce the most acceptable product, and adjust or somehow make the other machines work the same.

*Figure 2.9f*—This distribution also displays kurtosis. In this case, it has too sharp a peak. One of the assignable causes that could be operating to produce this distribution is that the measuring instrument lacks the necessary precision or discrimination. In other words, the characteristic cannot be measured precisely enough to distinguish more than a couple of different cells or measurements.

*Figure 2.9g*—Distributions that have an elongated tail on one side are said to display **skewness.** This distribution is described as positively skewed or skewed right (i.e., the tail is to the right). Positively skewed distributions are caused either by a measurement that could end in zero, such as a flatness measurement or total indicator readout (TIR), or by sorting. **Sorting** of product results when 100 percent inspection is performed on a product and the nonconforming product is taken out. A supplier that sorts product is a supplier that is unable to work to target. Costs for this supplier's product will likely remain the same or increase.

*Figure 2.9h*—Only sorting can be blamed for a negatively or left-skewed distribution. Buying from this supplier is probably not a good idea, unless the supplier will attempt to improve the process. Another supplier should be found that will provide the desirable characteristic in a more normally configured distribution.

Finally, the distributions shown in Figure 2.9g and h might indeed be normal if product were not sorted. The distribution in Figure 2.9h could be improved by merely centering the process, thus reducing costly inspection and scrap or rework.

EXAMPLE 2.3

Your company has contracted with a new supplier of motor shafts. The first parts received produced the data in Table 2.1 that resulted in the histogram in Figure 2.7. All shafts are within specification, but your manager wants to know whether or not the supplier is working to target (using SPC techniques) or to specification (could involve 100 percent inspection and sorting). The answer will determine whether the supplier can continually improve and drive down manufacturing costs. As discussed in Chapter 1, future competitiveness depends on the capability of the workforce to continuously improve processes.

The specification on the part is 25.5 ± 0.4 mm. What will your report show the new supplier's technique (SPC or sorting) to be? What could you recommend for the manager to suggest to the supplier concerning the process?

*Solution:* You are working with a QC team on the problem, and you first ask the incoming inspector, who selected the sample, if it was done randomly; for instance, were parts selected from the top, middle, and bottom of the lot's container? You next verify that measurements were taken with an instrument capable of a precision of 0.1 mm.

After assuring yourself of appropriate data collection, you draw a histogram (Figure 2.7). You see that the histogram is skewed negatively with a clifflike appearance on the right. A clifflike appearance on the left might indicate that zero-based measurements have been taken (e.g., TIR). The sharp drop-off is on the right, so a zero-based measurement is not the case in your study. The indication is that 100 percent inspection has occurred and the lot has been sorted. The company will never achieve the full capability of the process, quality will remain the same, and the cost of the parts can never become less.

Your quality team suggests that the supplier attempt to bring the process under control by using control charts (Chapter 4). After assignable causes have been detected and eliminated, the histogram of data collected on the process characteristic should appear normal. Once the process is in statistical control, and only then, can the process variability be reduced.

The distributions in Figure 2.10 are simplified representations of actual histograms. The simplification involves smoothing the edges of the histograms used previously and drawing a smooth curve through the midpoints of the cells of the histogram. The resultant graph is known as an *ogive,* pronounced "ohjive."

Returning now to the desirability of the normal curve, Figure 2.10a shows a process at first normal, but then changing over time. The centers of the distributions also change. This change in form occurs when assignable causes are operating. Can this kind of process be repeatable? No, it cannot.

Figure 2.10b depicts what happens over time when the process is normal and repeatable. *A normal process has no assignable causes*

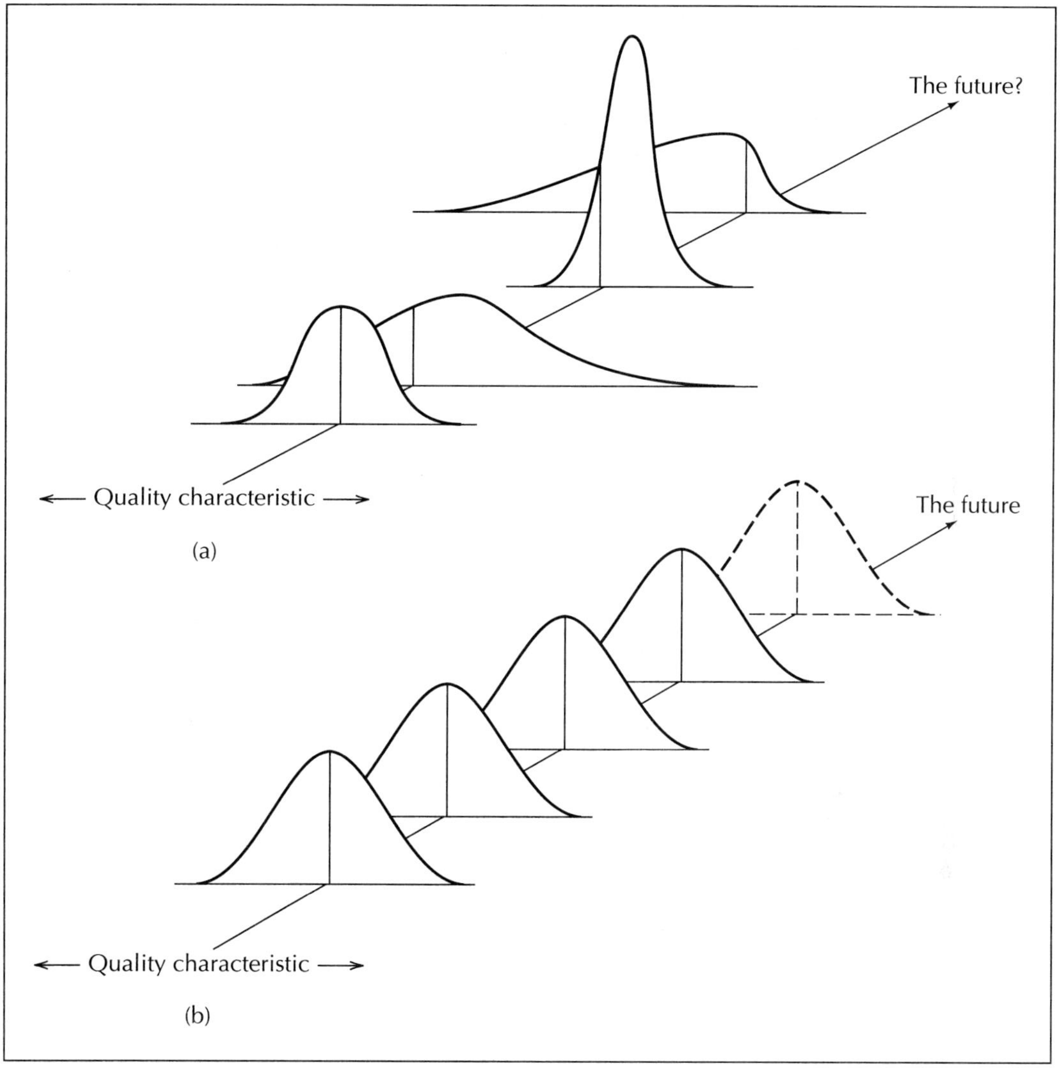


**FIGURE 2.10**
(a) These process distributions vary, making the process unpredictable; (b) these process distributions are normal, repeatable, and predictable.

*operating—only chance causes—and leads to increased repeatability.* When only chance causes affect a process, the natural randomness of the data usually yields the normal curve. This kind of process can be predicted in the future. It is interesting to note that *if normal and repeatable processes occur, variability in those processes can only be reduced by management acting to reduce the variability.* (This fact will be discussed at length in later chapters.) As an example, management may have to purchase a new machine, machining center, or batch-processing system.

## Section 2.4 Review Problems

1. The *normal distribution* is said to be ______________ shaped.
   a. rectangular
   b. circular
   c. columnar
   d. bell
2. Chance causes are due to ______________
   (random, nonrandom) variation and are relatively __________ (small, large).
3. Assignable causes are due to ______________
   (random, nonrandom) variation and are relatively __________ (small, large).
4. A distribution that is skewed positively
   a. has a tail to the right.
   b. has a tail to the left.
   c. has no tail.
   d. has a tail on the $Y$ axis.
5. Are assignable or chance causes operating when a process's distribution changes in form over time?

# 2.5 *Pareto's Principle*

The Pareto principle is named after the Italian economist Vilfredo Pareto. He found that approximately 80 percent of Italy's wealth was held by 20 percent of the population. Hence, the Pareto principle is often referred to as the 80/20 rule. Many others refer to it as "the vital few versus the trivial many." Industrial processes behave like the wealthy in Pareto's nineteenth-century study. That is to say, "Twenty percent of the process quality characteristics cause 80 percent of the problems with a product or service."

EXAMPLE 2.4

The data for one month's production shown in Table 2.7 were collected by a manufacturer of plastic film. The plastic film is used to wrap foods and must be clear or transparent (opacity), be of the correct thickness (gauge), be chemically homogeneous (carbon content), and not possess any other visible nonconformities. The data also include scrap due to "other causes." Notice also that some of the types of failures show zero for this month. They had resulted in scrap in previous months.

TABLE 2.7
**Scrap Amounts for Plastic Film, January**

| *Type of Failure* | *Amount of Scrap (kg)* | *Percent of Total Amount* |
|---|---|---|
| 1. Air bubbles | 500 | 0.7 |
| 2. Bubble breaks | 19,650 | 29.3 |
| 3. Carbon content | 0 | 0.0 |
| 4. Flatness | 3,810 | 5.7 |
| 5. Gauge | 27,600 | 41.1 |
| 6. Opacity | 0 | 0.0 |
| 7. Other | 600 | 0.9 |
| 8. Scratches | 3,840 | 5.7 |
| 9. Trim | 500 | 0.7 |
| 10. Wrinkles | 10,650 | 15.9 |
| Totals | 67,150 | 100.0 |

What types of failure should the manufacturer attempt to improve first?

*Solution:* The **Pareto diagram** in Figure 2.11a clearly depicts the data in Table 2.7. Note that the graph is like a histogram, a bar graph with frequency of occurrence scaled on the $Y$ axis. The only major difference is that the types of failures are graphed in descending order of occurrence. (You should verify this fact from Table 2.7.) A clear picture of troubleshooting priorities, or of processes that must be improved, results. Table 2.7 also shows that 20 percent of the processes—that is, bubble breaks (2) and gauge (5)—result in 70.4 percent of the failures. To reduce scrap significantly, gauge and bubble breaks must be reduced.

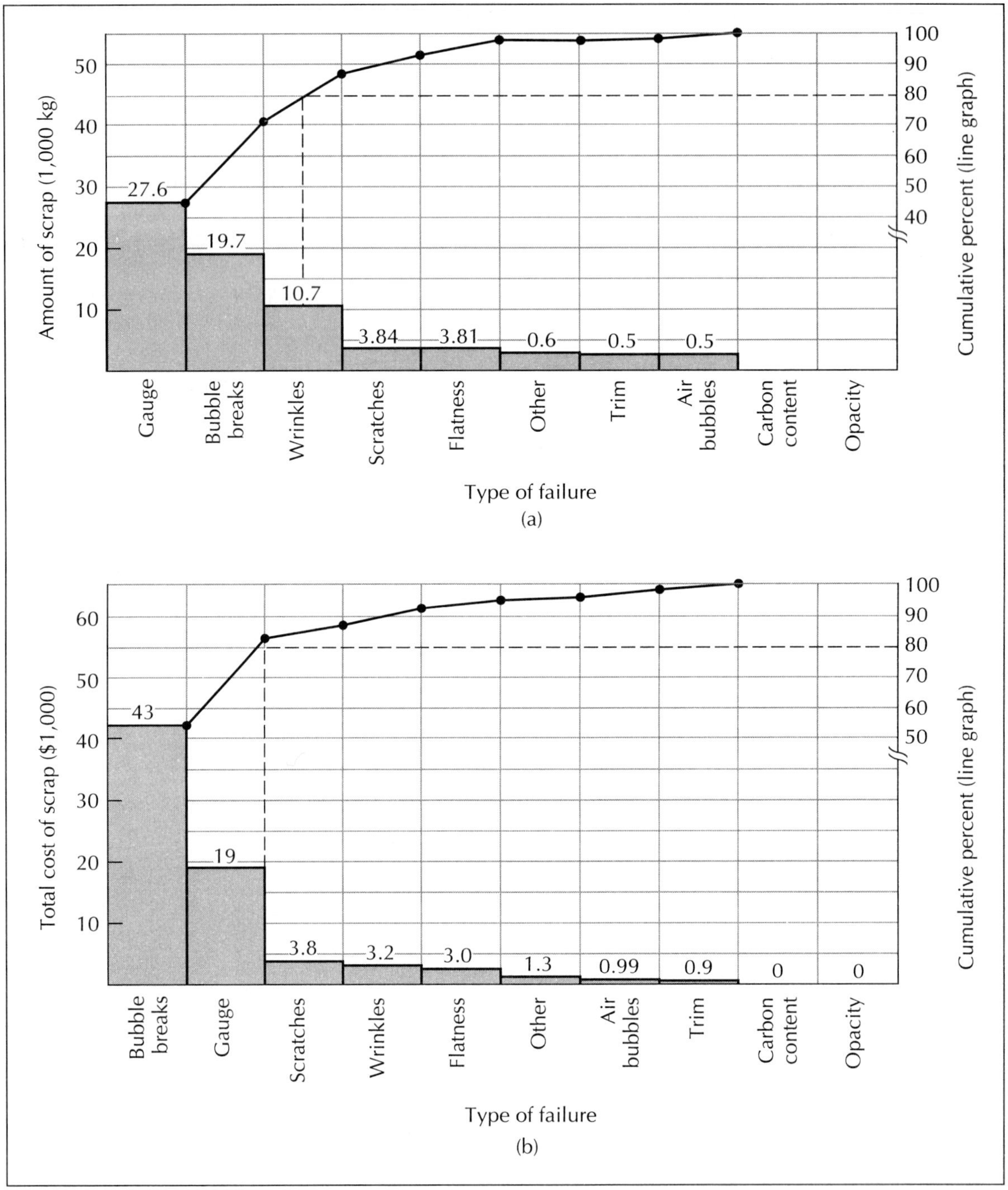


**FIGURE 2.11**
The Pareto diagram in (a) reflects amount of scrap due to type of failure. In (b) cost is tabulated for type of failure.

A summary of the steps necessary to construct Pareto diagrams follows:

1. Collect the data. Always use the same units of measure. In this case, the data are displayed in units of kilograms. In Figure 2.11b they are converted to U.S. dollar units.
2. Rank the quality characteristics by magnitude—of cost or scrap—from large to small.
3. A **cumulative frequency distribution** is a graph of the total frequency of all cells below the upper cell boundary of any individual cell (see the "Cumulative Percent" column in Table 2.8). Draw a line graph displaying the cumulative distribution in percent of total. The dashed lines on the Pareto diagrams indicate the point at which 80 percent of the problems are accounted for.
4. Complete the diagram using conventional graphing techniques.

To construct the Pareto diagram in Figure 2.11b, Table 2.8 was constructed. The types of failure are already ranked in descending order. Note that the ranking by cost varies from the ranking by amount of scrap. In some studies costs are proportional to the amount of nonconforming material. But sometimes, especially if rework is an option, a Pareto's main cause of loss will change between the amount of rework and the cost of the rework. In all cases, *it is best to work with the accounting department in determining cost reporting with Pareto analysis* (see Chapter 10).

TABLE 2.8
**Scrap Costs ($) for Plastic Film, January**

| *Type of Failure* | *Cost of Scrap ($/1000 kg)* | *Total Cost* | *Percent of Total Amount* | *Cumulative Percent* |
|---|---|---|---|---|
| 2. Bubble breaks | 220 | 43,230 | 57.0 | 57.0 |
| 5. Gauge | 70 | 19,320 | 25.5 | 82.5 |
| 8. Scratches | 100 | 3,840 | 5.1 | 87.6 |
| 10. Wrinkles | 30 | 3,195 | 4.2 | 91.8 |
| 4. Flatness | 80 | 3,048 | 4.0 | 95.8 |
| 7. Other | 220 | 1,320 | 1.7 | 97.5 |
| 1. Air bubbles | 220 | 990 | 1.3 | 98.8 |
| 9. Trim | 180 | 900 | 1.2 | 100.0 |
| 3. Carbon content | 0 | 0 | 0.0 | 100.0 |
| 6. Opacity | 0 | 0 | 0.0 | 100.0 |
| Total cost of scrap | | 75,843 | | |

In this example the "cost of scrap by type of failure" varies from the "amount of scrap by type of failure." This difference occurs because the scrapped rolls may be sold to discount dealers, and the price of the rolls of materials varies by the type of nonconformity.

The two obvious areas to work on are gauge and bubble breaks. These two categories are reversed between the cost and amount studies, but they explain almost 80 percent of the problems in both studies. The bottom line is that *it is usually easier as well as more cost-effective to reduce a large category than to reduce a small category.*

## *Section 2.5 Review Problems*

1. A bar chart showing the number of occurrences of different types of failures in a system or process is called a
   a. nonlinear function
   b. histogram
   c. normal curve
   d. Pareto diagram
2. Vilfredo Pareto determined that ______________ (20, 80) percent of the quality characteristics generating scrap or rework on any given product or service are caused by ______________ (20, 80) percent of the processes.
3. What are the two ways that the types of failures can be ranked in a Pareto analysis?

## *Summary*

- The effective quality technologist will construct and interpret line graphs; understand relationships such as positive, linear, and inverse; scale the $X$–$Y$ coordinate system; and identify dependent and independent variables.

- Scatter diagrams offer a simple and effective way to interpret the results of a controlled experiment. As one variable—the independent variable—is changed on the $X$ axis, the technologist will plot the dependent variable on the $Y$ axis.
- The histogram is a frequency plot on an $X$–$Y$ coordinate system. The quality technologist must know the basic steps in constructing the histogram: determining the range of the data, the cell interval, coding of the data, and scaling the $X$–$Y$ axis.
- The histogram provides vital information regarding the shape of the data distribution. The act of troubleshooting a process is greatly aided by examining these graphical shapes in terms of normality, skewness, or kurtosis.
- Working to target rather than specification is the only way to continuously improve a process. Normally distributed data from a measurable quality characteristic are an indication that the characteristic and process are capable of continuous improvement.
- The Pareto chart offers a simple tool for problem solving. By ranking the problem areas by cost or number of units affected, the QC technologist or manager can prioritize which problems to deal with.

## CASE STUDY · CHAPTER 2

Consider that you have been selected as the team leader to solve a problem concerning a manufacturing operation. Your company's product is an automobile component, a manifold, that your customer, the auto assembler (e.g., Ford, GM, Chrysler), must mount on an engine block. Standoffs (Figure 2.12) are supplied to you by a second-tier supplier. The standoffs are used to mate the manifold with the block. The problem is with the inner diameter (ID) of the standoffs. Your supplier, the manufacturer of the standoffs, is holding to specification, but the standoff's lower specification limit allows too close a fit over the bolts on the block for easy assembly. (The specification limits of the ID are 0.495 to 0.525 in., and the stud that the part is required to fit over is a ½-in. stud with a 0.495 in. major diameter.) Your customer, the assembler, is having difficulties mounting the manifold to the engine because of the tight clearance on some standoffs. Your customer wishes that the problem be rectified as soon as possible.

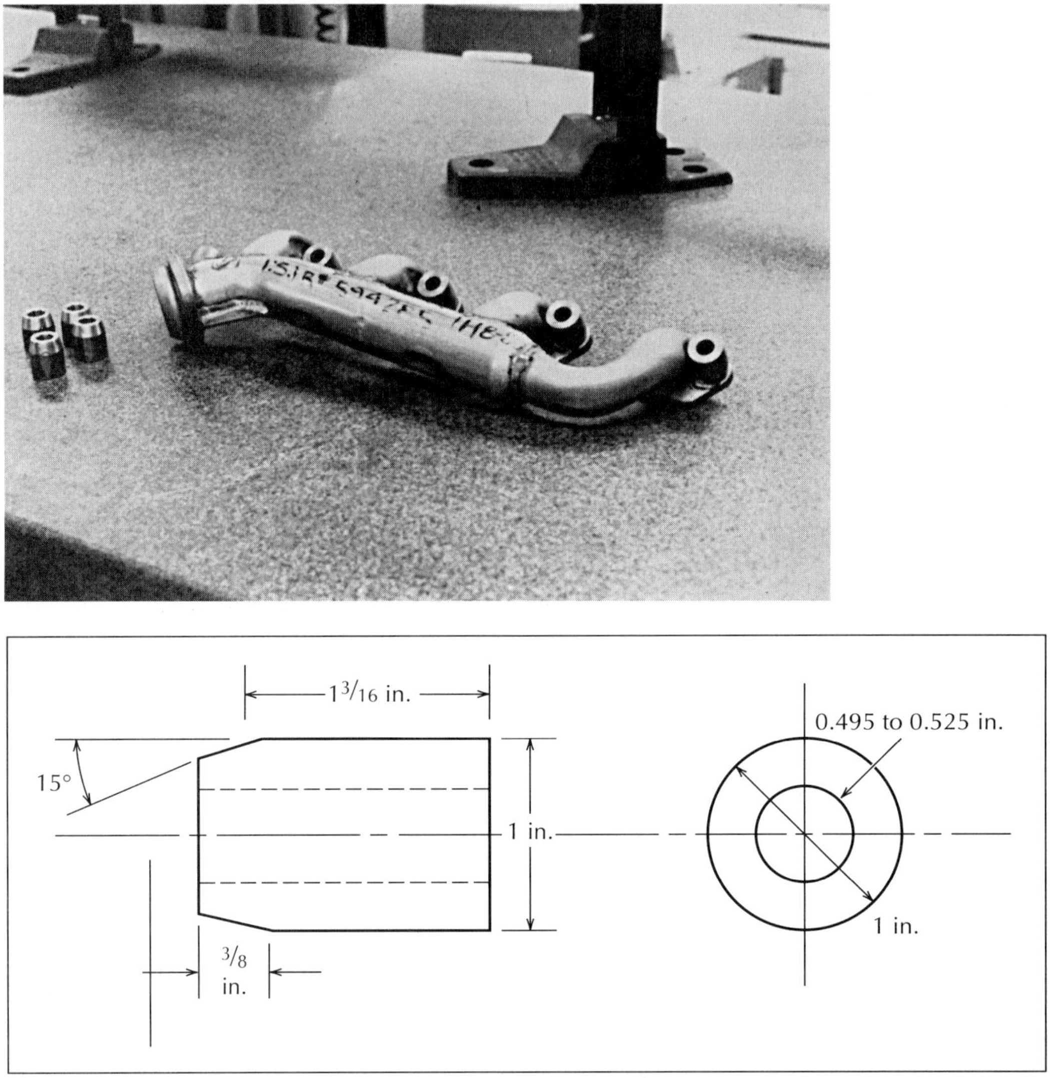


**FIGURE 2.12**
The standoffs—on left in photo—are used to guide the manifold onto studs on the engine block during assembly.

The quality assurance (QA) manager has just informed you of the assignment. He asks that you take a couple of days to prepare a plan with a list of needs to implement the plan. You know the QA manager to be very open to ideas, a student and practitioner of good quality management principles, and someone who will act on your ideas (give you empowerment). What will be your recommendations?

Effective problem solving is made easier with an outline of recommended activities that may be followed in reporting to the QA manager. As team leader, you will use the Shewhart cycle (Chapter 1) as your outline. Your plan should include such items as a time line, resources that are needed (gauges, flip charts, meeting room), how the data will be collected, how to define nonconforming parts (defective standoffs) operationally, and who should be involved in the project—such as operators, supervisors, sales and marketing personnel, purchasing, and maintenance.

Your proposal to the QA manager is as follows:

*Step 1: Project Selection.* Incoming inspection data on the inner diameter of the standoffs confirms the supplier to be in compliance with present specifications. The problem is that the specification limits are too wide, allowing the smaller diameter standoffs produced to be too tight a fit over the studs in the engine block for easy assembly. The project will involve a sample of standoffs from the supplier. Plant personnel will attempt a second drill operation on all standoffs in the sample—a reaming operation is too expensive and the surface smoothness of the hole is not critical to assembly. The drill size proposed is 33⁄64 in. (or standard metric, 13.2 mm), assuring an easy fit over the ½-in. (13-mm) studs during assembly.

The project team will consist of an inspector, a manufacturing engineer, the supervisor of the department installing the standoffs on the manifold, a purchasing agent, and a machinist. The quality training classroom will be scheduled for team meetings.

*Step 2: Project Implementation.* The team will prepare a flowchart of the proposed operation. The flowchart will include such process steps as follows: collecting the sample (random sampling techniques), moving the sample to the drill press, the drill operation, and inspection after drilling.

Inspection procedures will be rigorously defined. All personnel will agree on what constitutes the acceptable ID of the standoff and how it is to be gauged.

A control chart will be used to check the process, over time and during the drill operation, in order to ascertain the wear of the drill. For instance, how many parts may be drilled before a drill change is required?

Cause-and-effect diagrams will be used as needed.

*Step 3: Process Evaluation.* Final inspection will use a histogram to display the spread of the data. Capability of the process, with a reduced spread in the specification limits, will be determined from the histogram.

Cost data on the second drill operation will be collected.

*Step 4: Continuing Improvement.* Recommendations will be made by the purchasing agent (a member of the project team) to the supplier. The supplier will be given a tighter dimension on the ID (on the higher side of the present lower dimension) and asked to perform the second drill operation. The results of the study will show how much cost should be added for the extra drill operation.

This proposal, if accepted, should result in less scrap and rework for your company. Also, your company's customer—the assembler—will experience easier assembly of the manifolds. All of the steps proposed will be reviewed and modified as needed in the first team meeting.

The project, as outlined, should be accomplished within four to six months. Of course, it may be completed sooner if it becomes a high enough priority that other duties for personnel involved are reduced or eliminated.

## Chapter Review Problems

1. The following data were recorded during a pump test. Graph the data using conventional graphing techniques as explained in the text. Be sure to place the independent variable on the $X$ axis.

**Table for Problem 1**

| *Revolutions per Minute (rpm) of Pump* | *Gallons per Minute (gpm) of Pump Output* |
|---|---|
| 0 | 0 |
| 100 | 0.5 |
| 220 | 1.4 |
| 310 | 2.6 |
| 400 | 4.2 |
| 498 | 6.0 |

2. From the graph in problem 1, find the pump output at 250 rpm. Also, describe the function as linear or nonlinear.
3. Use your statistical calculator to verify the coefficient of correlation already calculated for the data in Example 2.1. First, design and complete a data table from the graph in Figure 2.6 (the first point is $X = 2.05$ and $Y = 0.47$). Your statistical calculator will also calculate the following: the slope of the function, the $Y$ intercept, and predicted values for $X$ or $Y$, given the other value. There are 25 points to enter, so be careful and enter the data "right the first time."

4. Repeat problem 3 by using the following equation for calculating the linear coefficient of correlation:

$$r = \frac{n(\Sigma xy) - (\Sigma x)(\Sigma y)}{\sqrt{n\Sigma x^2 - (\Sigma x)^2}\sqrt{n\Sigma y^2 - (\Sigma y)^2}}$$

To accomplish this formidable task, use the graph's even-numbered points only. Prepare and complete a table of all necessary quantities, such as the sum of the $X^2$ and $Y^2$ values. This exercise should result in a greater appreciation for the automatic functions on the statistical calculator.

5. List the steps involved in creating a quality histogram. Be sure to mention where repeated steps may be necessary.

6. Milk producers are regulated by container fill weights. Construct a histogram of the data in the table (weight in grams of 85 samples of product plus container). Determine whether or not the producer is sorting out overfilled containers.

**Table for Problem 6**

| | | | | | | | | | |
|---|---|---|---|---|---|---|---|---|---|
| 185 | 182 | 183 | 184 | 183 | 179 | 187 | 176 | 186 | 184 |
| 186 | 181 | 182 | 180 | 185 | 188 | 186 | 186 | 180 | 183 |
| 187 | 182 | 180 | 184 | 187 | 183 | 186 | 183 | 187 | 185 |
| 179 | 186 | 180 | 178 | 186 | 187 | 183 | 185 | 180 | 186 |
| 180 | 183 | 181 | 186 | 184 | 185 | 184 | 179 | 183 | 185 |
| 183 | 182 | 179 | 186 | 177 | 184 | 181 | 181 | 185 | 182 |
| 183 | 184 | 186 | 181 | 184 | 186 | 180 | 183 | 184 | 183 |
| 183 | 178 | 187 | 185 | 180 | 185 | 185 | 182 | 183 | 179 |
| 182 | 181 | 186 | 187 | 182 | | | | | |

7. A physician recognizes that blood pressure measurements are variable, and statistically determines the mean or average data and the variability of the data for each of her patients. The hospital she is employed by has a computer, and the blood pressure data are collected by the computer. A computer-constructed histogram results for each patient. From the data in the table—measurements of the systolic blood pressure of one of her patients—construct a histogram.

It is interesting to note that all data are in even numbers. Nurses are instructed to collect data in even numbers, because most manometer gauges used in collecting blood pressure have a

**Table for Problem 7**
**Diastolic Blood Pressure for David**
**(mm of Hg, $n$ = 100)**

| | | | | |
|---|---|---|---|---|
| 88 | 84 | 80 | 86 | 82 |
| 86 | 90 | 78 | 78 | 80 |
| 76 | 82 | 80 | 82 | 76 |
| 82 | 92 | 82 | 80 | 86 |
| 88 | 84 | 84 | 92 | 90 |
| 82 | 88 | 74 | 80 | 88 |
| 76 | 86 | 82 | 84 | 82 |
| 84 | $76^o$ | 84 | 78 | 84 |
| 86 | 78 | 74 | 78 | 84 |
| 80 | 80 | $74^o$ | 78 | 84 |
| $98^x$ | 82 | 84 | 84 | 88 |
| 82 | 84 | 88 | 82 | 84 |
| $76^o$ | 90 | 82 | $76^o$ | 80 |
| 78 | 82 | 86 | 88 | 90 |
| 90 | 82 | 84 | 84 | 88 |
| 86 | 88 | 90 | 92 | 88 |
| 84 | 80 | $88^x$ | 84 | $98^x$ |
| 80 | $94^x$ | 90 | $88^x$ | 84 |
| 82 | 80 | 80 | 78 | $80^o$ |
| 84 | 82 | 86 | 78 | 84 |

discrimination (the distance between two lines on the scale) of two units. This practice will affect the number of cells and the cell limits to achieve a quality histogram.

8. Several process distributions are offered in Figure 2.13. For each, determine the following: (a) Are assignable causes operating? (b) If so, what are your diagnoses of the causes? (c) If only chance causes are operating, what, if anything, needs to be done to improve the process? If assignable causes are operating, give an example of at least one possible assignable cause. Use Figure 2.9 and the associated discussion of the figure to answer these questions.

9. A fast-food service industry wishes to compare the percentage of fat content of the meat from two suppliers. From the data in the table (page 72) construct histograms for suppliers A and B, and determine which supplier has the better quality (lower fat content is better of course) and which supplier is more likely to improve the fat removal process.

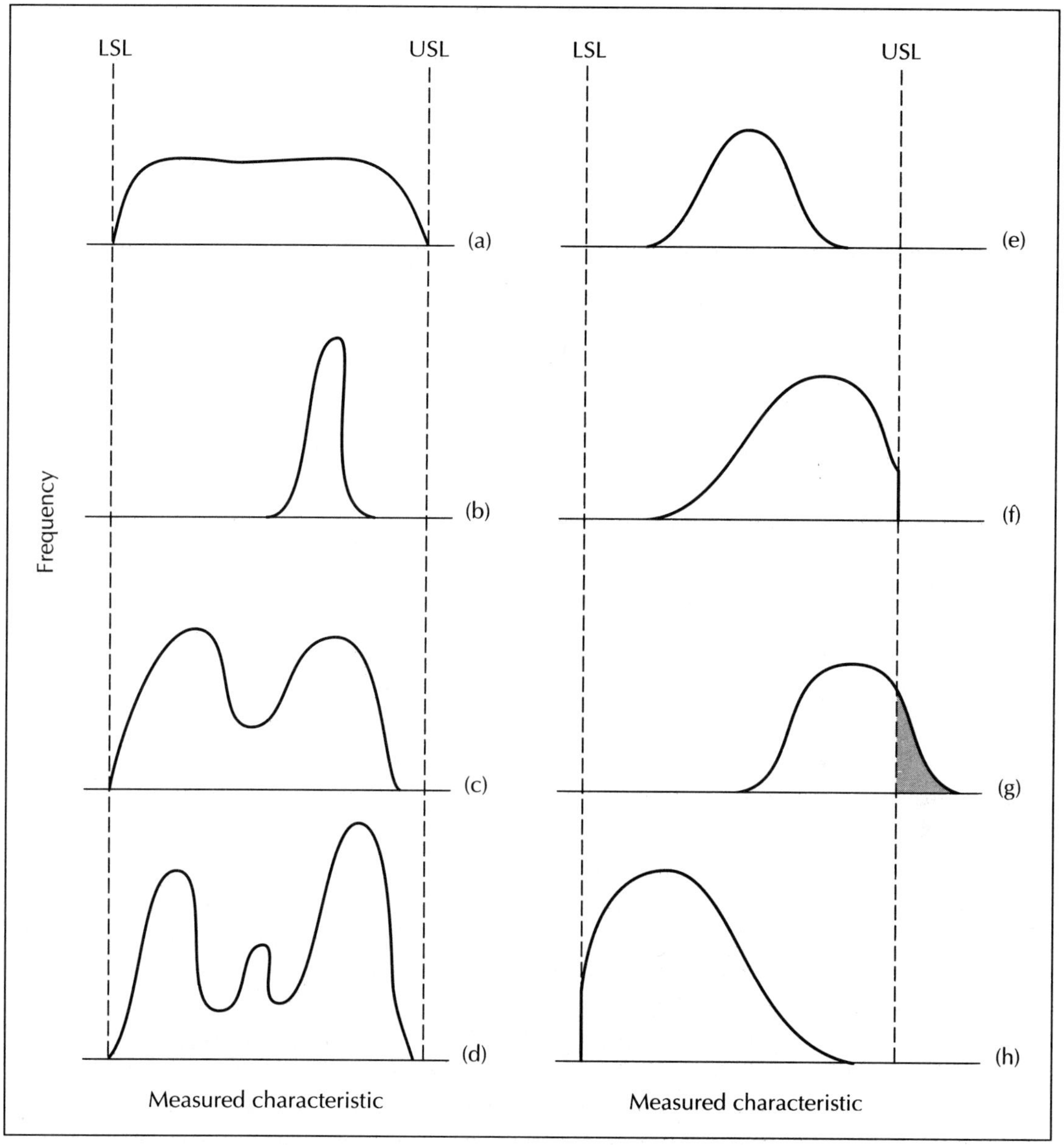


**FIGURE 2.13**
These smoothed-off histograms are to be used in answering the questions posed in problem 8. LSL = lower specification limit; USL = upper specification limit.

**Table for Problem 9**

| *Sample Measurement of Fat Content (%)* | *Frequency, Supplier A* | *Frequency, Supplier B* |
|---|---|---|
| 0.1 | 2 | 0 |
| 0.2 | 3 | 0 |
| 0.3 | 3 | 2 |
| 0.4 | 8 | 6 |
| 0.5 | 12 | 10 |
| 0.6 | 11 | 8 |
| 0.7 | 10 | 4 |
| 0.8 | 0 | 3 |
| 0.9 | 0 | 0 |
| 1.0 | 1 | 0 |

10. In your own words, describe in one or two paragraphs the importance of Figure 2.10.
11. Analyze the Pareto diagram in Figure 2.11. What is the top-priority problem based on the amount of scrap? What is the top priority based on the cost of the scrap or rework generated? Which problem would you choose to work on first? Justify your decision.
12. A milk producer found the following frequency of complaints on its skim milk: spoiled early, 10; top leaker, 20; out-of-date, 7; bottom leaker, 14; and all other causes, 5. Construct a Pareto diagram and interpret the results.
13. From an employee survey, a company categorized and compiled data describing the major obstacles to SPC implementation (Section 1.1, "1930—SPC"). The survey findings of what personnel felt were most serious were counted as follows: the overall organizational structure (too vertical), 43; lack of management commitment and presence, 22; the use of an unintegrated management-by-objectives (MBO) approach in upper-level management, 62; lack of feedback from the customer, 9; quality responsibility rests with QC department, 15; lack of cooperation from the union, 4; and lack of education and training of operators, 44. Construct a Pareto chart of the "Major Obstacles to Quality SPC Implementation."
14. From the Pareto diagram in problem 13, discuss in class the top three obstacles to SPC implementation and describe possible solutions.

# LABORATORY · CHAPTER 2
## Putting the Data into Consumable Form

Materials needed: Data from the Chapter 1 laboratory and graph paper.

Use the data collected in the experiment in Chapter 1 and draw the histogram. Write a conclusion describing the shape of the distribution and the actions that may be taken to obtain a more normal and less variable distribution.

# 3

# The Normal Distribution

**Central Tendency** 3.1
**Variability** 3.2
**Shape of Distributions** 3.3
**The Normal Curve** 3.4
**Working to Target** 3.5

## 3.1 *Central Tendency*

There are three methods for locating the center of a data distribution. These methods are referred to as measures of **central tendency.** Each method involves different techniques and may offer differing results. Only in the case of a normal distribution will all three measures be the same.

The focus of this text will be on sample statistics rather than population statistics. In statistical quality control work, the data collected from samples are used. The difference between a population and a sample is that a **sample** is a limited number of units taken from a large source. A **population** is the larger source from which the sample has been drawn. The sample is used to make predictions by calculating probabilities for the larger population.

The **mean** is the most frequently used measure of central tendency. (Some authors use the term "mean" when referring to a population and "average" when referring to a sample. This text will use the term "mean" for both.) The calculation is rather simple, and the formula is well known to most. It involves adding all of the numerical values in the distribution and dividing by the total number ($n$) of the values. The equation is as follows:

$$\overline{X} = \frac{X_1 + X_2 + X_3 + \cdots + X_n}{n} = \frac{\sum_{i=1}^{n} X_i}{n}$$

Where $\overline{X}$ (called $X$ bar) represents the mean; $X_1, X_2, X_3, \ldots, X_n$ represent specific data values; and $\sum_{i=1}^{n} X_i$ means the sum of all values of $X$, from the first value ($X_1$, or $i = 1$) to any number of additional data points ($X_n$, or $i = n$).

EXAMPLE 3.1

Find the mean age of the students attending a quality control class. Their ages are

18, 25, 45, 34, 23, 19, 28, 23, 31, and 17

*Solution:* The individual data values are first added, and the sum is then divided by $n$.

$$\overline{X} = \frac{18 + 25 + 45 + 34 + 23 + 19 + 28 + 23 + 31 + 17}{10}$$

$$= \frac{263}{10} = 26.3$$

And the mean or average of the data is found to be 26. Since ages are given only to the nearest year, the average age is accurate only to the nearest year.

The average seems to be somewhat high for this sample. The reason is that the mean of the data may be affected by extreme values (the 45-year-old in this case). The effect is most noticeable when dealing with small groups of data.

Another measure of central tendency is the median. It is calculated by ranking the data values from small to large. The **median** is the middle data value of the ranked distribution.

EXAMPLE 3.2 Find the median age of the students in Example 3.1.

*Solution:* We first rank the individual data values from small to large.

17, 18, 19, 23, 23,
25, 28, 31, 34, 45

In this case, as always when there is an even-numbered distribution, the middle of the group is between two data values. To complete the calculation, average the two—23 and 25—to find the median age to be 24:

$$\text{Median} = (23 + 25)/2 = 24$$

(For odd-numbered distributions simply select the number in the center of the distribution. For example, for 3, 5, 7, 9, 11, the median is 7.)

The *median* age of 24 seems to be more representative of this sample group than the *mean* value of 26. *The median is not affected by extreme values.*

The final measure of central tendency is the mode. The **mode** is the most frequently occurring data value.

EXAMPLE 3.3 Find the modal age, or the most frequently occurring age, of the students in Example 3.1.

*Solution:* The only age that appears more than once is 23. Therefore the modal age of the class is 23.

In this last example the mode is not particularly helpful. The mode is more appropriately used in describing larger distributions. Refer to the discussion of Figure 2.9d, Section 2.4, to better visualize this point.

To find the mean, median, and mode of grouped data requires an estimate, because all of the data in each cell are treated as the class mark or midpoint value of the cell. Example 3.4 demonstrates the methodology.

EXAMPLE 3.4 Find the mean, median, and mode of the grouped data in Table 3.1 from a sample of product. The data, the unthreaded diameter of ¼-inch screws,

TABLE 3.1
**Grouped Frequency Distribution of ¼-in. Screws, Unthreaded Shank Diameters (Coded from 0.2400 in.)**

| *Cell Boundaries* | | | |
|---|---|---|---|
| *Lower* | *Upper* | *Cell Midpoint* | *Cell Frequency* |
| 75.5 | 78.5 | 77 | 4 |
| 78.5 | 81.5 | 80 | 6 |
| 81.5 | 84.5 | 83 | 13 |
| 84.5 | 87.5 | 86 | 21 |
| 87.5 | 90.5 | 89 | 17 |
| 90.5 | 93.5 | 92 | 12 |
| 93.5 | 96.5 | 95 | 7 |
| | | | 80 |

have been taken from Table 2.6. Also see the histogram in Figure 2.8. Can you estimate the central tendency of the data from the histogram?

*Solution:* First, calculate the mean $\overline{X}$ by adding the products of $X_{mid}$ times the frequency of each cell. Then, divide by the total ($n$). Expressed mathematically, the equation is as follows:

$$\overline{X} = \frac{f_1X_1 + f_2X_2 + \cdots + f_nX_n}{n}$$

$$\overline{X} = \frac{\sum_{i=1}^{k} f_iX_{mid}}{n}$$

Where each $X$, in the first equation, is the class mark or midpoint of the cell. (The $X$s are depicted as $X_{mid}$ in the combined equation; $k$ is used rather than $n$ in the summation symbol of the combined equation to symbolize the number of cells.) Note that the exact value of the mean for the original data cannot be determined, but the mean for the grouped data will be very close.

The mean for the grouped data (refer to last two columns in Table 3.1) is

$$\overline{X} = \frac{(4 \times 77) + (6 \times 80) + (13 \times 83) + (21 \times 86) + (17 \times 89) + (12 \times 92) + (7 \times 95)}{80}$$

$$\overline{X} = \frac{6{,}955}{80} = 86.9$$

The mean of the uncoded data is 0.2487 inches after rounding to the significant figures of the original data. The data would not be rounded to the one significant figure of the denominator term (80), because the 80 is a **counting or statistical number** that does not affect the rounding of the data from the process.

As in the case of finding the $\overline{X}$ of the grouped data, finding the median of the grouped data also requires an assumption. If we *assume that the data are evenly distributed throughout the cell interval,* then the following equation may be used to approximate the median:

$$\text{Median} = L + \left(\frac{n}{f} \cdot i\right) = 84.5 + \left(\frac{17}{21} \cdot 3\right) = 86.9$$

To use the equation, first find the cell that must contain the median (refer to the last column of Table 3.1). In this case, the median value must be in the cell containing 21 data points, because the data below this cell added to the 21 units within the cell equals 48 units (more than one-half of the total in the sample of 80). Next, the number of data points below the median value ($n$) in the cell must be divided by the total number of data points within the cell ($f$). In this case the median value is the 40th unit. The first three cells contain 23 units, and $n$ is calculated by $40 - 23 = 17$. The ratio of $n$ to $f$ is then multiplied by the cell interval value of $i = 3$. The result is added to the lower limit ($L = 84.5$) of the median cell. This process is known as **interpolation.**

The median of the uncoded data is 0.2487 inches (after rounding to the significant figures of the original data), or 13 ten-thousandths less than the screw designation of ¼ inch. The median of the distribution, 0.2487, is equal to the mean of 0.2487, meaning that the distribution is as depicted in Figure 3.1c (page 85).

The mode is simply the cell having the greatest number of data points, or having the highest frequency. In this example the midpoint value of 86 is used. In this case, the distribution only has one mode ($f = 21$). The distribution is said to be **unimodal.** With the mode being slightly less than the mean, the distribution might be described as slightly negatively skewed (Figure 3.1a, page 85).

By far, the most important measure of central tendency is the mean. *Most statistical calculators will compute the mean or average of grouped distributions.* The products of the frequency and the class midpoints are entered and stored in memory. The calculator sums the products entered and computes the mean of the grouped data. Check your statistical calculator's owner's manual for instructions.

In conclusion, central tendency measures include the mean, median, and mode. The mean is the statistic used most frequently. Its calculation is understood by most business and industrial employees, and its value usually varies least from sample to sample. The main disadvantage of the mean is that it is affected by extreme values.

The median is not affected by extreme values and should be used as the preferred measure of central tendency to describe samples or populations that possess extreme values. Examples of distributions with extreme values are those describing income or age (Examples 3.1 and 3.2). The main disadvantages of the median include the difficulty in calculation (especially with grouped data) and the lack of sensitivity to extreme data points. In many statistical studies extreme data points needed to be included. Another distinct disadvantage is that company personnel might not understand the assumptions used and the characteristics of the median.

The mode is used with large samples or populations to describe the most frequently occurring cause or factor. Examples include number of children in the average family, most frequent cause of accidents, and most frequent type of nonconformance found on a particular business or industrial form. The main disadvantages of the mode is that some distributions are multimodal, or contain more than one mode. If multimodality is the case, the mode could not be used as a measure of central tendency.

## *Section 3.1 Review Problems*

1. What three different measures of central tendency may be used to describe a distribution of variable data?
2. State at least one strength and weakness for each of the measures of central tendency listed in problem 1.
3. Solve for the mean height of five men having the following heights:

   6′0″, 5′9″, 5′4″, 5′11″, 6′8″

4. Solve for the median height in problem 3.
5. Solve for the mean, median, and modal height of the people in your class. Solve for the same measures using age (you are free to lie about your age). Add in your instructor's age and recalculate. Add in (an estimate of) Deming's and/or Juran's age and recalculate. What happens to the various measures as extreme values are used?

# 3.2 *Variability*

Controlling the variability of a product is vital to the overall quality of that product. Product design that does not consider the variability of materials and manufacturing processes may—or will almost surely—result in processes that do not meet design specifications. For example, a plastics process that cannot produce parts to a given design strength could have too much variability in the temperature of the injection molder or excessive variability (lack of homogeneity) of the incoming material (plastic pellets).

All processes exhibit variability. If a process quality characteristic reveals little or no variability, it is due to the lack of sensitivity of the measuring instrument (Figure 2.9f with associated discussion). An investment in improved measurement equipment that can measure process variability may lead to improved process control and improved quality of product.

How is process or product variability described and studied statistically? The range as a measure of variability has already been discussed in Chapter 2. The equation for the range is $R = X_{\text{high}} - X_{\text{low}}$. The range is easy to calculate, and it is often quite adequate. One disadvantage of the range, however, is how greatly it is affected by extreme values. Take the simple data set consisting of the five numbers 1, 3, 5, 7, and 9. Using the range calculation, the variability of the sample is 8. However, if the 9 is changed to 12, the range becomes significantly larger (11), only because of a change in one datum. The most significant weakness of the range calculation is that data points in the middle of the sample are ignored. For instance, what if the numbers 3 and 7 in the example were both 5? The range would not change, even though the variability of the data would be less.

The variance and the standard deviation are more meaningful measures of variability. The same data used to calculate the range are used to demonstrate the necessary calculations. The sample **variance** ($s^2$) is defined as the measure of variability (dispersion) of a sample based upon the sum of the squared deviations from the arithmetic mean, divided by $n - 1$. (The quantity $n - 1$ represents the degrees of freedom of the sample, which results in an unbiased estimate of the population variance.) The equation is written as

$$s^2 = \frac{1}{n-1}\sum_{i=1}^{n}(X_i - \overline{X})^2$$

EXAMPLE 3.5

Find the variance and standard deviation for the sample data 1, 3, 5, 7, 9.

*Solution: Step 1:* Construct a table of the necessary intermediate calculations.

TABLE 3.2
**Preliminary Calculations for Variance and Standard Deviation**

| $X$ | $X - \bar{X}$ | $(X - \bar{X})^2$ |
|---|---|---|
| 1 | −4 | 16 |
| 3 | −2 | 4 |
| 5 | 0 | 0 |
| 7 | 2 | 4 |
| 9 | 4 | 16 |
| | 0 | 40 |

Note that the sum of the $(X - \bar{X})$ column is zero. This will always be the case if the mean is in the exact center of the distribution. The $(X - \bar{X})$'s are squared in order to make each of the results positive.

*Step 2:* To find the variance, divide by $(n - 1)$. The sum of the $(X - \bar{X})^2$ column results in 40, a number ten times greater than the actual data. The value $(n - 1)$ is used to bring the data closer to the range of the original data squared. This action is analogous to finding the mean of the data (by dividing by $n$) in Section 3.1. The final value of the variance is

$$s^2 = \frac{\sum_{i=1}^{n} (X_i - \bar{X})^2}{n - 1} = \frac{40}{4} = 10$$

This value approximates the range $(9 - 1 = 8)$ found earlier.

*Step 3:* Find the standard deviation ($s$) by taking the square root of the variance ($s^2$).

$$s = \sqrt{10} = 3.16$$

This result, obtained by taking the square root of the squared data, represents the variability of the data within the values found in the sample.

*Note that your statistical calculator will automatically calculate the standard deviation of a sample or population.* Consult your owner's manual for the order of operations. The difference between the population variance ($\sigma^2$) and the sample variance ($s^2$) calculation is the degrees of freedom $(n - 1)$ term. On some calculators the sample standard deviation is symbolized by an $s$ and on others it is symbolized by $\sigma_{n-1}$. Use your calculator in the statistical mode to confirm the value for the standard deviation $s$ of Example 3.5. The value of the variance $s^2$ may be found by simply squaring the standard deviation. If your answers do not agree with the example calculation, you are probably pressing the standard deviation of the population key ($\sigma$), rather than the standard deviation of the sample key ($s$ or $\sigma_{n-1}$).

EXAMPLE 3.6 Solve for the standard deviation and variance of the sample data in Example 3.4.

*Solution:* It is assumed that you have a statistical calculator that will automatically enter the frequency of each cell times the midpoint value of the cell. Use your calculator's owner's manual to find the necessary procedure. The result is

$s = 4.69$ (Note that the standard deviation of the population $\sigma$ is slightly smaller at 4.66.)

$s^2 = 22.0$

Just as the value of the mean for grouped data was an estimate (Example 3.4), the value of the standard deviation is also an estimate. In this case, using grouped data, each data point is assumed to be dispersed evenly across the cell.

The preceding examples demonstrate how the standard deviation, unlike the range, considers every data point or every cell midpoint. The **standard deviation,** then, describes the clustering or dispersion of the data about the mean. Each data point—or cell midpoint in the case of grouped data—is weighted by its proximity to the mean. These procedural differences make the standard deviation the preferred choice of measurement of variability. Understanding the concept of the standard deviation is critical to understanding process capability.

## *Section 3.2 Review Problems*

1. Calculate the sample standard deviation of the following set of five sample observations:

   8.76, 7.93, 8.43, 8.55, 8.01

2. Calculate the variance of the sample observations in problem 1.

3. Calculate the range of the sample observations in problem 1.
4. Calculate the population standard deviation of the sample observations in problem 1. Explain the difference between the population standard deviation $\sigma$ and the sample standard deviation $\sigma_{n-1}$.

## 3.3 Shape of Distributions

The shape of histograms has already been discussed in Chapter 2. The normal curve was briefly described, and a smooth curve, called an ogive, was used to represent a distribution and to show how the distribution may change over time (Figure 2.10, Section 2.4). The normal curve was described as the preferred shape of a distribution, one that would be most stable and predictable over time.

One of the most controversial problems in applied statistics is how important it is to confirm *normality* before interpreting statistical tools such as the control chart (Chapter 4). Several tests exist to determine normality. Two of the normality tests are graphical in nature. The first is to graph the histogram, round off the edges, and judge the resultant ogive (Figure 3.1). The other method is to graph the cumulative distribution by using normal probability paper. A normal distribution will graph as a straight line. Both graphical tests are nonnumerical and therefore require judgment. The author recommends simply using the graph of the histogram.

Using the graphical method, consider the characteristic of shape known as **skewness**—that property of a distribution displaying a large tail to one side. The three measures of central tendency (Section 3.1) may aid in visualizing skewness. When a distribution is skewed, the mean is shifted from the median and mode in the direction of the skewness. Figure 3.1 depicts how the three measures are related for negative skewness (Figure 3.1a), positive skewness (Figure 3.1b), and normality (Figure 3.1c; the median, mode, and mean are superimposed).

Two numerical methods exist for determining normality—the chi-square ($\chi^2$) test, and the skewness and kurtosis tests. Both tests require large numbers of sample data, the accuracy of many significant figures, and laborious calculations that can often go wrong. Most computer software uses the chi-square ($\chi^2$) statistic as well as skewness and kurtosis tests to determine normality. An example of calculating skewness and kurtosis follows.

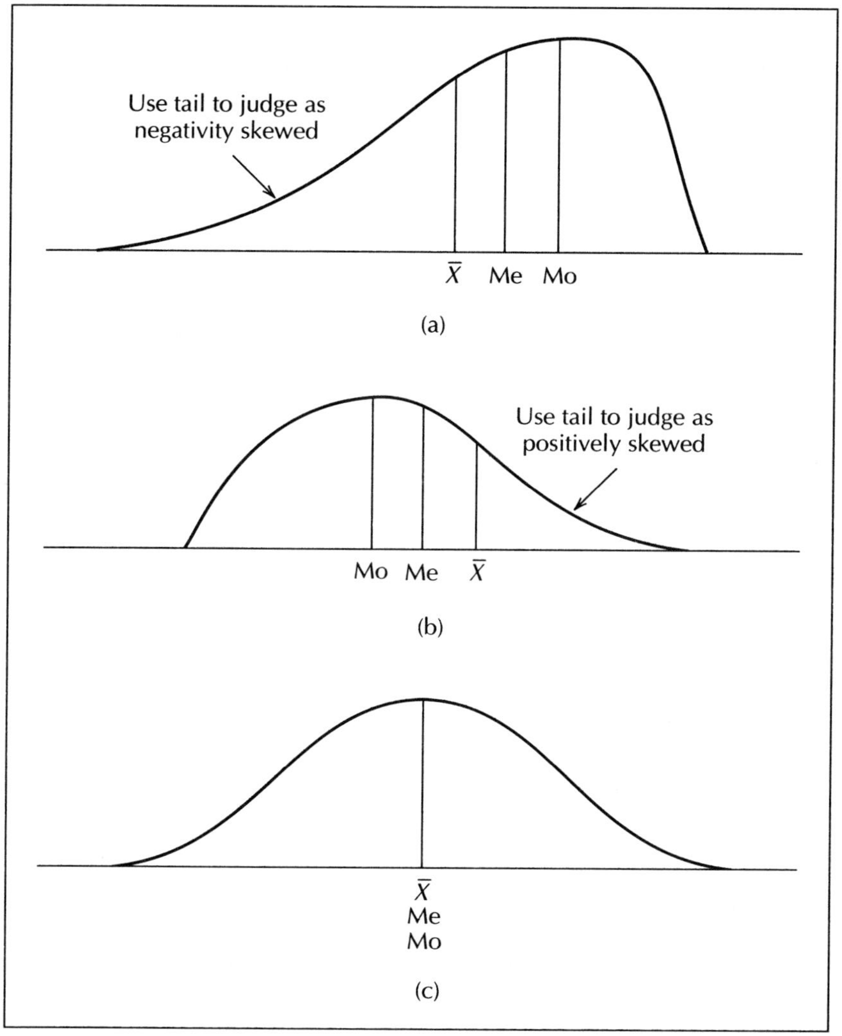


**FIGURE 3.1**
The mean ($\overline{X}$) pulls away from the median (Me) and the mode (Mo) in the direction of the skewness.

## *Calculating Skewness and Kurtosis*

The following equation may be used to calculate skewness:

$$g_1 = \frac{\sum_{i=1}^{k} f_i(X_i - \overline{X})^3/n}{s^3} = \frac{\sum_{i=1}^{k} f_i(X_i - \overline{X})^3}{ns^3}$$

Where $g_1$ represents skewness for a sample. The variable $k$ indicates the number of classes used in a calculation from grouped data.

**Kurtosis** ($g_2$) describes the flatness (**platykurtic** distribution, Figure 2.9c) or peakedness (**leptokurtic** distribution, Figure 2.9f) of a distribution. A normal curve is described as **mesokurtic.** The equation for $g_2$ is

$$g_2 = \frac{\sum_{i=1}^{k} f_i(X_i - \overline{X})^4}{ns^4}$$

Example 3.7 will demonstrate calculations for skewness and kurtosis.

EXAMPLE 3.7

Find the skewness and kurtosis of the distribution described by the grouped sample data in Figure 1.12. The mean ($\overline{X}$) of this distribution is 182.47. The standard deviation $s$ used for the calculation is 3.08.

*Solution: Step 1:* Construct a table of the necessary intermediate calculations. The following data table represents the distribution shown in Figure 1.12 (an example of fill weight).

TABLE 3.3
**Preliminary Calculations for Skewness and Kurtosis**

| *Class Mark* ($X$) | $f$ | $X - \overline{X}$ | $f(X - \overline{X})^3$ | $f(X - \overline{X})^4$ |
|---|---|---|---|---|
| 175.87 | 6 | −6.60 | −1,724.976 | 11,384.842 |
| 177.52 | 11 | −4.95 | −1,334.161 | 6,604.098 |
| 179.17 | 12 | −3.30 | −431.244 | 1,423.105 |
| 180.82 | 15 | −1.65 | −67.382 | 111.180 |
| 182.47 | 16 | 0.00 | 0.0 | 0.000 |
| 184.12 | 34 | 1.65 | 152.732 | 252.008 |
| 185.77 | 28 | 3.30 | 1,006.236 | 3,320.579 |
| 187.42 | 3 | 4.95 | 363.862 | 1,801.118 |
| | | | −2,034.933 | 24,896.93 |

*Step 2:* Complete the calculation of the skewness ($g_1$).

$$g_1 = \frac{\sum_{i=1}^{k} f_i(X_i - \overline{X})^3}{ns^3} = \frac{-2{,}034.933}{(125)(3.08)^3}$$

$$= -0.557 \approx -0.6$$

*Step 3:* Complete the calculation of the kurtosis ($g_2$).

$$g_2 = \frac{\sum_{i=1}^{k} f_i(X_i - \overline{X})^4}{ns^4} = \frac{24{,}896.93}{(125)(3.08)^4}$$

$$= 2.213 \approx 2.2$$

The information in Example 3.7 may be interpreted as follows:

- The skewness value ($g_1$) of $-0.6$ shows significant negative (left) skewness. Refer to Figure 1.12, the histogram of the data, to see that the distribution is definitely negatively or left-skewed. Significant probability of departure from normality is approximately $0 \pm 0.3$ for sample distributions of 100 to 200.
- The kurtosis value ($g_2$) of 2.2 displays moderate flatness. A value of 3 is the accepted value for the normal, mesokurtic, curve. If $g_2$ is greater than 3 ($>3$), the distribution is leptokurtic, and if $g_2 < 3$, the distribution is platykurtic. For sample sizes of 100 to 200, divergences from 3 on the order of plus or minus 0.6 are considered departures from normality. Since the 2.2 is lower than three by 0.8, the distribution is flatter than normal. Refer to Figure 1.12, the histogram of the data, to see that many of the cells in the distribution are about the same height (flat or platykurtic).
- The skewness and kurtosis calculations confirm the nonnormality of the distribution. Nonnormality is not desirable in the process, and the project team must continue to study methods to reduce assignable causes and approach normality.

Calculations of skewness and kurtosis are not considered to be reliable for sample sizes under 100 and 200, respectively. Multimodal distributions also offer invalid results. Therefore, in industrial or business settings, especially without the aid of a computer, it is recommended, and reasonable to assume, that a graphical judgment of normality will suffice. *It is further recommended that the quality team always check the histogram for normality of the actual distribution, so as to know when assignable causes are operating.* Undetectable assignable causes will hinder process improvement efforts.

## *Section 3.3 Review Problems*

1. When a distribution has a large tail to the left, it is said to be ______________ (negatively, positively) skewed.
2. A positively skewed distribution would have the mean shifted ______________ (upscale, downscale) from the mode.
3. A kurtosis value of 3 means that the distribution is
   a. normal
   b. leptokurtic
   c. platykurtic
   d. mesokurtic
4. The author recommends using ______________ to determine normality.
   a. normal probability paper
   b. the histogram
   c. the chi-square test
   d. skewness and kurtosis tests

# 3.4 *The Normal Curve*

The normal curve has been intensively studied since it was first discovered by Abraham De Moivre in the eighteenth century. The normal distribution was used extensively by the German mathematician and physicist Carl Friedrich Gauss in his studies of astronomy. Because of his work the distribution is sometimes known as the Gaussian distribution.

The bell-shaped curve or **normal curve** is the graph of the normal distribution. Some characteristics of the distribution include the following:

1. The mean, median, and mode occur at the same point.
2. Its graph is a bell-shaped curve with tails asymptotic (approaching the $X$ axis) on both sides.
3. The data tend to cluster around the mean with few values more than three standard deviations from the mean.

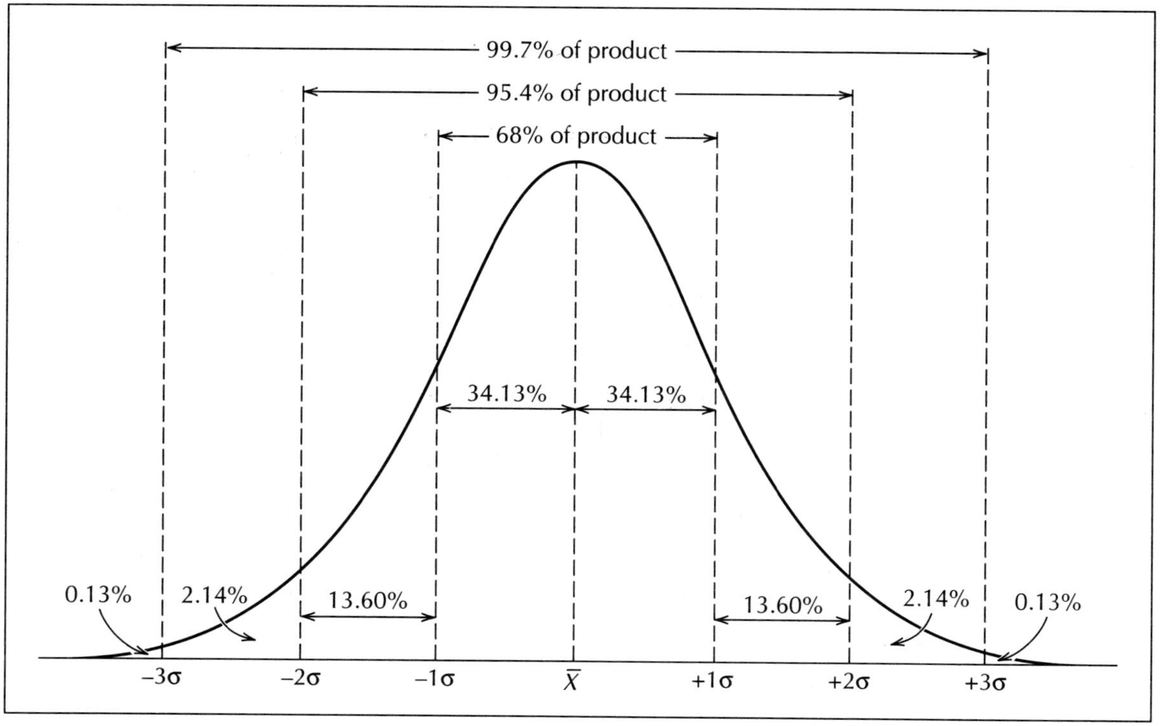


**FIGURE 3.2**
Selected areas under the ogive of the normal curve.

4. For any given mean and standard deviation there is only one normal distribution.

The area under the normal curve yields the probability of finding a given data point ($X$ value) in the distribution. As seen in Figure 3.2, the area enclosed by $\pm 1$ standard deviation ($\pm 1s$) represents approximately 68 percent of a product quality characteristic's distribution. A table of the "standardized, normal probabilities" is given in **Appendix A.** In order to use the normal table, the $X$ values must be converted to $Z$ values. The $Z$ values convert the distribution to be studied—regardless of its shape, spread, or central tendency (Figure 3.3)—into the standardized form, from which the probabilities of the table are

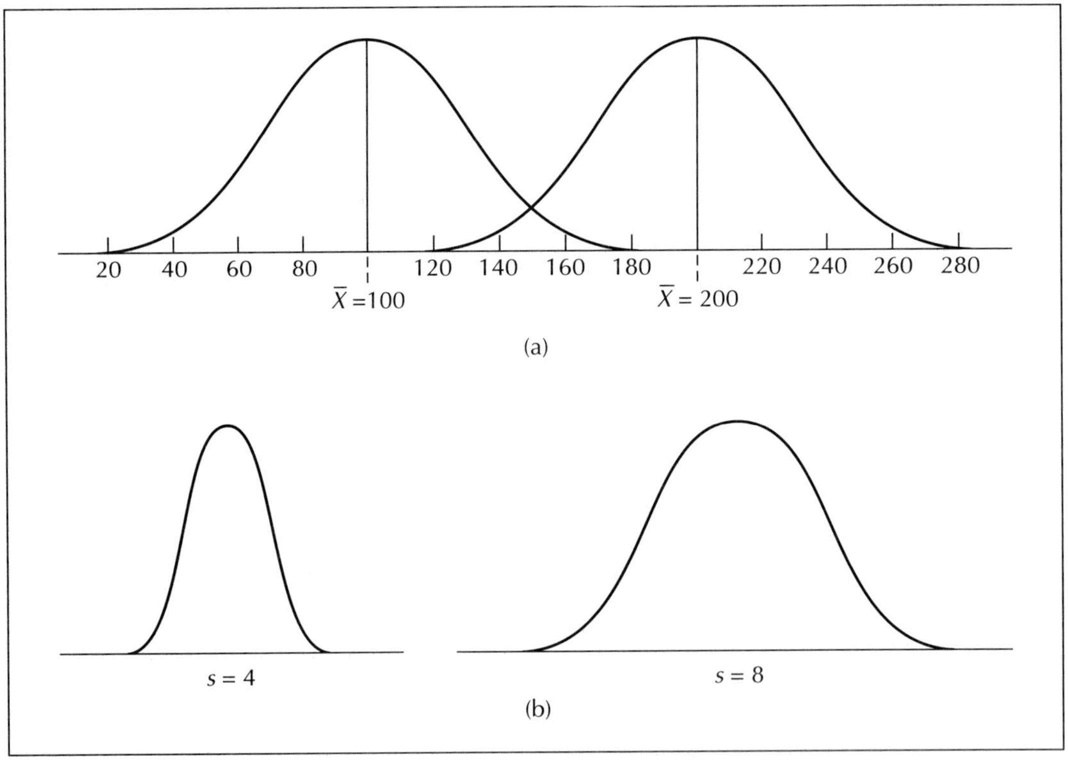


**FIGURE 3.3**
Normal curves possessing different means (a) and different standard deviations (b) may be transformed to the standard normal curve in order to use a single table of probabilities.

computed. The "Standardized, Normal Table" will be known as the **normal table** for simplicity.

To convert an $X$ value to a standardized $Z$ value, use the following equation:

$$Z = \frac{X - \overline{X}}{s}$$

It is more conventional to use

$$Z = \frac{X - \mu}{\sigma}$$

Sigma ($\sigma$), the standard deviation of the population, is slightly smaller than $s$, the standard deviation of a sample. This difference will not significantly affect the following calculations, and sample statistics are those used almost exclusively in troubleshooting industrial processes. Therefore, this text will use the sample statistical symbols in the equation

$$Z = \frac{X - \overline{X}}{s}$$

The use of the standardized normal curve and the table of standardized normal probabilities in Appendix A will be demonstrated by Examples 3.8 through 3.12.

EXAMPLE 3.8

A sample distribution (considered normal) of a critical part dimension has a mean $\overline{X}$ of 1.85 inches and a standard deviation $s$ of 0.330 inches. The lower specification limit (LSL) and upper specification limit (USL) are 1.41 and 2.83 inches, respectively. Find the probability of obtaining a part below the LSL of 1.41 inches.

*Solution:* Using the mean and standard deviation, compute the standardized $Z$ value of the data point.

$$Z = \frac{X - \overline{X}}{s}$$

$$Z = \frac{1.41 - 1.85}{0.330} = -1.33$$

The area covered by the normal table moves from left tail to right. In Figure 3.4, the area of the curve that lies between the left tail of the standardized normal distribution to the transformed $Z$ value may be visualized. The area

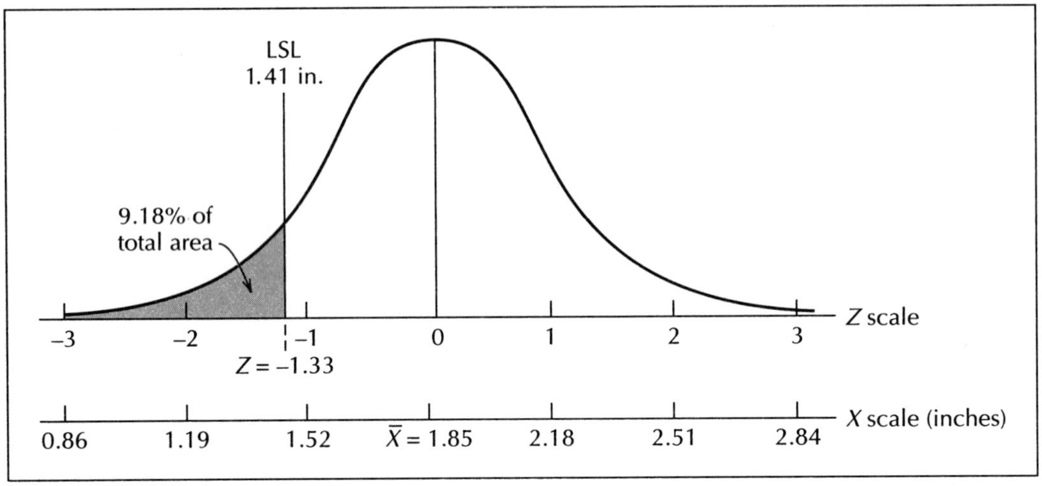


**FIGURE 3.4**
Data from Example 3.8 are transformed to the standardized normal distribution.

(probability of occurrence), from the table, is 0.0918. This value is found by locating −1.3 on the left-hand side of the table and moving across the table to the 0.03. (Note that the values across the top of the table are in reverse order for negative values of *Z*.) The probability of finding a part in this region of the distribution (less than the LSL of 1.41 in.) is, therefore, 0.0918, or 9.18 percent.

**EXAMPLE 3.9** For the distribution of Example 3.8, find the probability of obtaining a part measuring more than the USL of 2.83 inches.

*Solution:* First calculate the standardized *Z* value of the part.

$$Z = \frac{2.83 - 1.85}{0.330} = 2.97$$

The area or probability obtained—from the left-reading normal table—is depicted in Figure 3.5. The value of the area to the left of the *Z* value of 2.97 is found from the normal table to be 0.9985.

Note, however, that the area of interest is not from the left of the standardized normal distribution, but from the right. Since the table reads from the left, the area calculated above must be subtracted from a probability of 1.0 or 100 percent. Therefore, the area of interest is

$$100\% - 99.85\% = 0.15\%$$

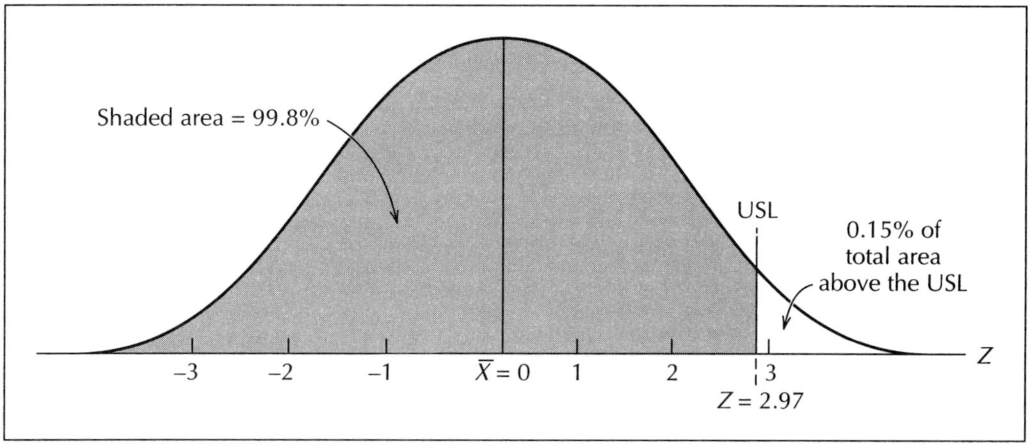


**Figure 3.5**
Data from Example 3.9 are transformed to the standardized normal distribution.

Therefore, there is a 0.15 percent probability of finding a part greater than the USL of 2.83 inches. In other words, the probability is 0.15 chances in 100 or 15 chances in 10,000!

EXAMPLE 3.10 What is the probability of finding a part that is within the specification limits for the distribution described in Example 3.8?

*Solution:* Now we are interested in the area between the LSL and USL. The drawing of the area of interest is left for the student. The calculation involves taking the area from the left of the standardized, normal distribution to the USL, and subtracting that area from the smaller area described from the left of the $Z$ distribution to the LSL. The calculation becomes

$$0.9985 - 0.0918 = 0.907$$

And the probability of finding a part dimension between the two values is 90.7 percent. This calculation reveals a scrap and rework rate of almost 10 percent, a quite significant internal quality cost (Chapter 10).

EXAMPLE 3.11 The process may be improved by moving the mean or center of the distribution. For a drilling operation, the process should be shifted to the left. Undersized drilling can be reworked, but oversized drilling creates scrap. On the other hand, a grinding operation creates scrap by undersizing parts. Assume that this process is a grinding operation. Note that almost 10 percent of the parts must be scrapped because they are undersized (material cannot very efficiently be added to parts). Less than 1 percent of the parts may be reworked. Shifting the process center to the right will create less scrap and more rework, definitely a more competitive strategy.

Shift the process center so that only 2 percent of the product needs to be scrapped.

*Solution:* In order to find the new center of the distribution, the new mean must be solved for. Using the $Z$ transformation equation, solve for $\overline{X}$. Also refer to Figure 3.6.

$$Z = \frac{X - \overline{X}}{s}$$

$$X - \overline{X} = Zs$$

$$-\overline{X} = -X + Zs$$

$$\overline{X} = X - Zs$$

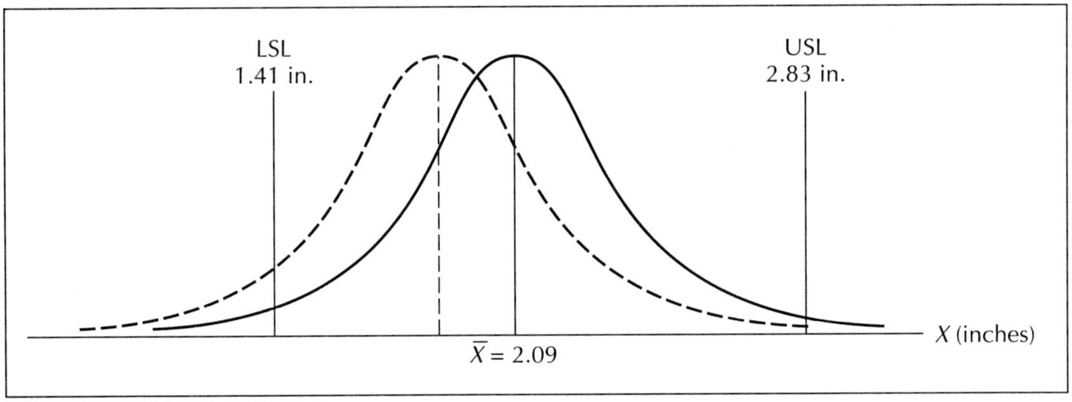


**FIGURE 3.6**
The original process (dashed distribution) has been recentered in order to reduce scrap.

Note that the measurement value $X$ used in the resulting equation is the LSL of 1.41 inches, because we desire approximately 2 percent of the parts (rather than over 9 percent) to be below the lower specification limit. To find the value for $Z$, enter the table's rows and columns to find the value closest to 2 percent. The closest value is 0.0202 at a $Z$ value of −2.05. (Interpolation could be used to find the exact $Z$ for 2 percent, but in the process centering of the machine this would surely be judged as "overkill.") Since $s$ is already known, the equation may be solved for $\overline{X}$.

$$\overline{X} = 1.41 - (-2.05)(0.330) = 2.09 \text{ inches}$$

The process distribution, after centering, is shown in Figure 3.6.

A further check reveals another benefit of moving the process center. This exercise is left for the student.

One final note before leaving the normal curve and its resultant probabilities. If a process or product quality characteristic is not normal but known to be unimodal and operating in a random manner, the percent of product that will fall outside $t$ standard deviations of the mean will always be less than

$$\% \text{ product lying outside of } t \text{ standard deviations of } \overline{X} < 100/2.25t^2$$

**EXAMPLE 3.12** Automobile batteries have a mean life of 46 months with a standard deviation of 6.5 months. The distribution is nonnormal. Find the minimum percentage of batteries having a life of 40 to 52 months.

*Solution:* First calculate $t$, the number of standard deviations included between 40 and 52 months.

$$t = \frac{\text{specific range of interest}}{s} = \frac{46 - 40 \text{ or } 52 - 46}{s} = \frac{6}{6.5} = 0.923$$

$$\text{and maximum \% of product falling outside} < \frac{100}{2.25(0.923)^2} = 52$$

So, the total proportion of batteries falling within 40 to 52 months must be at least 100 − 52 = 48%. What would the percentage be if the distribution were normal? Use the normal table at $Z = \pm 0.92$ to find the percent of product with a life between 40 to 52 months. (See Review Problem 4.) The difference is large enough in this case to use the inequality.

It should be mentioned at this point that data distributions take many forms, the normal distribution representing only one of many. Appendix B shows other distributions that data take and their usual applications. The **Poisson distribution (Appendix D)** will be used in Chapter 6. This will help to increase your awareness of other types of statistical distributions.

## Section 3.4 Review Problems

1. Approximately what percentage of the area under the normal curve is included within ±3 standard deviations about the mean (a six-sigma spread)?
   a. 68.0
   b. 85.7
   c. 95.3
   d. 99.7
2. Using the normal table in Appendix A, find the probabilities associated with the following values of Z: 1.98, 0.22, −2.85, −3.0, −0.22. Convert the probabilities to percent.
3. Locate the following probabilities (expressed as percent) on the normal table. Convert the probabilities to Z values.

   20.05%, 0.060%, 99.89%, 57.93%, 50.00%

4. Refer to Example 3.12. What is the proportion of automobile batteries having a life of 40 to 52 months, when using the statistics of the normal table?
5. Do you feel there is a significant difference between your answer in problem 4 and the answer in Example 3.12? Is the proportion using the normal curve greater or less? Why?

# 3.5 Working to Target

This chapter has brought focus to the desirability of the normal distribution in process control. For many processes, a normal frequency distribution is the best evidence that a system of chance (random) causes are operating. **Chance causes** are often referred to as **unassigned causes,** because they are difficult to detect or identify.

The following are examples of chance causes:

1. The machine does not allow for tight control. Some backlash and hysteresis exist in all machines. (Figure 3.7a shows the ideal movement of a machine part that is being sequenced.

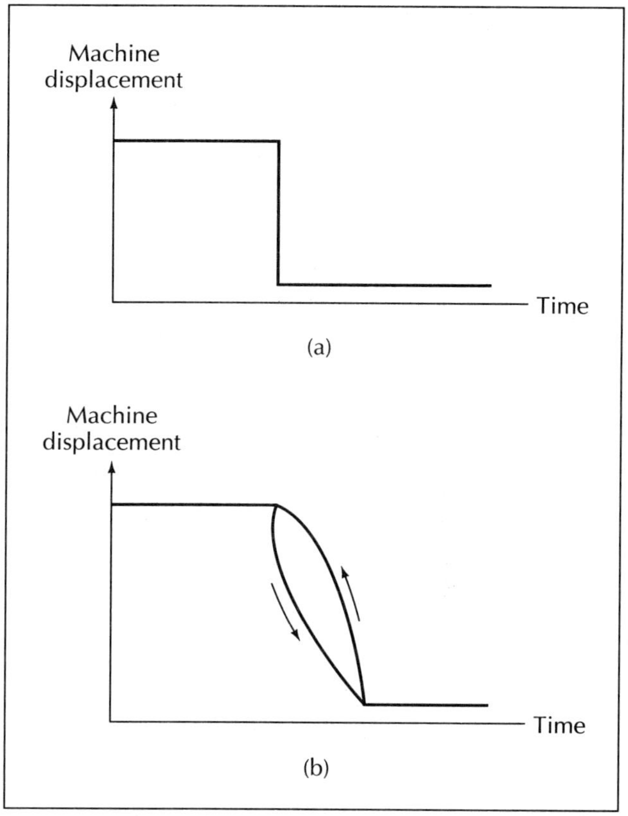


**FIGURE 3.7**
Backlash, also known as hysteresis, is the property of the switching action in a machine.

Figure 3.7b shows what actually happens, with the arrows indicating movement of the machine part in opposite directions.)

2. Machinery varies between established maintenance cycles (preventive or predictive maintenance systems only—otherwise, maintenance is only accomplished after obvious problems with the machine or process occur).
3. Materials vary, even when sourcing from the same supplier.

Without assignable causes operating, the normal distribution often results. Only chance causes are operating. Under these conditions

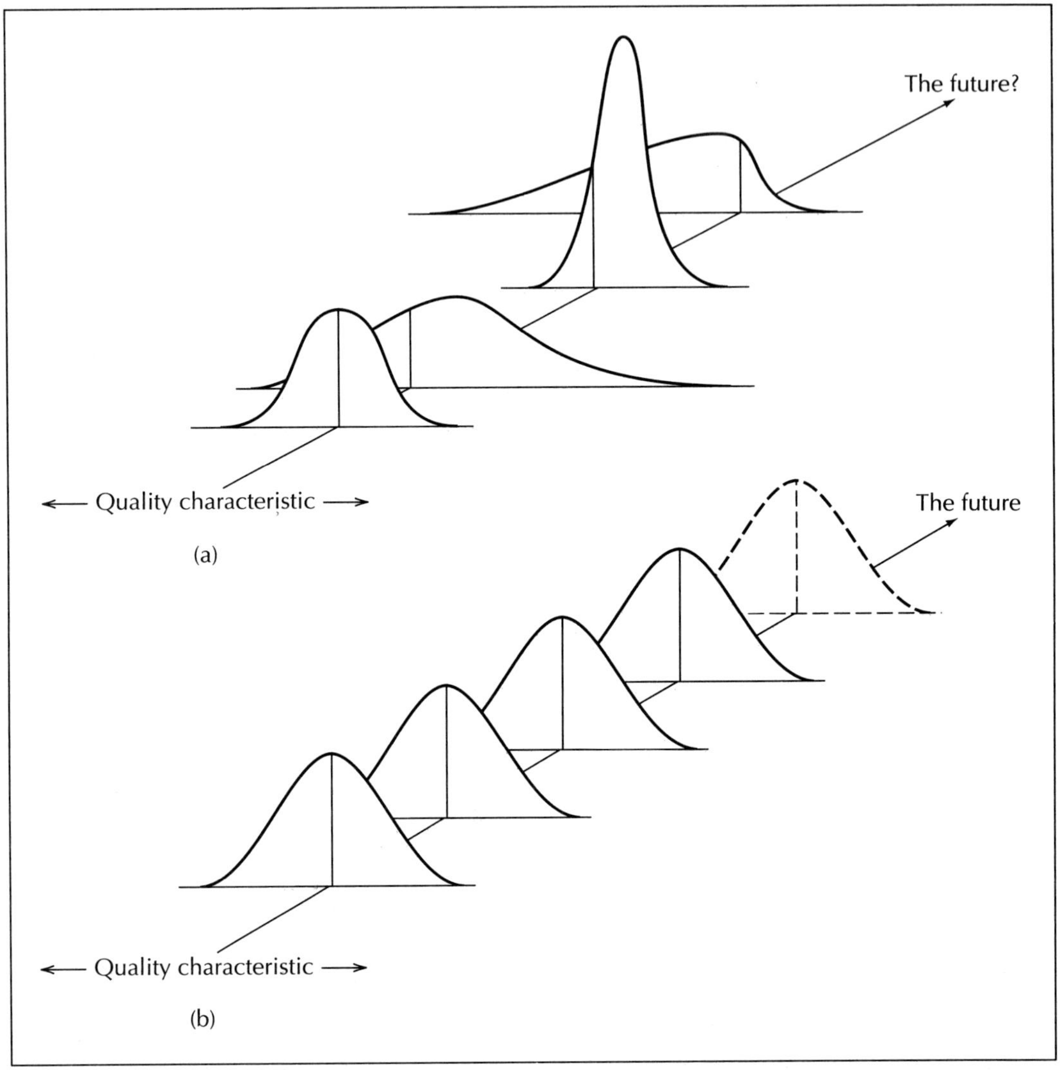


**FIGURE 3.8**
Assignable causes must be identified and removed in order to achieve a predictable process over time (a). With only chance causes operating and assignable causes removed, a process delivers product with predictable quality characteristics (b).

the process will remain stable and predictable over time. The process must undergo dramatic change to further reduce variability.

What if the product's quality characteristic does not form a normal distribution after all assignable causes have been eliminated? For instance, right-skewed distributions always result when a measurement is truncated at zero (e.g., runout, Figure 2.9g, Section 2.4). *In most cases dealing with nonnormality, the normal table can still be effectively used to predict process centering and scrap and rework.* The statistical calculations of the mean and standard deviation apply in any case. Only the interpretations differ if the distribution is not normal. Even with these differences in interpretation, most industrial processes can use the analysis from the normal table, discussed in Section 3.4, to verify the efficacy of attempts to improve the process.

Figure 3.8a shows what happens when assignable causes are operating. The process is changing shape, variability, and centering. The product quality characteristic is not predictable, and improvements to the process are difficult if not impossible to verify.

Examples of assignable causes are as follows:

1. The material is changing; for example, the single supplier cannot control variations or there is more than one supplier.
2. The methods are changing; for example, the length of a heat-treat operation is constantly changing.
3. Personnel are changing; for example, reluctance to hire and train full-time employees has required that untrained, temporary workers run the process.
4. Production runs begin too soon after startup—cold starts.
5. Maintenance is not performed correctly or in a timely manner.

Other examples of assignable or unassigned causes may be created by remembering MMMMEP, the five main branches in the cause-and-effect diagram (Figure 1.9, Case Study—Chapter 1). Figure 3.8b depicts the distribution when assignable causes have been eliminated.

EXAMPLE 3.13 A television manufacturer operated two plants with differing quality philosophies. One plant worked to target, and the other worked to specification limits. For one quality characteristic, that of color density, the product quality distributions are overlaid as shown in Figure 3.9.

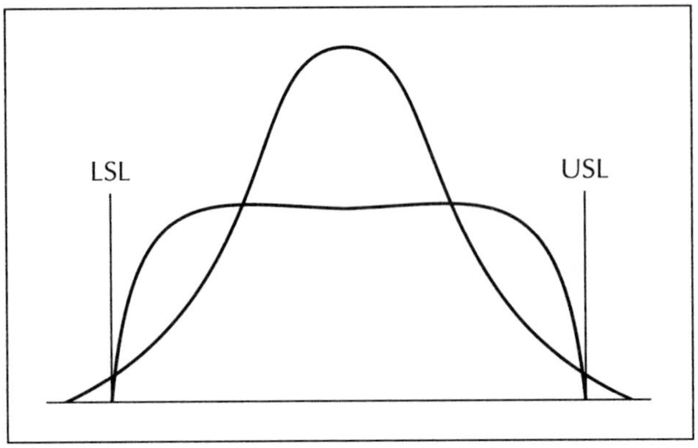


**Figure 3.9**
Working to target versus working to specification limits.

The flatter distribution is due to sorting of product that lies beyond specification limits. Sorting always leads to a flatter distribution because the tails of the more normal curve have been eliminated. Note that the plant producing the normal distribution does have some TV sets that are out of specification regarding color density.

Even though the plant with the normal distribution produced more nonconforming product, the losses computed (e.g., lack of sales and warranty costs) were much less. Why did the more normal distribution best satisfy the customer?

*Solution:* In attempting to find the reasons for this situation, company officials explored the differences between the two plants and their products (see discussion of benchmarking, Chapter 10). One of the most interesting findings was that customers could easily spot differences in color density. The customers were selecting the television sets from the plant having most of the sets near the target value, especially when the sets from both plants were viewed side by side in the showrooms.

The study forced the conclusion: *Working to target is superior to working to specification limits.* Not only does working to target reduce inspection (no sorting occurred in the plant working to target), but it also produces goods that are closer to customer expectations. Pleasing the customer increases sales and decreases returned product and other warranty costs.

*When working to target we attempt to reduce the variation to the target value. Continuous quality improvement results.*

## Section 3.5 Review Problems

1. Backlash or hysterisis usually is an example of a/an ______________________ (chance, assignable) cause.
2. Eliminating ______________________ (chance, assignable) causes produces the normal distribution.
3. In Example 3.13, the plant with the normal distribution produced more nonconforming product. Why did that plant sell more product?

## Summary

- A distribution is described in three ways—central tendency, variability, and shape.
- The mean is the most frequently used of the three values of central tendency. Its equation is

$$\overline{X} = \frac{X_1 + X_2 + X_3 + \cdots + X_n}{n} = \frac{\Sigma X_i}{n}$$

- The variability of a distribution may be described by the range and the standard deviation.
- The range, obtained by simply subtracting the highest data point from the lowest, will be used for constructing $\overline{X}$ and $R$ charts in the next chapter.
- The standard deviation considers all data points in a distribution.
- The best way to calculate the standard deviation is to use your statistical calculator.
- The standard deviation will be used to calculate process capability in Chapter 4.
- Determining the shape of a distribution is one of the best methods to find assignable causes for inferior product or process quality characteristics.
- Viewing the process histogram will usually suffice in determining the normality of a distribution.

- For large samples the following equations of skewness and kurtosis are useful:

$$g_1 = \frac{\sum_{i=1}^{k} f_i(X_i - \overline{X})^3}{ns^3}$$

$$g_2 = \frac{\sum_{i=1}^{k} f_i(X_i - \overline{X})^4}{ns^4}$$

- The normal table was used to determine percent of product out of specification. The statistics of the normal curve was found to be useful in selecting the best center for a process. The following equations were used:

$$Z = \frac{X - \overline{X}}{s}$$

$$\overline{X} = X - Zs$$

- The value of working to target to assure continuous quality improvement was discussed.
- Eliminating assignable or easily identifiable causes aids us in working to target.
- The next chapter, variable charting, will aid us as we attempt to work to target.

## CASE STUDY · CHAPTER 3

At the reception desk of a large luxury hotel (Figure 3.10a) the following "times to register" were recorded:

| *Time to Register (min)* | *Frequency* |
|---|---|
| 3 | 7 |
| 4 | 10 |
| 5 | 17 |
| 6 | 22 |
| 7 | 16 |
| 8 | 11 |
| 9 | 3 |
| 10 | 1 |
| Total | 87 |

(a)

**FIGURE 3.10**
Lobby of large luxury hotel (a) and the histogram of time to register (b, page 104).

Currently, about 5 percent of the guests of the hotel are delayed by more than 8 minutes. The hotel manager has asked the quality department to give the registration personnel a realistic target (an average time) to work toward. The target should be established so that no more than 2 percent of the guests have to wait more than 8 minutes to register. See the histogram of times to register in Figure 3.10b.

The quality department at the hotel has already done extensive studies on the waiting-time-to-register problem, since this is one of the largest sources of complaints from hotel guests. Therefore, quality personnel already know that assignable causes for delays—such as insufficient staffing or desk clerks using the phone while guests wait to register—have been eliminated. The general distribution of "time to register" is relatively normal because assignable causes have been eliminated. Normality means that the target or center of the distribution can be changed, and the percent of customers waiting more than 8 minutes to register can be predicted.

The data from the most recent sample, as shown in Figure 3.10b, support the claim that almost 5 percent of the guests are detained more than 8 minutes (4/87 = 4.6%). In order to shift the mean to improve the wait time so only 2 percent of the guests must wait more than 8 minutes, the normal table may be used. The first step is to locate the closest value to 0.02 within the table. The associated $Z$ value is 2.04. The $s$ of the data is calculated to be 1.61. The final calculation (from Section 3.4) is

$$\overline{X} = X - Zs = 8 - (2.05)(1.61) = 4.70$$

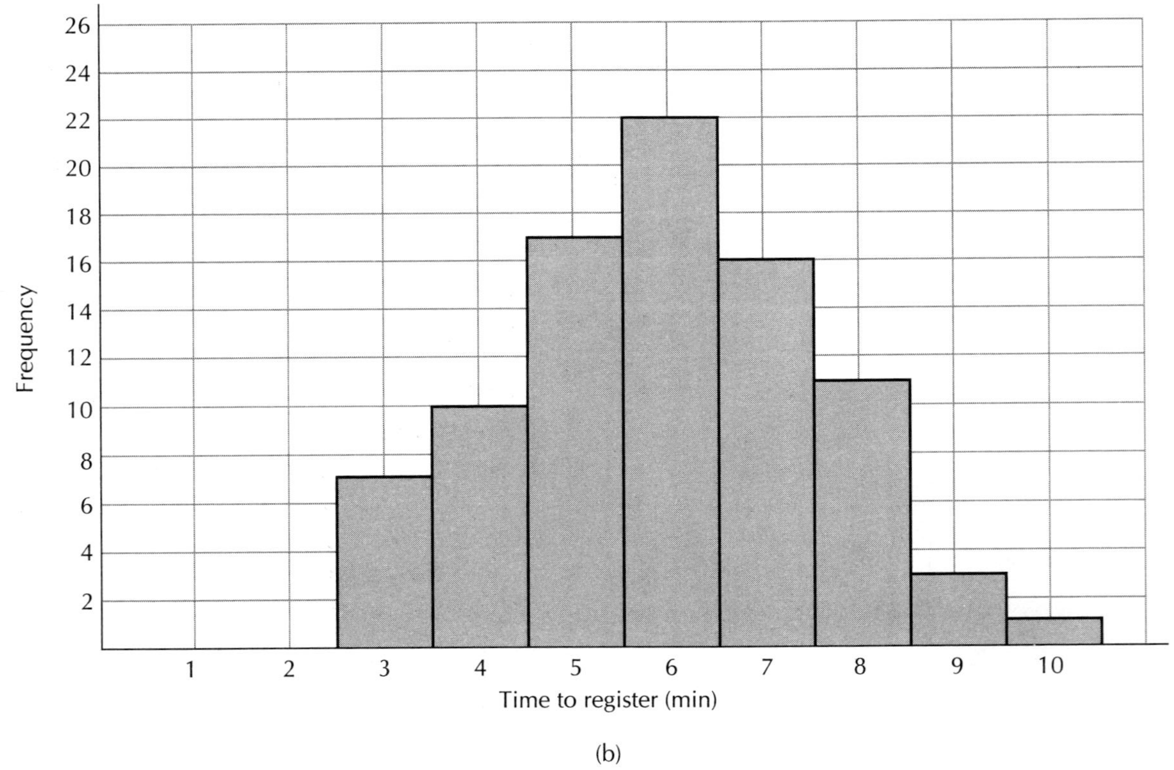


**Figure 3.10** *(continued)*

The quality department recommends that the desk manager attempt to serve everyone within an average of 5 minutes. The previous average was approximately 6 minutes.

## Chapter Review Problems

1. Find the mean and median height of the following five individuals:

   4′11″, 6′4″, 5′7″, 6′0″, 6′2″

2. What is the standard deviation of the heights of the individuals in problem 1?
3. One advantage of the **International** or **metric system (SI)** is that units are not mixed as they are in the U.S. customary system (the units in problem 1). What is the mean and standard deviation of the heights of the following individuals?

   1.66 m, 1.73 m, 1.93 m, 1.70 m, 1.83 m, 1.88 m

4. Solve for the median of the data in problem 3.
5. Find the mean, standard deviation, and mode of the following ungrouped data. Use your calculator to calculate the $\overline{X}$ and $s$, by directly keying in the cell values and frequencies.

| *Rockwell Hardness C-Scale* | *Frequency* |
|---|---|
| 51 | 3 |
| 52 | 17 |
| 53 | 135 |
| 54 | 1,100 |
| 55 | 495 |
| 56 | 11 |
| 57 | 1 |

6. Find the mean and standard deviation of the data in the Case Study—Chapter 3. Compare it with the information given in the Case Study.
7. Examine the shape of the histogram in Figure 3.10b. After referring to Section 3.5, make a statement concerning the amount of assignable causes operating (almost none, a few, many). Do you think

the hotel has worked on the problem of waiting times to register? Why, or why not?

8. The following data were collected on a critical quality characteristic, surface finish of a bearing:

| *Surface Finish (μ in.)* | *Frequency* |
|---|---|
| 43 | 17 |
| 54 | 20 |
| 65 | 27 |
| 76 | 32 |
| 87 | 28 |
| 98 | 16 |
| 109 | 13 |
| 120 | 14 |
| 131 | 12 |

Solve for $\overline{X}$ and $s$.

9. In problem 8, solve for the skewness ($g_1$) and the kurtosis ($g_2$). What conclusions can you draw from the results?

10. Explain what the standardized, normal ($Z$) values mean in terms of an actual distribution (Section 3.4). For instance, what does the mean of any distribution equal in terms of $Z$?

11. A sample distribution from a precision drill operation is normal. The distribution has a $\overline{X}$ of 2.05 mm and an $s$ of 0.35 mm. What is the percent of product below the LSL of 1.5 mm?

12. Using the information in problem 11, what is the percent of product above the USL of 3 mm?

13. What percent of product in problems 11 and 12 will be within specification limits?

14. Shift the mean of the drilling process to 2.25 mm to center the process. Solve for the percent of product within specification. Which way should the mean be shifted from center in order to increase rework and reduce scrap?

15. A sample distribution of the fill weights of a dairy product is considered to be a normal distribution. If the distribution has a $\overline{X}$ of 62 oz. and an $s$ of 0.5 oz. what is the percent of product below the LSL of 60.5 oz.?

16. Using the conditions specified in problem 15, what is the percent of product above the USL of 63.5 oz.?
17. What percent of product in problems 15 and 16 would be found to be within specification limits?
18. Move the center of the process in problems 15 and 16 to insure that less than one-tenth of one percent of product is below the LSL. What is the value of the new $\overline{X}$?
19. Using the actual data from the Case Study—Chapter 3, find the percent of service times below 5 minutes and above 8 minutes.
20. Draw a normal curve of the recentered distribution in Example 3.11, as fitted to the specification units. Calculate the percent out of specification after recentering. Did recentering the process result in less scrap or rework? If so, why? Which was changed more by moving the $\overline{X}$—scrap or rework?
21. It is often necessary to calculate a weighted average when averages from data occur with different frequencies. A student's grade point average (GPA) is a good example, because the cumulative GPA depends on the number of credit hours used to calculate it. One application of the weighted average is to estimate the change in your GPA after acquiring a number of new credit hours: If your cumulative GPA is 3.24 with 97 credit hours, how much will it change if you earn a quarter average of 4 (all A's) with a load of 18 credit hours? Use the following equation:

$$\overline{X} = \frac{\sum_{i=1}^{n} w_i \overline{X}_i}{\sum_{i=1}^{n} w_i} = \frac{(3.24 \times 97) + (4)(18)}{97 + 18}$$

22. Calculate your new cumulative grade point average after predicting your average for this quarter.
23. See Laboratory—Chapter 3. The rule of thumb given for gauge error is 20 percent. Explain what this means, in one paragraph.
24. The accuracy of a measuring instrument is linked to NIST by (see Laboratory—Chapter 3)
    a. standardization
    b. traceability
    c. R&R studies
    d. capability

# LABORATORY · CHAPTER 3
# Gauge Capability Study (Short R&R Study)

*Materials needed:* Measurement tools, 20 to 30 parts (depending on the number of operators in the study), and short R&R forms, as used in the following example.

Before collecting data, the capability of the measuring device must be confirmed. There are three characteristics of interest—accuracy, repeatability, and reproducibility.

**Accuracy** is a systematic error or bias in a set of observed values. More specifically, accuracy is the average number of readings by a gauge measuring a traceable part such as a gage block. Accuracy cannot be determined without **traceability** to some standards group. In the United States, traceability is to the **National Institute of Standards and Technology** (NIST). NIST, in turn, is traceable to the International Bureau of Weights and Measures in Paris, France.

**Repeatability and reproducibility** (R&R) represent the random errors present in measurement. R&R studies involve the measurement of actual parts with the gauge of interest. First we will define the two terms.

- **Repeatability**—The ability of the gauge (gauge includes any measuring instrument) to repeat a measurement on a specific type of part, with the measurement being performed by a single operator (often referred to as gauge variation).
- **Reproducibility**—The ability of the gauge to repeat a measurement with more than one operator (operator variation).

The distributions for a typical R&R study are shown in Figure 3.11. The top distribution, much narrower than the bottom two, represents the reproducibility of a gauge between two operators. The spread of the means of the measurements taken from different operators is often referred to as operator variation. The repeatability distributions are for each of the operators being studied. $R_1$ and $R_2$ represent the range of the data taken by operators 1 and 2, respectively. Repeatability is often referred to as gauge variation, because we have eliminated the effects of more than one operator.

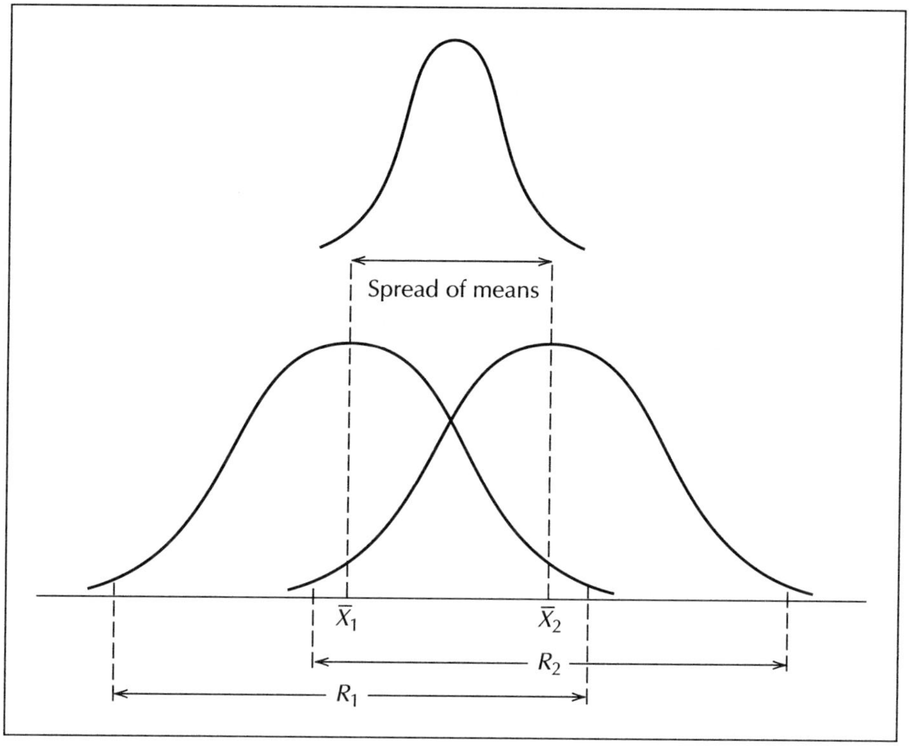


**FIGURE 3.11**
Drawings of the distributions produced by two operators performing an R&R study.

Performing the R&R study (short method):

1. Select two operators (or students in the quality laboratory) and one gauge.
2. Select five production parts and wash, deburr, and number them. Discuss the best method to measure the quality characteristic of interest. (For example, if measuring an inner diameter, it is best to mark the part at the point the measurement will be taken. Actual practice in performing this type of measurement requires measuring two points 90° from each other.) This action reduces error involving within-part variation. The discrimination or least count of the gauge must also be decided.

For instance, will the smallest divisions be used as a least count, or will the operators use one-half of the least count?

3. Each operator, using the same gauge, measures the parts in a *randomly decided order*, and without seeing the other's readings. The data are recorded on the form in Figure 3.12.

| Parts | Operator A | Operator B | Range \|A – B\| |
|---|---|---|---|
| 1 | 0.5128 | 0.5124 | 0.0004 |
| 2 | 0.5129 | 0.5131 | 0.0002 |
| 3 | 0.5129 | 0.5128 | 0.0001 |
| 4 | 0.5124 | 0.5122 | 0.0002 |
| 5 | 0.5130 | 0.5130 | 0.0000 |
| | | Sum of ranges ① | 0.0009 |

The study is conducted by two operators using the same five parts which were selected at random and identified 1 through 5. Each operator gages or measures each part and records data. R&R error is calculated as follows:

Average range ② = Sum of ranges ① ÷ 5 = 0.0009 ②

R&R error ③ = 4.33 × 0.0009 ② = 0.0039 ③

Specification tolerance range ④ = 0.525 (High Limit) – 0.505 (Low Limit) = 0.020 ④

% R&R error ⑤ = 0.0039 ③ ÷ 0.020 ④ × 100 = 19.5% ⑤

**FIGURE 3.12**
The completed R&R form using the short method.

The range of the data collected by the operators is calculated by subtracting the readings on each part. Notice that the absolute (no minus signs) value of the subtraction is recorded. After finding the sum of the ranges [1], the average range [2] is found by dividing by five (the number of measurements). Next, the R&R error [3] is determined by using a statistical factor of 4.33. (The factor 4.33 determines the total range possible in a situation involving the measurement of a great number of parts.) The R&R error is divided by the specification tolerance range [4] to calculate the percent R&R error [5].

The rule of thumb for deciding whether a gauge is capable or not is to use the percent of R&R to specification range, as follows:

- If R&R is 20 percent or less of the specification range, the gauge is acceptable (some quality practitioners use as low as 10%).
- If R&R is greater than 20 percent of the specification range, repair the gauge and reevaluate.
- If R&R is still greater than 20 percent of the specification range, the gauge simply will not work and an alternative measurement method must be found.

This short method of the *gauge capability study* is unable to separate repeatability and reproducibility. To separate the two factors, the long method must be used. See Laboratory—Chapter 4.

# 4

# Variable Control Charts

**Preparation for Charting** 4.1
**Constructing $\overline{X}$ and $R$ Charts** 4.2
**Chart Analysis** 4.3
**Process Capability** 4.4

## 4.1 *Preparation for Charting*

The control chart was originally developed to aid in the study of process variability. All processes vary, and the statistically based control chart offers the quality technologist one of the best methods to determine when to take action on the process. With a properly completed control chart the technologist or manager can avoid overcontrol or overreaction to a process change.

**Overcontrol** is a major cause of increased costs in manufacturing and service industries. Knowing when to control a process—and when not to!—eliminates scrap and rework. Every one dollar of scrap or rework eliminated leads to an estimated saving of six dollars in hidden costs. Hidden costs are increased costs, such as inventory control of scrap and material handling of rework (see the discussion of quality costs in Chapter 10).

The "gas mileage" example (from Section 1.1, "1930") is an example we can all relate to (see Figure 4.1). An example of overcontrol is to have a tune-up at point number three, when such effort and expense are not necessary. The chart shows that the lower gas mileage at point 3 is not due to the need for new spark plugs and the like, but is due to the natural variation in the driving experience. For instance, the poorer gas mileage at point 3 may be attributed to a short-term road construction problem on the driver's way to work. To have a tune-up at this time would result in overcontrol and waste. The control chart can eliminate overcontrol and waste.

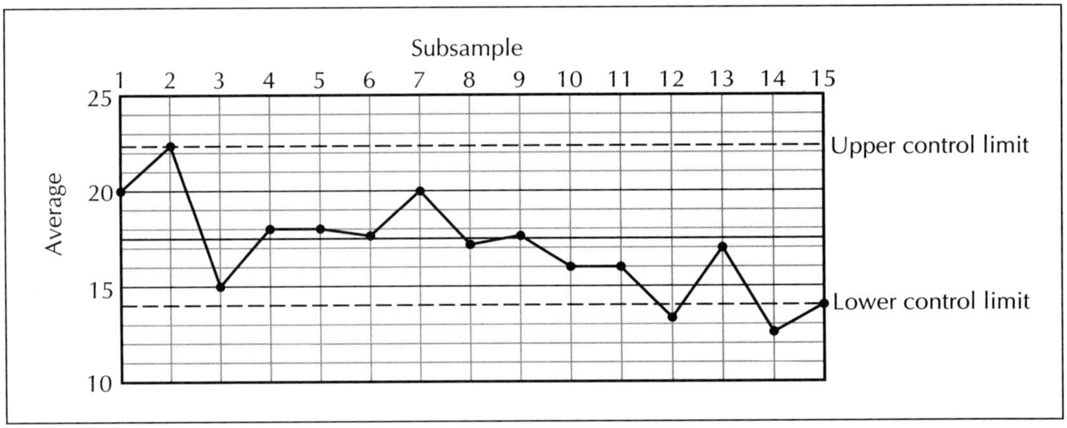


**Figure 4.1**
The control chart tracks miles per gallon.

One of the greatest advantages of the control chart is its ability to differentiate "assignable causes" from "chance causes." Assignable and chance causes have already been described in Section 2.4, "Troubleshooting with Histograms." These important concepts are redefined here. An **assignable cause** is one that produces a large variability, so large that a cause may be "assigned" to that variability (see points 12 and 14, which are beyond the control limits or natural limits of the chart, Figure 4.1). The chart has revealed, between points 12 and 14, that there is an assignable cause. The points are extending below the control limits. There is a clear need for a tune-up at this time.

A chance cause is a more minor change. It is due to the inherent variability of the process that chance alone can cause. **Chance causes** are more difficult to detect and usually cannot be eliminated without large changes in the *system* producing the quality characteristic of interest. In this case, large system changes might involve the elimination of the highway construction problems discussed in the example, or the cost of putting a new, more fuel-efficient carburetor on the car. In a company, large system changes can only be accomplished with the approval of upper management.

One interesting story from the papermaking industry that vividly illustrates the problems of overcontrol describes a process that had far too much variability. After a control chart study was implemented, it became apparent that the process was most stable when the operators were away from the machine! Overcontrol in this industry had cost the company millions before management discovered by means of the control chart that "tweaking of the process" was the culprit.

*Tweaking* resulted in assignable causes operating. The control chart discovered the large number of assignable causes by showing the process over time and finally revealing that the cause was too much, rather than too little, *pride of workmanship.* The solution became a system problem, that of upper management providing appropriate training on control charts and showing operators how overcontrol was detrimental. *With control charts the operators knew best when to make adjustments.*

## Actions Prior to Implementation

Before the control chart can be implemented, several actions must have already taken place. Most of these actions were summarized in the Case Study—Chapter 1. *Please review the case study, especially the components of the Shewhart cycle, before continuing.*

Specific actions to precede the construction of a control chart include the following: The process or product quality characteristic to be studied must be identified; the capability of the measuring equipment must be determined; resources must be allocated; and the appropriate personnel must be prepared to conduct the study. Each of these considerations is discussed in the following paragraphs.

***Identifying the Project*** The Case Study—Chapter 1 discussed how methods such as the examination of historical data and Pareto analysis could be used to select the control chart study to be conducted. Consideration should also be given to the probability of success with the project. Probability of success is an especially important consideration when first implementing statistical process control (SPC) in the plant. Always select a "doable" project!

The quality characteristic must be identified beyond question. If it is a product characteristic, it could involve the diameter of the holes bored, the fit between two parts, the texture of a material, or the time of service in a restaurant. Identifying the quality characteristic is instrumental in selecting the measurement equipment.

***Measurement Capability*** After the quality characteristic has been identified, the measurement equipment used to gather the data must be identified. The equipment must be capable of the accuracy and repeatability necessary to show a suitable range of data points.

***Allocation of Resources*** The decision to perform a control-chart study is also a decision to spend money. The appropriate management personnel must be informed and the necessary funds allocated for purchasing equipment and funding additional personnel time. Quality costs should be tracked during the study. Therefore, the accounting department should be involved.

***Selection of Personnel*** Control charting of a process or product usually involves several departments. A typical "study team" might be composed of an industrial engineer or engineering technologist, an operator, a supervisor, and someone from purchasing. The team members should be granted enough time to meet and to implement the project. Discussion should take place as to who is to analyze the data and coordinate the preparation of the report. A chairman and recorder should be selected by the team members.

In summary, the four steps that must be accomplished before control charting are

- project identification;
- determination of capability of measuring equipment;
- allocation of resources;
- selection of personnel.

Once these steps have been accomplished, we are prepared to implement the charting of the selected quality characteristic.

You will probably find the **Glossary of Symbols** printed inside the back cover to be useful as you work through the control chart examples.

## *Section 4.1 Review Problems*

1. Overcontrol is major cause for increased ________________ (quality, costs) due to scrap and rework. Overcontrol can be reduced by the use of ______________________________ ______________________________.
2. Every one dollar of scrap or rework eliminated leads to an estimated ______________ dollars in hidden costs.
   a. 2
   b. 4
   c. 6
   d. 8
3. Tweaking the process, one form of overcontrol, would be considered a/an ____________________________ (chance, assignable) cause.
4. List the actions that should take place before implementation of a control chart.

# 4.2 *Constructing $\overline{X}$ and R Charts*

Construction of the $\overline{X}$ and $R$ chart involves many of the principles already discussed in Chapter 2, "Graphical Analysis of Data." The $X$ and $Y$ axes must be scaled using the proper units. The $X$ axis, always containing the independent variable, will usually be in units or periods of time. The $Y$ axis, the dependent variable, will have the units of the quality characteristic of interest, such as strength, length, diameter, or weight. The main difference between the $\overline{X}$ and $R$ chart and the line graph is that the points in $\overline{X}$ and $R$ charts are joined by connecting lines so that trends and shifts may more readily be seen.

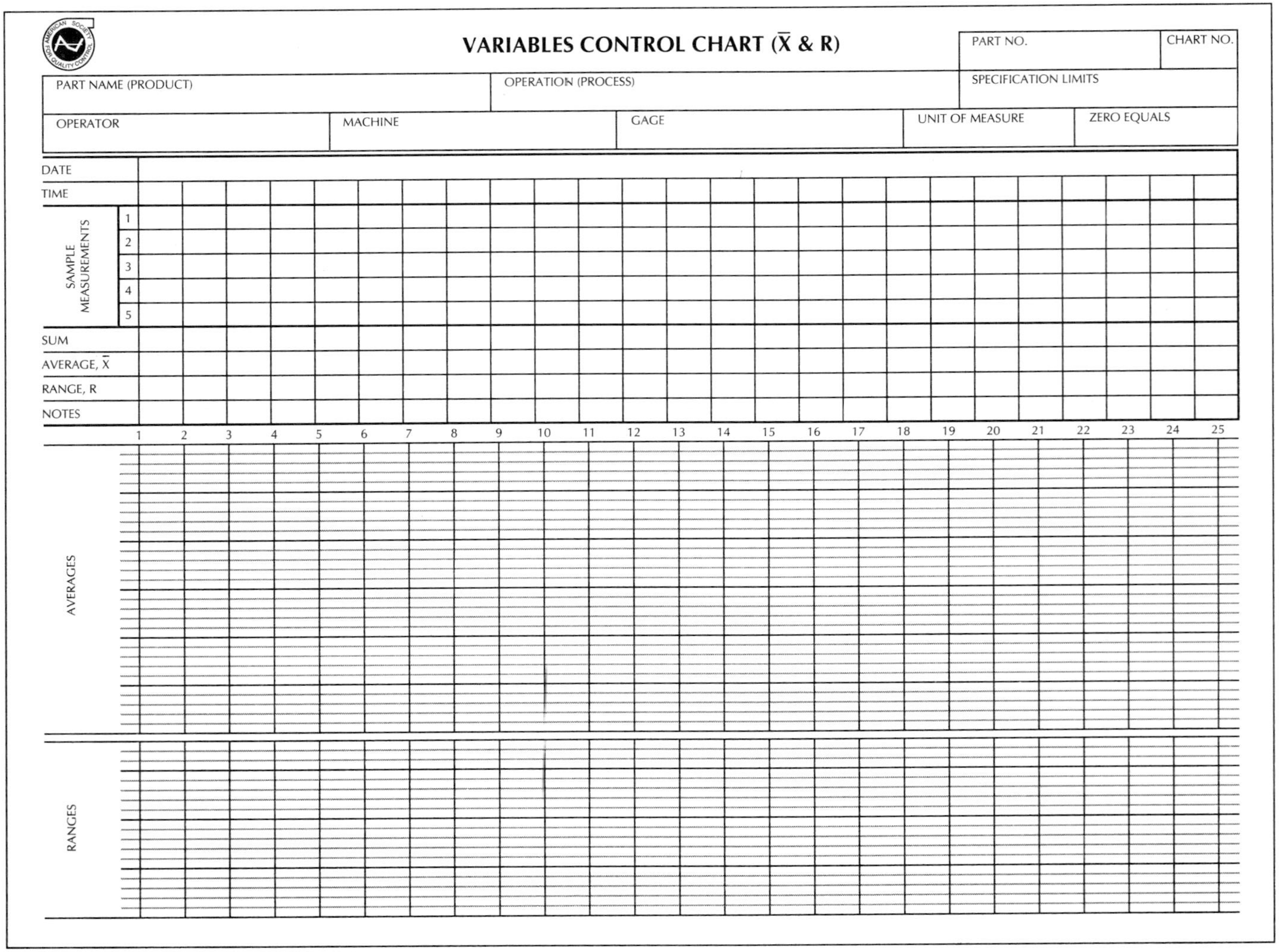


**FIGURE 4.2**
A control chart form distributed by the ASQC. (*Source:* Reprinted with the permission of the ASQC.)

*The best way to learn how to construct a chart is to do it.* The form used is the form developed and distributed by the ASQC and shown in Figure 4.2.

Begin constructing the chart by completing all header information. Properly completing this important information cannot be stressed too much. A great deal of data is useless to others because header information is inaccurate or incomplete. The chart's header is completed as shown in Figure 4.3.

From the header information the technologist or manager may be certain of the exact product the study concerns and of the quality characteristic that is being charted. The operators names are recorded. They usually are trained to perform the charting, reducing the need for additional inspectors and giving the operators the chance to control their own processes. The type of gauge being used (in this case a scale for weighing) and the specific machine are also noted.

The data may be coded to allow faster recording of the data. In this particular case the "zero equals" block could have a 100 entered (zero = 100), and the first data point, 187.0, would then be entered as 87.0. (However, this procedure has not been followed in our example. In order to clarify the example as much as possible, the data are not coded.)

The back of the form has a space for notes. This space is used to record gauge serial numbers, a change in operators, and assignable causes that were identified during the charting process. These assignable causes will be linked to out-of-control points found on the chart.

Most important, you should think of charting as an interactive and creative enterprise. Whoever is doing the charting should be intimately linked with the process as it is running and feel free to note any and all thoughts concerning the potential assignable causes revealed by the chart.

Before moving on with the charting example, it is more meaningful to examine the actual process. Please note that the chart under construction is the one illustrated in Figure 1.10, a part of the fill-weight process discussed in the Case Study—Chapter 1. As a continuation of that discussion, let's take a look at some of the equipment being used in the process. Figure 4.4a shows the product entering the four fill cylinders located below the filler bowl (the hopperlike object in the top-center of the photograph). Figure 4.4b is a closeup of the fill cylinders with the screw adjustment (fill adjustment) visible on one of the

**VARIABLES CONTROL CHART ($\bar{X}$ & R)**

| PART NO. | CHART NO. |
|---|---|
| N/A | One |

| PART NAME (PRODUCT) | OPERATION (PROCESS) | SPECIFICATION LIMITS |
|---|---|---|
| 170 g (6 oz.) yogurts | Fill | 182.0 to 190.0 |

| OPERATOR | MACHINE | GAGE | UNIT OF MEASURE | ZERO EQUALS |
|---|---|---|---|---|
| Quality team | Filler bowl and valve assy. | Top-loading scale | Grams | 0 |

DATE: 6-22-91 (A.M.)

| TIME | | 9:05 | 9:10 | 9:15 | 9:20 | 9:25 | 9:30 | 9:35 | 9:40 | 9:45 | 9:50 | 9:55 | 10:00 | 10:05 | 10:10 | 10:15 | 10:20 | 10:25 | 10:30 | 10:35 | 10:40 | 10:45 | 10:50 | 10:55 | 11:00 | 11:05 |
|---|---|---|---|---|---|---|---|---|---|---|---|---|---|---|---|---|---|---|---|---|---|---|---|---|---|---|
| SAMPLE MEASUREMENTS | 1 | 187.0 | 176.2 | 187.0 | 187.0 | 176.3 | 184.0 | 181.3 | 183.1 | 184.3 | 176.9 | 181.0 | 183.3 | 178.2 | 181.5 | 183.6 | 184.5 | 184.9 | 175.0 | 182.1 | 185.2 | 184.9 | 187.6 | 187.7 | 186.0 | 185.7 |
| | 2 | 179.6 | 179.5 | 182.9 | 179.6 | 180.0 | 184.2 | 183.0 | 184.2 | 177.0 | 181.0 | 183.3 | 184.5 | 181.3 | 183.5 | 184.8 | 178.5 | 177.9 | 182.2 | 183.9 | 179.1 | 185.1 | 185.6 | 185.8 | 186.0 | 185.7 |
| | 3 | 181.6 | 181.7 | 182.9 | 183.7 | 182.9 | 183.0 | 184.2 | 176.9 | 181.2 | 183.2 | 184.4 | 178.4 | 183.5 | 184.7 | 178.0 | 180.3 | 182.3 | 183.7 | 185.1 | 181.6 | 179.9 | 185.6 | 185.8 | 185.6 | 187.6 |
| | 4 | 184.0 | 184.0 | 184.0 | 184.1 | 184.1 | 184.2 | 177.0 | 180.1 | 183.2 | 184.3 | 177.6 | 180.1 | 184.6 | 178.6 | 182.4 | 183.4 | 183.7 | 189.8 | 179.3 | 183.9 | 182.0 | 176.7 | 176.9 | 185.6 | 187.6 |
| | 5 | 185.3 | 185.3 | 185.4 | 185.4 | 185.5 | 183.9 | 180.6 | 183.1 | 184.4 | 177.3 | 180.6 | 183.3 | 178.7 | 182.5 | 183.6 | 184.6 | 185.1 | 179.6 | 182.0 | 185.2 | 183.7 | 185.7 | 185.9 | 185.5 | 188.2 |
| SUM | | | | | | | | | | | | | | | | | | | | | | | | | | |
| AVERAGE, $\bar{X}$ | | 183.5 | 181.3 | 184.4 | 184.0 | 181.8 | 183.9 | 181.2 | 181.5 | 182.0 | 180.5 | 181.4 | 181.9 | 181.3 | 182.2 | 182.5 | 182.3 | 182.8 | 182.1 | 182.5 | 183.0 | 183.1 | 184.2 | 184.4 | 185.7 | 187.0 |
| RANGE, R | | 7.4 | 9.1 | 4.1 | 7.4 | 9.2 | 1.2 | 7.2 | 7.3 | 7.4 | 7.4 | 6.8 | 6.1 | 6.4 | 6.1 | 6.8 | 6.1 | 7.2 | 14.8 | 5.8 | 6.1 | 5.2 | 10.9 | 10.8 | 0.5 | 2.5 |
| NOTES | | | | | | | | | | | | | | | | | | | | | | | | | | |

**FIGURE 4.3**
Accurate header information is vital. (*Source:* Reprinted with the permission of the ASQC.)

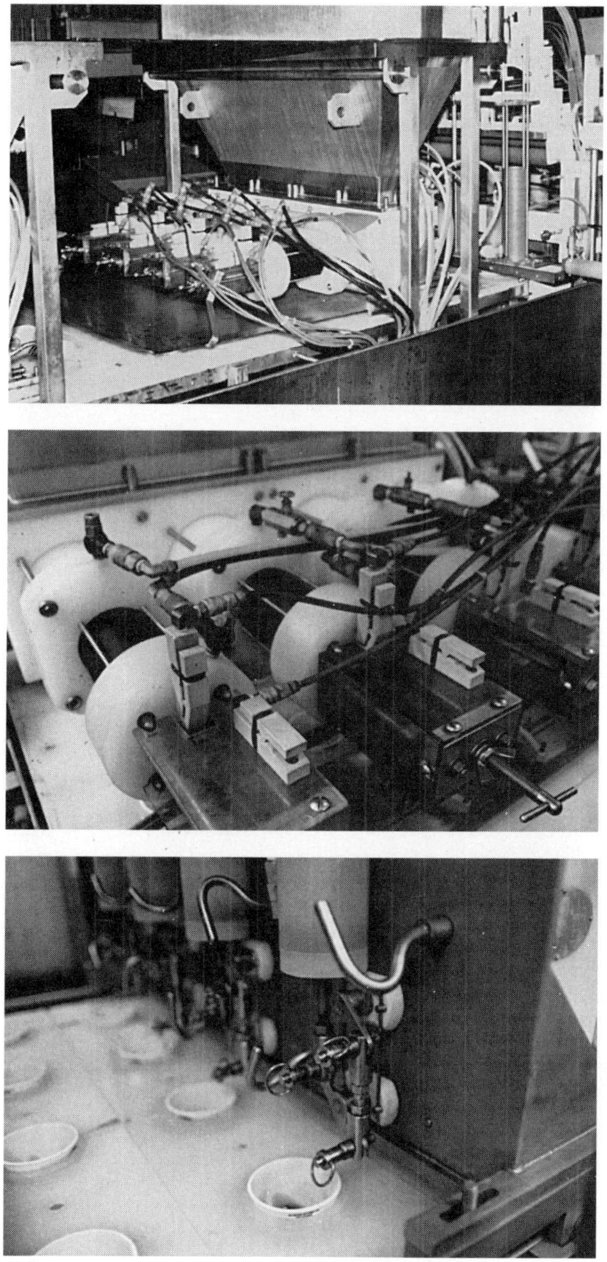

FIGURE 4.4
The machinery used to fill the yogurt cups.
(*Source:* Reprinted by permission of Tamarack Farms Dairy of Newark, Ohio.)

cylinders. Figure 4.4c shows the other end of the fill cylinders (the fill end) with the yogurt cups positioned underneath.

The data portion of the form is completed next (return to Figure 4.3). On the form are blocks provided for up to five measurements to a subsample. A **subsample** contains the data in one column of the chart, a sample of product taken during one time period. (A subsample is often referred to as a **subgroup.**) It is the subsample that yields the average and the range calculations. The subsample number is symbolized by the lowercase letter $n$, and the number $n$ of the observations remains constant for each particular study. In this example, the subsample size is five ($n = 5$), and the data are accurate to four significant figures and one-tenth of a gram.

The next step in constructing the chart is to calculate the $\overline{X}$s or averages of each subsample and the ranges $R$ (Figure 4.3). The lowest number has a square drawn around it, and the highest number is circled, to help you with the range calculation. The form provides a block for the total of each subsample, but it is not necessary to complete this calculation when using a statistical calculator. *At this point, confirm some, if not all, of the* $\overline{X}$ *and* R *calculations.* Here are the calculations for subsample 1:

$$\overline{X} \text{ of subsample 1} = (187.0 + 179.6 + 181.6 + 184.0 + 185.3)/5$$
$$= 183.5$$

$$R \text{ of subsample 1} = 187.0 \text{ (circled)} - 179.6 \text{ (in square)} = 7.4$$

The averages portion of the chart will be completed first. A scale for the $Y$ axis might be determined from inspecting the subsample averages ($\overline{X}$s), but it is best to calculate the control limits first. After calculating the control limits, we will use their spread in determining the $Y$ axis scale factor. Use the "Control Limits" section of the "Calculation Work Sheet" to calculate the control limits for the averages (Figure 4.5). Also see inside front cover.

The subsamples noted in the left-hand column of the "Control Limits" section are all of the subsamples from 1 to 25. The right-hand column will be used later in the study. In Figure 4.5 the calculations are shown for the averages of the range ($\overline{R}$) and of the $\overline{X}$ (the average of the averages is known as the X–double bar: $\overline{\overline{X}}$). The $k$ in the denominator of the equations for the $\overline{R}$ and $\overline{\overline{X}}$ symbolizes the number of subsamples (25). (The symbol $n$ has already been used to symbolize the number within each subsample, 5.)

**Calculation Worksheet**

*Control Limits*

Subgroups included 1–25

$\bar{R} = \frac{\Sigma R}{k} = \frac{169.8}{25} = 6.8$ =

$\bar{\bar{X}} = \frac{\Sigma \bar{X}}{k} = \frac{4570.5}{25} = 182.8$ =

*OR*

$\bar{X}'$ (Midspec. or Std.) $= 186$ =

$A_2\bar{R} = 0.577 \times 6.8 = 3.9$ × =

$UCL_{\bar{X}} = \bar{\bar{X}} + A_2\bar{R} = 186.7$ =

$LCL_{\bar{X}} = \bar{\bar{X}} - A_2\bar{R} = 178.9$ =

$UCL_R = D_4\bar{R} = 2.114 \times 6.8 = 14.4$ × =

*Factors for control limits*

| $n$ | $A_2$ | $D_4$ | $d_2$ | $\frac{3}{d_2}$ |
|---|---|---|---|---|
| 2 | 1.880 | 3.267 | 1.128 | 2.659 |
| 3 | 1.023 | 2.574 | 1.693 | 1.772 |
| 4 | 0.729 | 2.282 | 2.059 | 1.457 |
| 5 | (0.577) | (2.114) | 2.326 | 1.290 |
| 6 | 0.483 | 2.004 | 2.534 | 1.184 |

**FIGURE 4.5**
The calculation work sheet and the factors used in the calculations. (*Source:* Reprinted with the permission of the ASQC.)

The factor $A_2$ is found on the table "Factors for Control Limits" (Figure 4.5). The number circled (0.577) is selected from the row for an $n$ of 5. These statistically derived factors, when multipled by the average range, are used to determine the three standard deviations that the control limits must be from the center line ($\bar{\bar{X}}$) of the process. Plus and minus three standard deviations capture 99.7 percent of the

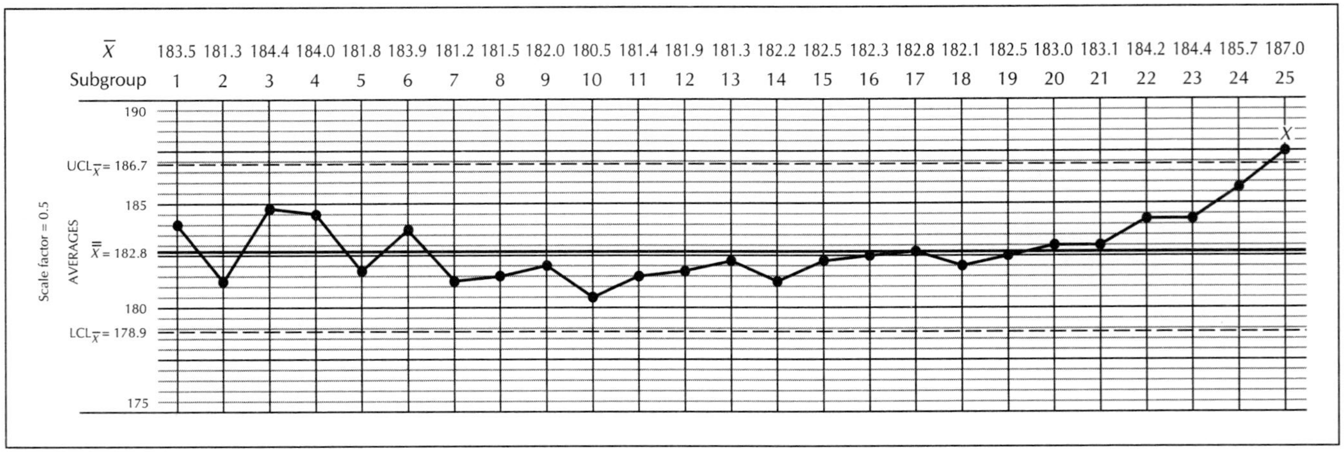


**FIGURE 4.6**
The completed "Averages" section of the control chart (scale factor = 0.5 units).

data if the distribution is normal (Section 3.4). It will be shown later that *by using subsample averages instead of each of the individual data points, the averages will tend toward a normal distribution even if the individual data are not normal.* Appendix C contains a more complete table of the factors used in control charting and capability determination. (Understanding the derivations of these statistical factors is not necessary in constructing $\overline{X}$ and $R$ charts and will not be shown in this text.)

The next step is to calculate the control limits (Figure 4.5). Notice that the $UCL_{\overline{x}}$ and the $LCL_{\overline{x}}$ symbolize the upper and lower control limits for the $\overline{X}$s (not the $\overline{\overline{X}}$s). The $UCL_R$ symbol does not have a bar over the subscript $R$ because the control limit is for each range point ($R$), not for the average range ($\overline{R}$). The lower control limit for $R$ is zero unless the subsample $n$ is at least seven. The equation for the $LCL_R$ is $D_3\overline{R}$, where the statistical factor $D_3$ is zero for $n < 7$ (Appendix C).

Turning to the "Averages" section of the chart (Figure 4.6), the chart's scale factor is determined to be 2.5 units per boldly illustrated line (0.5 units per line). This scale factor *allows the space between the control limits to occupy about two-thirds of the space available.*

$$186.7 - 178.9 = 7.8 \text{ units}$$
$$7.8 \text{ units}/20 \text{ minor divisions} \approx 0.5 \text{ units}/\text{line}$$

Also, see that the scale factor note is placed on the chart. Attempt to use multiples or submultiples of five for the scale factor, since there are five lines to each bold line.

Place the control limits (dashed lines) and the center ($\overline{\overline{X}}$) line (solid) on the chart for averages. Do not concern yourself with not getting the center line exactly in the middle of the chart or on a bold line. Control limit lines as well as the line for averages are more clearly seen if they are not superimposed on a bold line. *Use dashed lines for control limit lines and solid lines for averages.*

Each subsample average $\overline{X}$ is plotted on the chart in Figure 4.6. The scale factor note on the left-hand side helps you to plot each point correctly. Confirm the correct plotting of a few points. The lines connecting the $\overline{X}$ points show the movement of the process between the subsample averages. These lines help you to spot trends as well as other out-of-control signals.

The $R$ chart (titled "Ranges" on the form) is shown in Figure 4.7. Since there are three bold lines on the chart form, a scale factor of one unit per line will place the $UCL_R$ to occupy about two-thirds of the

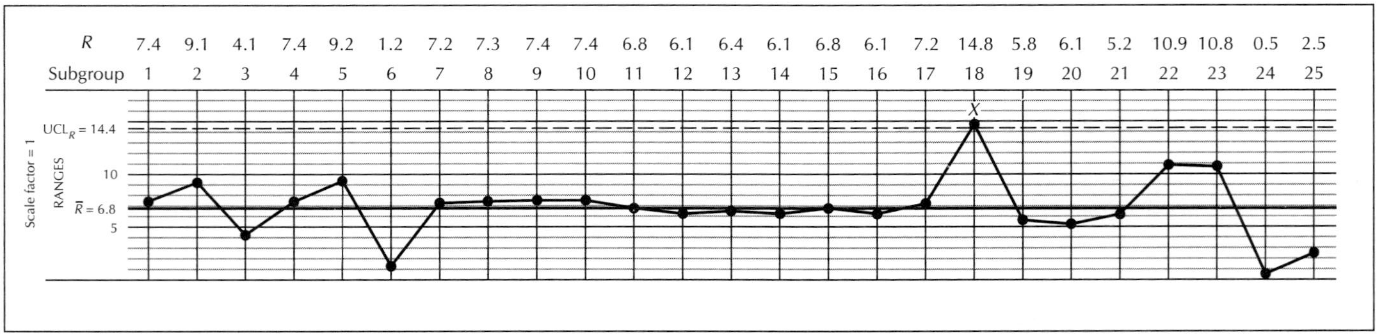


**FIGURE 4.7**
The completed "Ranges" section of the control chart (scale factor = 1 unit).

space available. The same conventions for the drawing of the lines are used as before on the "Averages" chart. Again, you should confirm the placement of the $\overline{R}$ and control limit lines and the correct plotting of the points for $R$.

You perhaps noted that the $\overline{\overline{X}}$ value of 182.8 grams and the specification limits are greater than the 170 g yogurts we are striving for (refer to header information in Figure 4.3). The added weight for the 170 g yogurts is due to the weight of the container (the tare). Also, it is necessary to explain that grams instead of ounces were selected as the unit of measurement, because a gram is a smaller and more sensitive measurement. Another consideration is that measurements in ounces are traditionally made in fractions of an ounce—for example, 1⁄16 oz. The metric or International System of measurement always uses decimal fractions, which are much easier to deal with.

You will also note on the completed $\overline{X}$ and $R$ form shown in Figure 4.8 that the points out-of-control due to being beyond the control limits are marked with an $X$. These points are more than three standard deviations from the center lines, and the probability of their being there is slight to nonexistent—approximately three chances out of 1,000. These points are considered obvious process shifts, and often an assignable cause is operating. If an assignable cause is identified for the two points marked—more on this later—then it is all right to throw out the subsamples associated with the points and recalculate the control limits. We now come to the use of the right-hand side of the "Control Limit" portion of the work sheet (Figure 4.9).

**EXAMPLE 4.1** Confirm the values for the new control limits shown on the work sheet in Figure 4.9.

*Solution:* First we must throw out subsamples 18 and 25. Throwing out these subsamples and adding the remaining subsample ranges result in the sum of 152.5. Since the number of subsamples has declined to 23, then the new denominator ($K$) is 23.

$$\overline{R} = 152.5/23 = 6.6$$

The averages total 4,201.6 after throwing out the out-of-control points. So that number must also be divided by 23 to yield

$$\overline{\overline{X}} = 4{,}201.6/23 = 182.7$$

The calculations show very little effect on the control limits for this study. Some studies show a much greater effect when out-of-

**VARIABLES CONTROL CHART ($\overline{X}$ & R)**

| PART NO. | CHART NO. |
|---|---|
| N/A | One |

| PART NAME (PRODUCT) | OPERATION (PROCESS) | SPECIFICATION LIMITS |
|---|---|---|
| 170 g (6 oz.) yogurts | Fill | 182.0 to 190.0 |

| OPERATOR | MACHINE | GAGE | UNIT OF MEASURE | ZERO EQUALS |
|---|---|---|---|---|
| Quality team | Filler bowl and valve assy. | Top-loading scale | Grams | 0 |

DATE: 6-22-91 (A.M.)

| TIME | 9:05 | 9:10 | 9:15 | 9:20 | 9:25 | 9:30 | 9:35 | 9:40 | 9:45 | 9:50 | 9:55 | 10:00 | 10:05 |
|---|---|---|---|---|---|---|---|---|---|---|---|---|---|
| SAMPLE MEASUREMENTS 1 | 187.0 | 176.2 | 187.0 | 187.0 | 176.3 | 184.0 | 181.3 | 183.1 | 184.3 | 176.9 | 181.0 | 183.3 | 178.2 |
| 2 | 179.6 | 179.5 | 182.9 | 179.6 | 180.0 | 184.2 | 183.0 | 184.2 | 177.0 | 181.0 | 183.3 | 184.5 | 181.3 |
| 3 | 181.6 | 181.7 | 182.9 | 183.7 | 182.9 | 183.0 | 184.2 | 176.9 | 181.2 | 183.2 | 184.4 | 178.4 | 183.5 |
| 4 | 184.0 | 184.0 | 184.0 | 184.1 | 184.1 | 184.2 | 177.0 | 180.1 | 183.2 | 184.3 | 177.6 | 180.1 | 184.6 |
| 5 | 185.3 | 185.3 | 185.4 | 185.4 | 185.5 | 183.9 | 180.6 | 183.1 | 184.4 | 177.3 | 180.6 | 183.3 | 178.7 |
| SUM | | | | | | | | | | | | | |
| AVERAGE, $\overline{X}$ | 183.5 | 181.3 | 184.4 | 184.0 | 181.8 | 183.9 | 181.2 | 181.5 | 182.0 | 180.5 | 181.4 | 181.9 | 181.3 |
| RANGE, R | 7.4 | 9.1 | 4.1 | 7.4 | 9.2 | 1.2 | 7.2 | 7.3 | 7.4 | 7.4 | 6.8 | 6.1 | 6.4 |
| NOTES | | | | | | | | | | | | | |
| | 1 | 2 | 3 | 4 | 5 | 6 | 7 | 8 | 9 | 10 | 11 | 12 | 13 |

| TIME | 10:10 | 10:15 | 10:20 | 10:25 | 10:30 | 10:35 | 10:40 | 10:45 | 10:50 | 10:55 | 11:00 | 11:05 |
|---|---|---|---|---|---|---|---|---|---|---|---|---|
| SAMPLE MEASUREMENTS 1 | 181.5 | 183.6 | 184.5 | 184.9 | 175.0 | 182.1 | 185.2 | 184.9 | 187.6 | 187.7 | 186.0 | 185.7 |
| 2 | 183.5 | 184.8 | 178.5 | 177.9 | 182.2 | 183.9 | 179.1 | 185.1 | 185.6 | 185.8 | 186.0 | 185.7 |
| 3 | 184.7 | 178.0 | 180.3 | 182.3 | 183.7 | 185.1 | 181.6 | 179.9 | 185.6 | 185.8 | 185.6 | 187.6 |
| 4 | 178.6 | 182.4 | 183.4 | 183.7 | 189.8 | 179.3 | 183.9 | 182.0 | 176.7 | 176.9 | 185.6 | 187.6 |
| 5 | 182.5 | 183.6 | 184.6 | 185.1 | 179.6 | 182.0 | 185.2 | 183.7 | 185.7 | 185.9 | 185.5 | 188.2 |
| SUM | | | | | | | | | | | | |
| AVERAGE, $\overline{X}$ | 182.2 | 182.5 | 182.3 | 182.8 | 182.1 | 182.5 | 183.0 | 183.1 | 184.2 | 184.4 | 185.7 | 187.0 |
| RANGE, R | 6.1 | 6.8 | 6.1 | 7.2 | 14.8 | 5.8 | 6.1 | 5.2 | 10.9 | 10.8 | 0.5 | 2.5 |
| NOTES | | | | | | | | | | | | |
| | 14 | 15 | 16 | 17 | 18 | 19 | 20 | 21 | 22 | 23 | 24 | 25 |

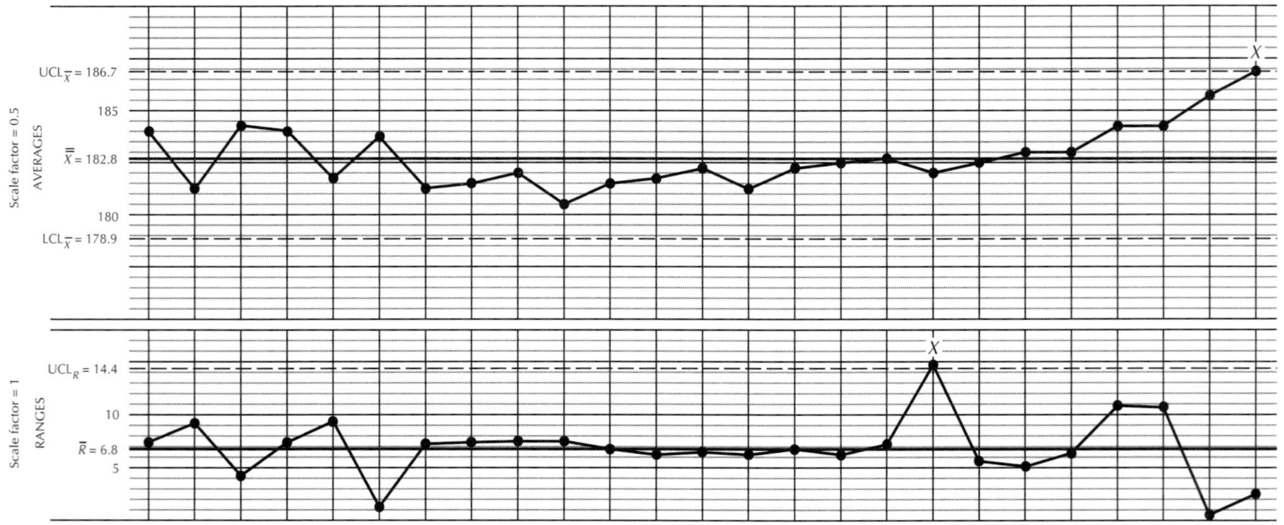


**FIGURE 4.8**
The completed $\overline{X}$ and $R$ chart.

**Calculation Worksheet**

*Control Limits*

| | | | | |
|---|---|---|---|---|
| Subgroups included | 1-25 | | 1-25 (excluding OOC points 18 and 25) | |
| $\overline{R} = \frac{\Sigma R}{k} =$ | $\frac{169.8}{25}$ | $= 6.8$ | $\frac{152.5}{23}$ | $= 6.6$ |
| $\overline{\overline{X}} = \frac{\Sigma \overline{X}}{k} =$ | $\frac{4570.5}{25}$ | $= 182.8$ | $\frac{4201.6}{23}$ | $= 182.7$ |
| $\overline{X}'$ (Midspec. or Std.) | | $= 186$ | | $=$ |
| $A_2\overline{R} =$ | $\times$ | $=$ | $0.577 \times 6.6 = 3.8$ | |
| $UCL_{\overline{X}} = \overline{\overline{X}} + A_2\overline{R}$ | | $=$ | $182.7 + 3.8 = 186.5$ | |
| $LCL_{\overline{X}} = \overline{\overline{X}} - A_2\overline{R}$ | | $=$ | $182.7 - 3.8 = 178.9$ | |
| $UCL_R = D_4\overline{R} =$ | $\times$ | $=$ | $2.114 \times 6.6 = 13.9$ | |

**FIGURE 4.9**
After eliminating two out-of-control points, the control limits are recalculated.

control points are eliminated. The new control limits are used as baselines in charting future studies and in the calculation of process capability (yet to be addressed). *Out-of-control points must not be eliminated unless the assignable cause is identified.*

It is usually better to just rerun and rechart the process after assignable causes have been corrected. Throwing out several out-of-control points often reveals other out-of-control conditions, thus revealing other assignable causes.

# *Section 4.2 Review Problems*

1. Average ($\overline{X}$) calculations on a control chart come from
   a. all samples
   b. each subsample ($n$)
   c. the highest and lowest data points
   d. the average of the averages

2. To measure the variability of the data on an $\overline{X}$ and $R$ chart, one would use the
   a. range
   b. standard deviation
   c. mode
   d. median
3. For 25 subgroups, an $n$ of 4, $\overline{R}$ of 12, and $\overline{\overline{X}}$ of 65, find the $UCL_{\bar{x}}$, $LCL_{\bar{x}}$, $UCL_R$, and $LCL_R$.
4. Given an $n$ of 4, $\overline{R}$ of 0.08, and $\overline{\overline{X}}$ of 0.043, find the $UCL_{\bar{x}}$, $LCL_{\bar{x}}$, $UCL_R$, and $LCL_R$.
5. Explain the difference between the literal factors $n$ and $k$. See the Glossary of Symbols printed inside the back cover.

# 4.3 *Chart Analysis*

## *Out-of-Control Signals*

**Out of control** means that a process has changed from a system of chance causes to a system of chance and assignable causes. When assignable causes are present, the process is no longer predictable (Figure 2.10). It is important to plant personnel in a manufacturing facility to know that their processes are consistent. Service industry personnel also need to know whether or not their services will be consistent. Both groups wish to consistently deliver quality products, and the control chart can confirm they are doing so.

There are only three types of out-of-control signals that you must be concerned with. When analyzing a chart all three should be considered. All three apply to both the average ($\overline{X}$) and range ($R$) charts. The three out-of-control signals are

- point out;
- rule of sevens;
- rule of thirds.

(Inexperienced evaluators usually review the $\overline{X}$ chart first and sometimes ignore the $R$ chart altogether. *Experienced evaluators will first consult the* R *chart.* Indeed, some practitioners only use a range chart in the beginning stages of bringing a process into control. The range chart shows the changing spread of the process and is a more sensitive measure of process consistency.)

You have already seen two out-of-control signals due to a **point out** (Figure 4.8). The first, on the range chart, signifies some type of gross error or instability on the part of the operators or equipment. The second appears at the end of a trend (defined later) on the average chart. (Students often ask about points on the line. If a plotted point is on a control limit line, it is usually best to give the process the benefit of the doubt and not count it as out of control.)

The point out should not be difficult to spot. It is often the only out-of-control condition considered. Actually, there is much more that can be learned from the $\overline{X}$ and $R$ charts. The laws of probability show us that there are signals within the control limits that tell us the process has shifted.

The **rule of sevens** offers two out-of-control (OOC) signals:

- seven points in a row above (Figure 4.10a) or below the $\overline{\overline{X}}$;
- seven points in a row going up (Figure 4.10b) or going down.

In Figure 4.10a the last seven points are below the centerline. What is the probability of seven points in a row, all being above or below the centerline, when only random or chance causes are operating? The answer lies in realizing that the probability that any point will be above or below the centerline is 50 percent, the same as the probability of a head or a tail on the toss of a coin. It follows that the probability of having seven things happen in a row with a probability of 0.5 on each measurement would be

$$P(\text{7 points in a row above or below the } \overline{\overline{X}}) = 0.5^7 = 0.0078 = 0.78\%$$

There is a 0.78 percent chance of this being a system of chance causes. The interpreter of this chart would be advised to find a reason for the out-of-control condition.

EXAMPLE 4.2

You are using a control chart to monitor the time of service in a large restaurant. Every one-half hour the average time of service to three customers is recorded on the chart. The centerline and control limits have been set in previous chart studies. Between 1 P.M. and 5 P.M. every day you notice that all the points recorded are above the $\overline{\overline{X}}$ on the chart. What will you tell the manager of the restaurant?

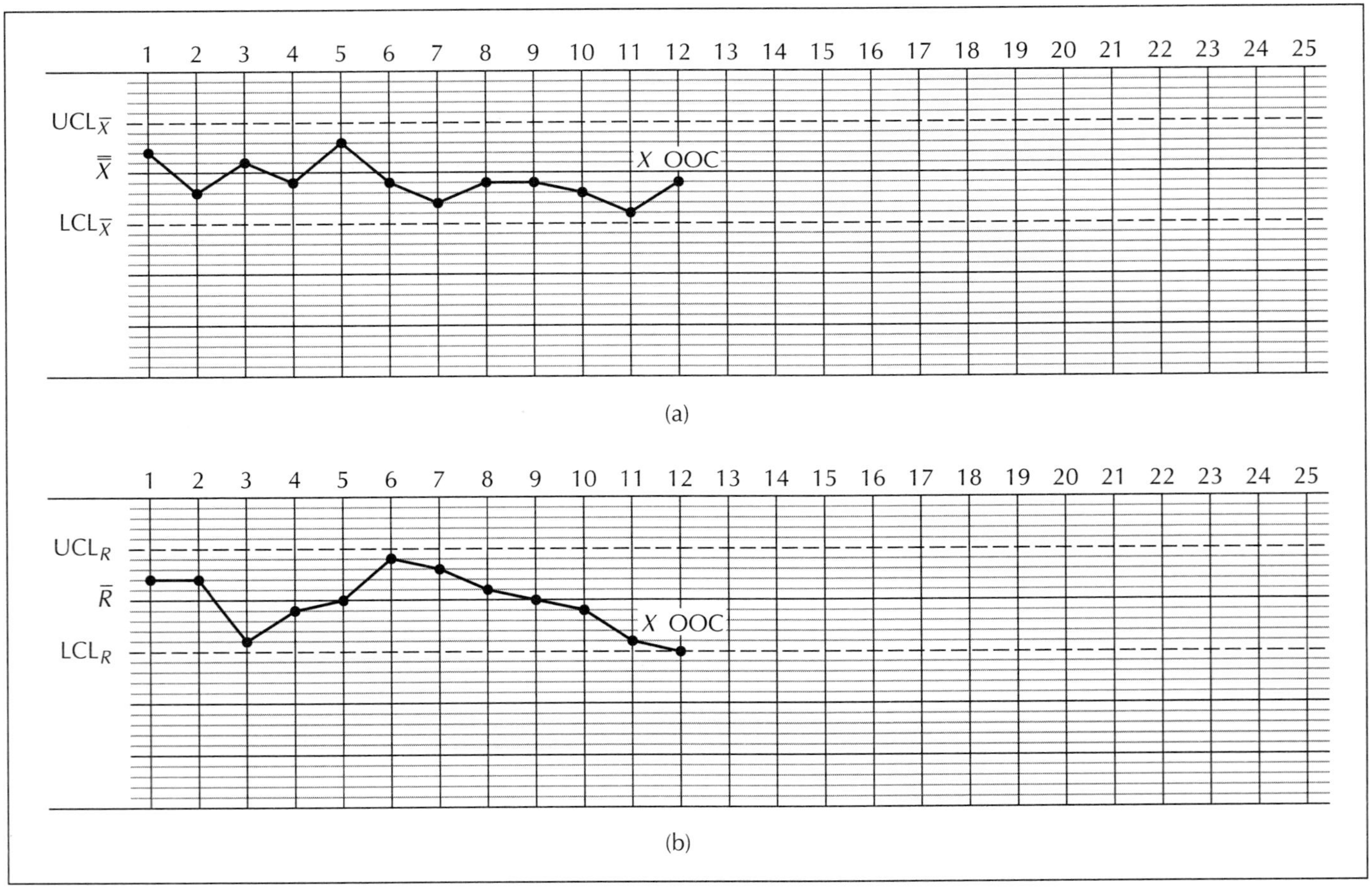


**Figure 4.10**
Two violations of the rule of sevens.

*Solution:* During the four-hour period you plot eight subsample averages. The probability of all of the points being above the line is less than 0.4 percent ($0.5^8$). After some investigation and discussions with the waiters, you inform the manager that, because of breaks after the noon hour (assignable cause), the staff is not capable of meeting customer needs. You suggest to the manager that evening shift servers be brought in early to allow for staff breaks after the noon hour.

Patterns of points going up or down are known as **trends** (Figure 4.10b). With only chance causes operating, what is the probability of one point being above or below its neighboring point? Again, the answer is 50 percent. So, in a system of chance causes, the probability of seven points in a row being consistently either above or below the last one is the same as that calculated for seven points above or below the centerline. The 0.78 percent probability of this being a system of chance causes, again, is a strong indication that an assignable cause is acting to shift the process. The chart interpretation must be that an assignable cause is operating, and it would be advantageous to all that it be identified.

Trends happen often in manufacturing. The usual cause is tool or die wear. Take, for instance, the die on a press operation that is used to shear parts. Naturally the die will wear a slight bit on each stroke of the press. Eventually the die will be so worn that the parts are going to be larger than desirable. (The inner diameter [ID] of a hole will become less for a tool like a die punch.) Some manufacturers watch these trends on a chart and know exactly when to change dies. A new die can be readied for installation, and the production team can install the die in a timely manner, so the expensive press may be utilized a greater percent of the time, continuing the production of high-quality parts. See also Section 5.3, ''Trend Charts.''

The rule of thirds is the third and final out-of-control signal to identify. The chart in Figure 4.11 shows a violation of the rule of thirds. We will use the normal curve drawn on the side of the chart to understand this violation of the rule of thirds.

With chance causes operating, how many of the points on the chart should be near the centerline (Zone 1)? Reflecting back to the percent probabilities under the normal curve presented in Section 3.4, your answer is ''about two-thirds.''

The **rule of thirds** simply recognizes that for chance causes alone to be operating there should be about two-thirds of the points within

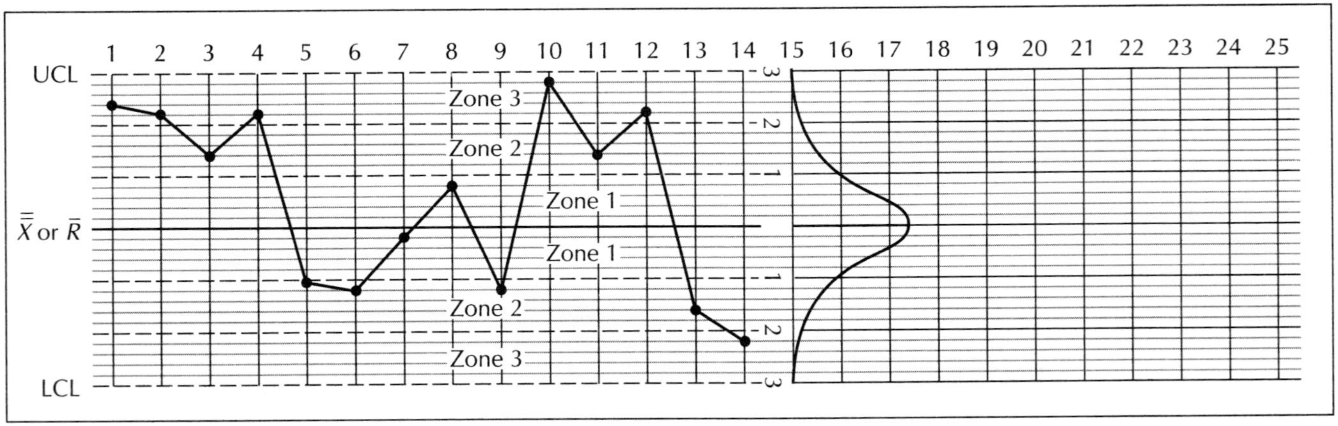


**Figure 4.11**
A violation of the rule of thirds.

plus and minus one standard deviation and the other one-third near the control limits.

There are too many points near the control limits (Zone 3) on the chart in Figure 4.11. The actual fraction in the two-to-three-standard-deviation region is almost one-half. If this were an *R* chart involving a machine process in manufacturing, you might conclude that there is a lot of variability in the backlash and that the machine needs maintenance.

(Several different machines being studied with one chart may also produce too many points near the control limits. A histogram of the distribution from a chart of the averages [$\overline{X}$s] with points near the control limits is bimodal. Other causes for the bimodal distribution as seen on the average chart are multispindle machines, such as screw machines; batch processes where periodic mixing is involved; and cold startups when the machine backlash is maximum.)

EXAMPLE 4.3

You are being interviewed for a quality technologist position by the quality manager at a brake-manufacturing company. The brake cylinder bores must be smooth for the piston to work smoothly and not wear out too fast (see a previous example of this process in Chapter 2, Example 2.1). So, smoothness of bore is an important product quality characteristic. The chart from the process shows all of the averages ($\overline{X}$s) to be very close to the centerline ($\overline{\overline{X}}$), which disobeys the rule of thirds.

The quality manager says, "Isn't this a great chart?" You know that the chart shows out of control, but you are not sure that the manager knows, and you certainly don't wish to embarrass your interviewer.

*Solution:* You cautiously answer, "It does look *awfully* good," giving the manager a puzzled look. The manager catches your drift, smiles, and says, "Yeah, it looks too good!"

The manager is impressed with the way you handled the test and warms up by asking what you think could have caused the out-of-control condition. You answer by pointing out that the control limits could have been set some time ago and the process is running much better than before. In this case, the manager should recommend a new chart be constructed if further improvement is desired. You also note that the inspectors may be editing the data. **Editing the data** is a kind way of saying that measurements are not done randomly; for example, inspectors take several measurements and record only the best readings, and the result is better-than-expected smoothness-of-bore measurements.

When does an out-of-control condition become an assignable cause? Only when there has been a cause identified for the out-of-control condition. Simply to throw out points and to recalculate control limits makes the chart useless. The value of the chart is to identify the assignable causes. The removal of an assignable cause will lead to process improvement if the cause is not good for the process, or the assignable cause must be continually implemented if the cause is good for the process.

"Assignable causes ain't always bad." The use of incorrect English, in this case, is meant to emphasize the importance of objectively dealing with out-of-control conditions. If we are monitoring time of service, for example, and the chart shows out of control to the low side of the average chart, then we want to know the assignable cause so it may be implemented. "Lower is better" in this case, and shorter time of service is desirable.

## *Averages and Individuals*

Before continuing our study, we need to see why subsample averages are used to calculate control limits rather than using all of the individual data in the calculations. There are two major reasons. The reasons are that, when compared to individual data, averages

1. tend to form a more normal distribution;
2. display less variability.

Both phenomena can be demonstrated by using a deck of cards. The distribution of a deck of cards is flat. Think of the ace as the number one and the jack, queen, and king as the numbers 11, 12, and 13, respectively. Since there are four cards of each value, the distribution (or shape of the histogram of the individual values) is flat as shown by the histogram in Figure 4.12a.

Use subsamples of five to find the distribution of the averages. (You should enjoy doing this five-card stud experiment yourself.) Random sampling may be accomplished by shuffling the deck before each sample of five is drawn. If the shuffle is well done—if the sample is truly a random representation of the deck of 52 cards—the average of each subsample will tend to gravitate toward the center. The histogram in Figure 4.12b shows the result of one such experiment using 50 "hands" of five cards. (Yours will be different of course.)

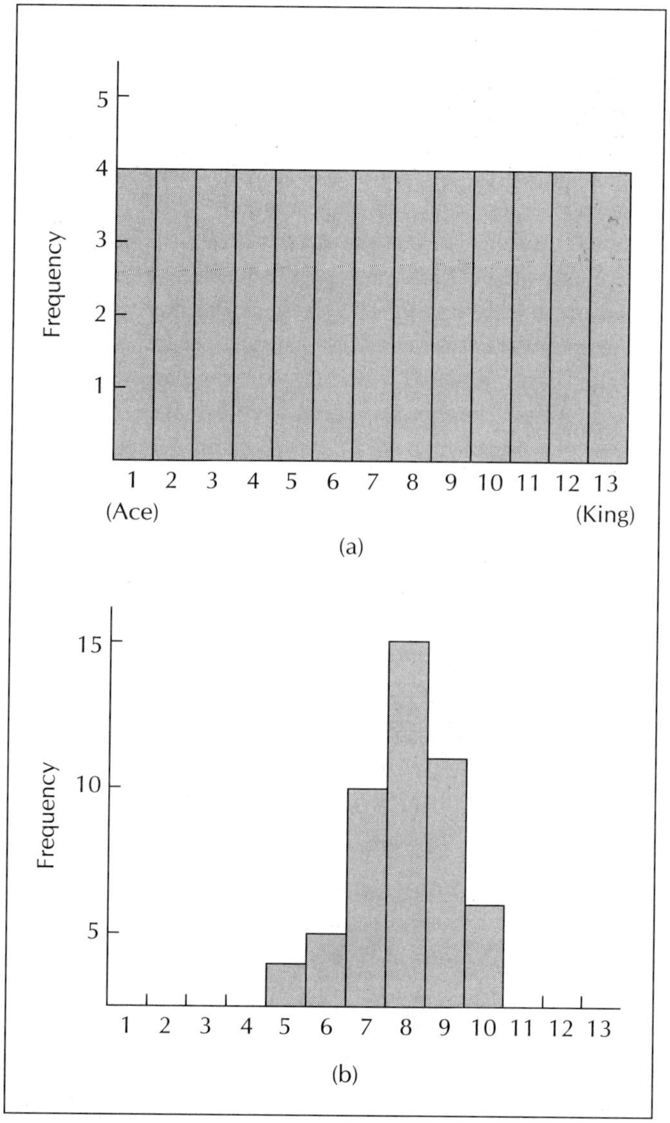


**FIGURE 4.12**
Distribution of a deck of cards: (a) histogram of individuals; (b) histogram of fifty averages (five to a subsample).

From the experiment and Figure 4.12b, we can see that averages tend to cluster about the middle or tend toward a normal distribution (reason number 1). The control chart statistics, like the interpretation of the standard deviation zones on the chart (Figure 4.11), depend on assuming that the distribution of the averages is normal.

This discussion leads to an important statistical theorem known as the central limit theorem. The **central limit theorem** states:

> No matter how the population distribution is shaped, average values ($\overline{X}$s) computed from samples of size $n$ drawn from that population will tend toward a normal distribution as the sample size $n$ increases.

The equation that describes the central limit theorem is

$$\sigma_{\bar{x}} = \frac{\sigma}{\sqrt{n}}$$

where sigma ($\sigma$) is the standard deviation of the individuals,
sigma-sub $\overline{X}$($\sigma_{\bar{x}}$) is the standard deviation of the averages, and
$n$ is the subsample size.

What would be the effect on the standard deviation of the averages $\sigma_{\bar{x}}$ as subsample size increases? Since $n$ is in the denominator, it lowers the standard deviation or spread of the averages (reason number 2). Figure 4.12 (for $n = 5$) shows us that the range of the individuals is 1 through 13 ($R = 12$). The range of the averages is only 5 through 10 ($R = 5$).

**EXAMPLE 4.4** What happens to the width of the control limits on a chart as the subsample size is increased?

*Solution:* As the $n$ increases in the equation for the central limit theorem, the standard deviation of the averages ($\sigma_{\bar{x}}$) must decrease. Since the standard deviation is directly related to the dispersion, the dispersion of the averages and the control limits for the averages must also decrease. Example calculations are shown for a fixed standard deviation of the individuals (25).

| $n = 3$ | $n = 6$ | $n = 12$ |
|---|---|---|
| $\sigma_{\bar{x}} = \frac{\sigma}{\sqrt{n}} = \frac{25}{\sqrt{3}} = 14.4$ | $\sigma_{\bar{x}} = \frac{25}{\sqrt{6}} = 10.2$ | $\sigma_{\bar{x}} = \frac{25}{\sqrt{12}} = 7.2$ |

The decreased dispersion of the averages compared to the individuals means that the averages are less sensitive to a process change. In other words, if the averages change, then you will be more sure that the process has changed. To give you some idea of the probabilities involved, consider that a chart using a subsample average of four ($n = 4$) will show out of control when it *shouldn't* only 16 percent of the time. On the other hand, an individual chart ($n = 1$, discussed in Chapter 9) will show out of control when it *shouldn't* 84 percent of the time.

## Section 4.3 Review Problems

1. To identify a trend, one would use the rule of ________________.
2. Excessive looseness or backlash in a machine would most likely disobey the rule of __________________________.
3. How many of the following conditions reflect an out-of-control condition on a control chart: 32 percent of the points are greater than one sigma; 95% of the points are within one sigma; 95% of the points are beyond one sigma; and 4 successive points are moving up.
   a. one
   b. two
   c. three
   d. four
4. As the subsample number decreases, the spread of the control limits __________________________ (increases, decreases).
5. What is the difference between the spread of individual data values and of average subsample values?

# 4.4 *Process Capability*

There is a great difference between control limits and specification limits. **Control limits** are calculated when using the machines, materials, and personnel involved in the actual process. **Specification limits** come from design parameters. These design specifications may have nothing to do with the actual machines, materials, and other parts of the processes that produce the product. Design for manufacturing is becoming increasingly important to research and development departments (Chapter 10). **Design for manufacture** is a system that better links the capability of the processes that produce the product with the overall design and specifically the product's specifications.

So, being out of control does not mean being out of specification. In fact, many in-control processes make products that are out of specification. *Control limits should never be thought of as specification limits.*

**Capability** ($C_p$) calculations change the control limit values so the actual process distribution may be compared with the specifications. To understand this relationship we will first calculate the capability for the fill-weight process using the revised control limits shown in Figure 4.9 (right-hand column). If a process is obviously out of control, capability is meaningless, so all assignable causes should be identified and the related out-of-control points removed before calculating capability with the revised control limits.

Figure 4.13 is from the back of the ASQC's control chart form. The title "**Limits for Individuals**" describes the work sheet as a system of equations that transform the narrower process control limits to the wider limits of the individual data. Examining the original data from the fill-weight example, you will note that individual data deviate far more than the subsample averages, a phenomenon discussed in Section 4.3 (the central limit theorem). The first calculation (the $\overline{\overline{X}}$ and $\overline{R}$ values are simply carried over from Figure 4.9) is

$$3\hat{\sigma} = \frac{3}{d_2}\overline{R} = \frac{3}{2.326} \times 6.6 = 8.51$$

The $d_2$ is another statistical factor taken from Figure 4.5b. (Note that $3/d_2$ is also given as 1.290 for $n = 5$.) This factor divided into the number 3 (representing three standard deviations) transforms the range value into the approximation—three sigma-hat ($3\hat{\sigma}$)—of the population ($\hat{\sigma}$ is the ratio of $\overline{R}/d_2$). The hat (ˆ) on the sigma indicates the standard deviation of the individual data (which is only a sample from the population) is an estimate of the standard deviation of the population. The result is 8.51.

The upper limit for the individuals (do not confuse with the upper specification limit) is found by adding 8.51 to the $\overline{\overline{X}}$ of 182.7: $UL_X = 182.7 + 8.51 = 191.2$. And the lower limit for the individuals $LL_X$ is found to be 174.2. To summarize, the estimated range of weight for the entire population of 170 g (6 oz.) yogurts, including the weight of

**Limits for individuals**

*Compare with specification or tolerance limits*

$\bar{\bar{X}} = 182.7$

$\frac{3}{d_2}\bar{R} = \frac{3}{2.326} \times 6.6 = 8.51$

$UL_X = \bar{\bar{X}} + \frac{3}{d_2}\bar{R} = 191.2$

$LL_X = \bar{\bar{X}} - \frac{3}{d_2}\bar{R} = 174.2$

USL $= 190.0$

LSL $= 182.0$

USL − LSL $= 8.0$

$6\sigma = \frac{6}{d_2}\bar{R} = 17.0$

**FIGURE 4.13**
The ASQC form "Limits for Individuals" provides equations to calculate process capability. (*Source:* Reprinted with the permission of the ASQC.)

the container (tare), is 174.2 to 191.2 g. This naturally occurring process weight range ($UL_x - LL_x = 17$) is then compared with the specifications desired by the food-processing company (see USL and LSL on the form). (The specifications taken from the header information in Figure 4.8 are transferred to Figure 4.13.) It is readily apparent that the variability of the process is still too high. The range of the specification is the lower spec limit minus the upper spec limit

$$\text{USL} - \text{LSL} = 8.0 \text{ g}$$

The estimated $3\hat{\sigma}$ is multiplied by 2 at the bottom of Figure 4.13 because we are interested in the three-sigma deviation on each side of the centerline. The result of doubling the $3\hat{\sigma}$ is shown on the last line of the work sheet as 17.0 grams. This value will include 99.7 percent of the actual data points in the population, which the original sample of 125 ($5 \times 25$) was taken from. (If the distribution of the individuals is not normal, the spread of the individuals in the population will be greater than 17.0 grams.)

Using the 17 grams (estimated process spread) and the 8.0 grams (desired process spread), we can calculate the capability of the process $C_p$ by

$$C_p = \frac{\text{Upper spec} - \text{Lower spec}}{6\hat{\sigma}} = \frac{8}{17} = 0.47$$

The capability of the process is 47 percent, which means that a significant amount of the product is beyond specification limits when the process is centered. (Another way to think of this is to invert the original fraction to find that the natural process occupies approximately two times the specification spread: $17/8 \approx 2$). It is obvious that the process is not capable of meeting specification and that nonconforming product is being produced.

## *The Capability Index $C_{pk}$*

A more stringent capability index is $C_{pk}$. Used by some quality control practitioners, it considers the center of the process in calculating process capability. The equations used are

$$C_{pk} = \frac{\text{USL} - \bar{\bar{X}}}{3\hat{\sigma}} \quad \text{or} \quad \frac{\bar{\bar{X}} - \text{LSL}}{3\hat{\sigma}}$$

The term $3\hat{\sigma}$ is used to calculate the capability of one-half of the process distribution at one time. *The minimum of the two calculations is used as the* $C_{pk}$ *value.* Example 4.5 will illustrate.

EXAMPLE 4.5 Solve for the $C_{pk}$ of the fill-weight process.

*Solution:* The calculations are as follows:

$$\frac{\text{USL} - \bar{\bar{X}}}{3\hat{\sigma}} = \frac{190 - 182.7}{8.51} = 0.858$$

$$\frac{\bar{\bar{X}} - \text{LSL}}{3\hat{\sigma}} = \frac{182.7 - 182.0}{8.51} = 0.0823$$

And the minimum value of 0.0823 is selected as the $C_{pk}$. The off-centered process is far from the minimum desired capability of one (100%).

## Continuing Improvement

The project team met and reviewed the original chart (Figure 4.8). The range chart revealed a central area between subsamples 7 and 17 with less variability than the rest of the chart. The average chart also demonstrated less variability in the middle of the run. It was decided to begin the fill operation sometime after the startup. Startups usually are more variable and often violate the rule of thirds. Near the end of the run, the range chart shows a point out, and the average chart displays a

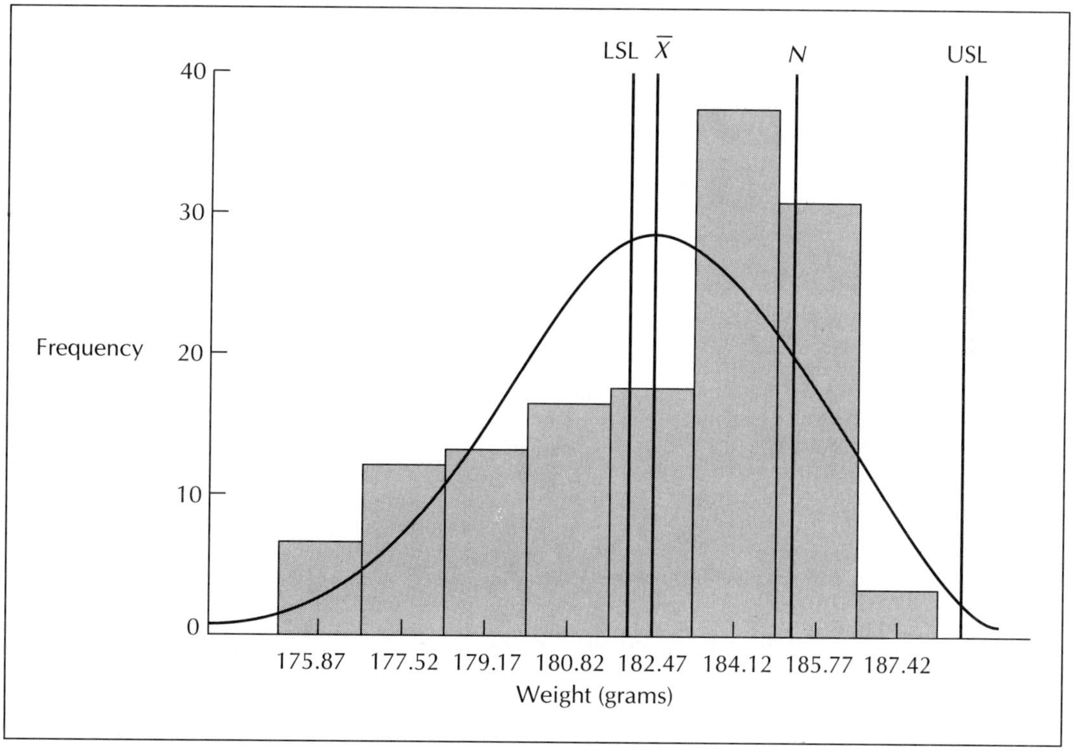


**FIGURE 4.14**
The histogram is not normal, and assignable causes remain.

trend. (Recall that two points—points 18 and 25—were identified as due to assignable causes and thrown out.) Also, the team examined the histogram of the fill process. The histogram, shown in Figure 4.14, is nonnormal. This finding means that the previous capability determinations were only rough estimates and the assignable causes had not been removed.

It was suspected that the pressure head in the filler bowl may have been at fault. A level control in the filler bowl is designed to measure and control the level to within one inch. Maintenance was brought in to check the level control. The level control required calibration, an assignable cause that affected the variability at the end of the run.

Figure 4.15 shows the results of the team's efforts. You should compare this chart with the original in Figure 4.8. The range shows no out-of-control points and, most importantly, the average range ($\overline{R}$) has decreased. The averages have reduced spread in the control limits, with no out-of-control trend signal, but with a centerline ($\overline{\overline{X}}$) at approximately the same place as before ($\overline{\overline{X}} = 182.7$).

With the reduced $\overline{R}$ of 2.4, the following calculations are used to determine the new capability (note that $\hat{\sigma}$ is approximately 1):

$$\hat{\sigma} = \frac{\overline{R}}{d_2} = \frac{2.4}{2.326} \cong 1$$

$$6\hat{\sigma} = 6 \times 1 = 6$$

$$C_p = \frac{\text{USL} - \text{LSL}}{6\hat{\sigma}} = \frac{190 - 182}{6} = 1.33$$

The project team celebrated the new process capability $C_p$ of 1.33. The process variation only occupied about three-fourths of the distance between specification limits. (The inverse of $C_p$ yields the number 0.75 or ¾).

Figure 4.16a shows the distribution after the process was brought into control. (Compare this ogive with the histogram in Figure 4.14) A great deal of product is still out of specification because the process is off center. The team's final step is to center the process. Since the process variability is smaller than the specification limits, the centering can be adjusted so that fill weight will be within the specification limits. The team adjusts the screw adjustments on the fill cylinders (Figure 4.4) to improve the centering of the fill weights. The result, with the new $\overline{\overline{X}}$ at 186 g, is shown in Figure 4.16b.

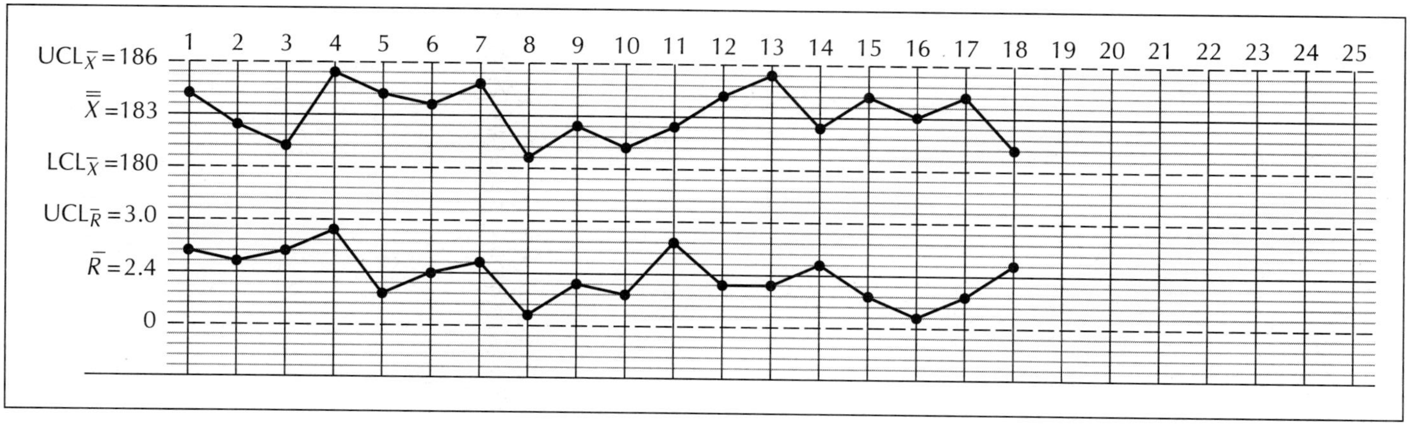


**Figure 4.15**
The project team has brought the fill weights into control.

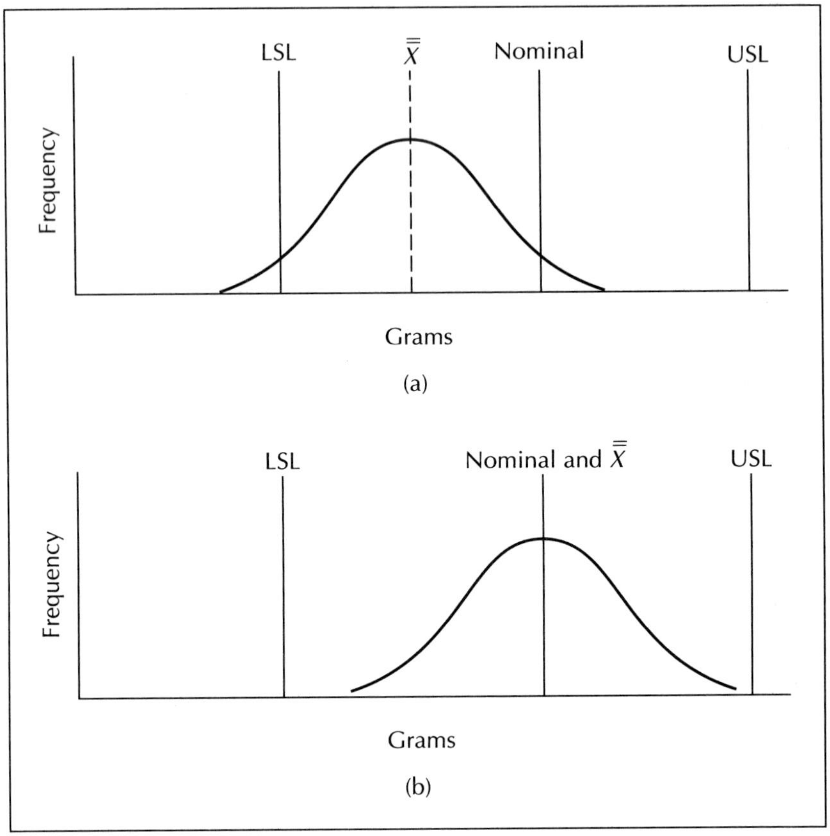


**FIGURE 4.16**
**Distribution for the fill process (a) after the process was brought into control; (b) after the process was centered.**

EXAMPLE 4.6 Solve for the $C_{pk}$ of the fill weight process before and after centering. The standard deviation will be considered to be 1.

*Solution:* The calculations are as follows:

$$C_{pk} = \frac{USL - \bar{\bar{X}}}{3\hat{\sigma}} \quad \text{or} \quad \frac{\bar{\bar{X}} - LSL}{3\hat{\sigma}}$$

$$= \frac{190 - 183}{3} \quad \text{or} \quad \frac{183 - 182}{3}$$

$$= 2.33 \text{ or } 0.333 \text{ (the minimum, 0.333 is selected)}$$

And, after centering

$$C_{pk} = \frac{190 - 186}{3} \quad \text{or} \quad \frac{186 - 182}{3}$$

The new $C_{pk}$ is 1.33.

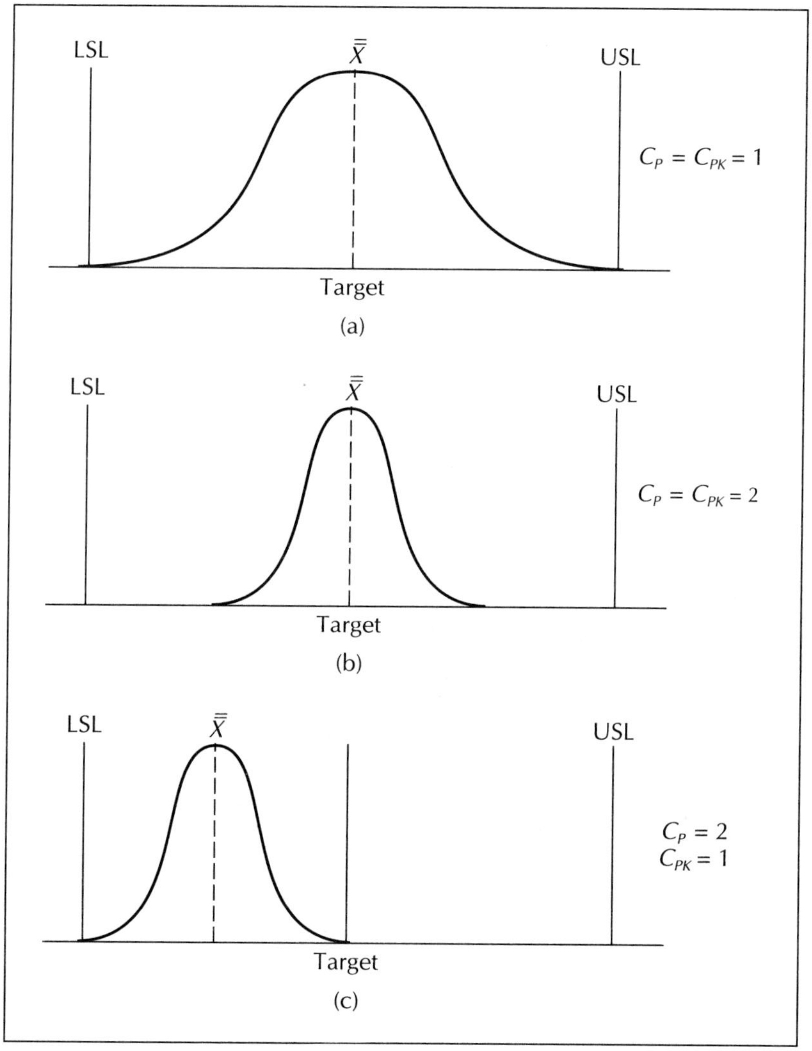


**FIGURE 4.17**
$C_p$ compared to $C_{pk}$.

The variable $C_{pk}$ approaches the value of $C_p$ by the improved centering of the process. To better understand the relationship between $C_{pk}$ and $C_p$, see Figure 4.17. Figure 4.17a shows the process centered with a $6\hat{\sigma}$ variability equal to the width of the specification limits ($6\hat{\sigma}$ = Upper specification limit − Lower specification limit). Both $C_{pk}$ and $C_p$ equal 1 in this case. Figure 4.17b shows the process centered with a variability one-half the width of the specification limits ($\hat{\sigma} = \frac{1}{2}$ [Upper specification limit − Lower specification limit]). Both $C_{pk}$ and $C_p$ have increased to 2. Figure 4.17c shows the process variability to remain the same, but the center has moved between the target value and the lower specification limit. The value of $C_p$ has not changed, because it does not depend on centering but only on variability; $C_{pk}$ exactly fills the area between target and the lower specification limit, decreasing its value to 1. (If the left tail of the distribution is below the lower specification limit LSL, $C_{pk}$ is less than 1.)

One final note: The capability calculations ($C_{pk}$ and $C_p$) are estimates. The estimates become more uncertain if the distribution of the individuals is not normal. Even though the chart for averages tends toward normality, the shape of the histogram for the individuals may not be normal. Many industrial process are not normal, causing the $C_{pk}$ and $C_p$ calculations to be far from perfect. However, it is the feeling of most quality practitioners that capability calculations transform control chart values to a reasonable estimate of the quality of the product population. A warning has already been stated in Chapter 3 that project teams should graph distributions, strive for normality, and attempt to prevent assignable causes from causing nonnormality. As much as normality is desired, however, many processes in the real world will not be normal, even when assignable causes have been removed.

## Section 4.4 Review Problems

1. A control chart is used to study the smoothness of bore of a brake cylinder. The variable $\overline{\overline{X}}$ is 21 microinches, and $\overline{R}$ is 4.8 microinches. The subsample size $n$ is four. If the unilateral tolerance is + 25 microinches (−0 microinches), what is the capability $C_p$? What is the capability $C_{pk}$?
2. A process that is in statistical control will
   a. produce product to specification
   b. consistently pass inspection

c. consistently produce product within statistical control limits
d. both a and c are true

3. A control chart is used to
   a. determine the stability of a process
   b. meet specifications of a product
   c. accept or reject a lot of material
   d. aid management in feasibility studies

4. Which of the following would be considered the most desirable for actual process spread versus product specification?
   a. USL − LSL = $6\sigma$
   b. USL − LSL > $6\sigma$
   c. USL − LSL < $6\sigma$
   d. USL = $3\sigma$

# Summary

- The construction and use of the control chart form the basis of this chapter.
- The control chart is the heart of **statistical process control (SPC),** a problem-solving tool developed by Walter Shewhart in the 1930s (Table 1.1).
- SPC eliminates sporadic (assignable) process changes, so we may work on the chronic chance (or system) causes.
- Prior to charting, the process must be identified and the personnel and resources must be allocated to complete the project successfully.
- Measurement equipment must be checked for accuracy, and repeatability and reproducibility (R&R) studies must be accomplished.
- Control charting involves the calculation of the chart centerlines ($\overline{\overline{X}}$ and $\overline{R}$), and the control limits $UCL_{\overline{X}}$, $LCL_{\overline{X}}$, and $UCL_R$).

$$\overline{X} = \frac{\Sigma X}{n} \ (n = \text{number in subsample})$$

$$R = X_k - X_l \ (\text{within each subsample})$$

$$\overline{R} = \frac{\Sigma R}{k} \ (k = \text{number of subsamples})$$

$$\overline{\overline{X}} = \frac{\Sigma \overline{X}}{k}, \qquad UCL_{\overline{X}} = \overline{\overline{X}} + A_2\overline{R}, \qquad LCL_{\overline{X}} = \overline{\overline{X}} - A_2\overline{R}$$

$$UCL_R = D_4\overline{R}$$

- The chart's scale factor should permit the control limits to occupy approximately two-thirds of each chart (20 minor divisions on the ASQC chart).
- The process control chart is checked for out-of-control conditions. The three out-of-control conditions are point out, rule of sevens, and rule of thirds. Any assignable causes are eliminated, and the process is brought into control.
- A histogram of the data should be normal for capability to be accurately determined, but nonnormal distributions can usually be estimated within 5 percent.
- Process capability ($C_p$) is calculated by

$$C_p = \frac{\text{Upper spec. limit} - \text{Lower spec. limit}}{6\hat{\sigma}}$$

- Process centering may be easily accomplished with $C_p \geq 1.3$. The process may then be moved within the specification limits to center the $\overline{X}$ over the desired or nominal value.
- $C_{pk}$ is calculated by

$$C_{pk} = \frac{USL - \overline{\overline{X}}}{3\hat{\sigma}} \quad \text{or} \quad \frac{\overline{\overline{X}} - LSL}{3\hat{\sigma}}$$

  With the least value being the $C_{pk}$.
- If capability is less than one ($C_p < 1$ or $C_{pk} < 1$), the project team works to eliminate system causes (e.g., improved maintenance, procedures, etc. [MMMMEP]) to further reduce the variability, or to center the process.
- The capability indices $C_{pk}$ and $C_p$ are optional. The statistics of the standard normal curve (Z values) may be used to determine the capability of the process and to track improvement on the process (Chapter 3).

## CASE STUDY · CHAPTER 4

You are a supervisor with a company that supplies weather stripping for car doors (referred to in the industry as door gaskets). As supervisor you have been selected to serve on a quality team to improve the cutoff length of the door gaskets. Door gaskets that are too long or too short lead to immense problems in trying to fit doors to a car on a rapidly moving assembly line.

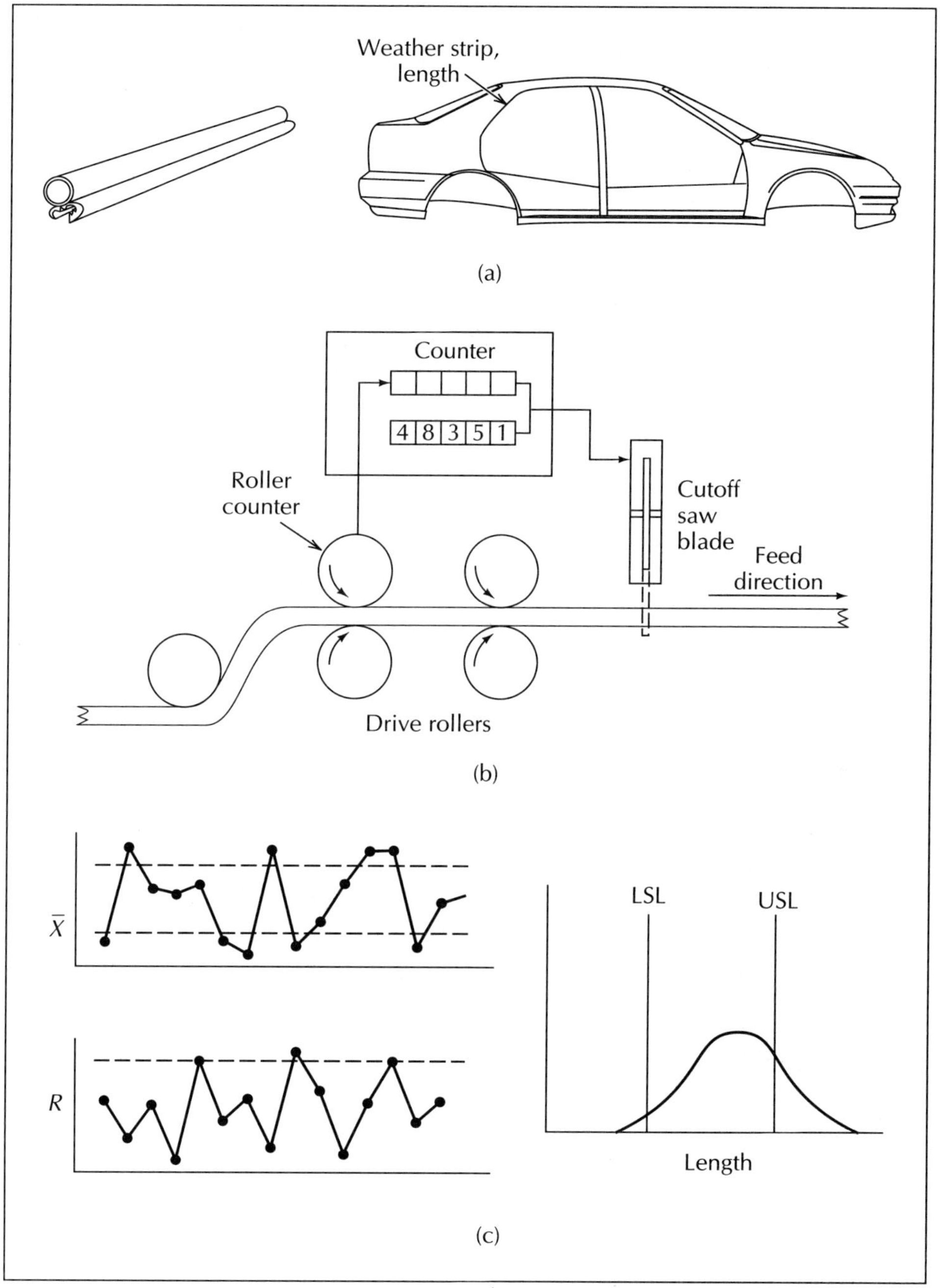


**FIGURE 4.18**
Weather strip placement (a), equipment used in the process (b), and the statistical results (c).

The rubber material has always been difficult to cut to proper length on the automated equipment used for the process. Short lengths represent the specific problem. A drawing of the automated equipment and the associated control chart (outline of pattern only) are provided to the team to facilitate problem solving (Figure 4.18).

The chart shows the process to be consistently out of control on both average and range charts. (Process capability cannot be determined with the process out of control.) The drawing of the process is a tipoff to the assignable cause for variation. As the door gasket feed rate varies, the roller counter slips on the material because of backlash between the feed rollers and the roller counter. Your recommendation to the committee will be for engineering to find a new way to measure cutoff length. What will you do to verify that any of the engineering's departments modifications work?

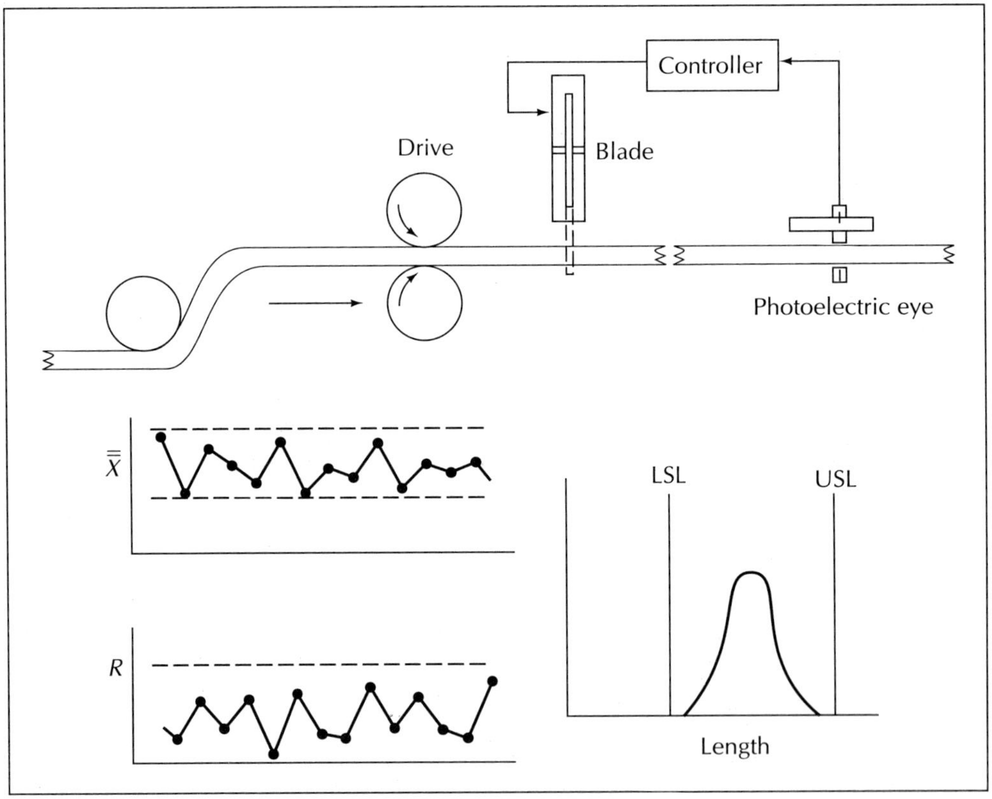


**Figure 4.19**
The process is in control and capable, with the new equipment.

You recommend the committee continue the control chart study it has already implemented. Any changes in the measurement system should be verified by the chart. If the system is in control, the distribution of individual data should also be compared to the original distribution to see if process capability has been improved and to confirm that the improvement is all that is necessary.

Engineering's improvement is depicted in Figure 4.19. The new counter is a photoelectric device. The noncontact measurement is made after cutoff, and the feedback through the process controller is the actual cutoff length information.

The control charts and the new distribution reflect an in-control process. The distribution of the individuals shows the process spread to occupy about one-third of the specification width. The coefficient $C_p$ is excellent at approximately three. All that remains is to adjust the process controller to better center the process. The lengths are still too short.

The quality team reports to management that scrap has been reduced and customers (auto assemblers) will have fewer complaints. More orders should follow. The project team rewards itself with a little party after work.

## *Chapter Review Problems*

1. The following measurements are from a machined bore difficult to hold within tolerance. The specifications are 5.252 to 5.256 inches. Since short runs are used to produce the parts, a small subsample size is used ($n = 2$). Prepare a variable control chart, $\overline{X}$ and $R$, of the data, using all of the proper conventions discussed in the text.

| *Sample* | *Observed Diameters* | | *Sample* | *Observed Diameters* | |
|---|---|---|---|---|---|
| 1 | 5.256 | 5.254 | 11 | 5.255 | 5.256 |
| 2 | 5.253 | 5.256 | 12 | 5.255 | 5.255 |
| 3 | 5.256 | 5.254 | 13 | 5.255 | 5.256 |
| 4 | 5.251 | 5.254 | 14 | 5.257 | 5.255 |
| 5 | 5.255 | 5.254 | 15 | 5.255 | 5.253 |
| 6 | 5.251 | 5.256 | 16 | 5.253 | 5.255 |
| 7 | 5.257 | 5.255 | 17 | 5.254 | 5.255 |
| 8 | 5.256 | 5.257 | 18 | 5.254 | 5.254 |
| 9 | 5.256 | 5.254 | 19 | 5.256 | 5.254 |
| 10 | 5.252 | 5.256 | 20 | 5.252 | 5.254 |

2. Identify the out-of-control points in problem 1. Also discuss any out-of-control conditions due to the rule of thirds. If this is a multiple spindle machine and each spindle drills one of the bores in

the subsample, what could be the reason for the instability as demonstrated by the rule of thirds? What study could you recommend for further improvement?

3. Not counting the points above the control limits, there are at least two out-of-control signals on the fill-weight example (Figure 4.8). Can you identify them?

4. A major airline collects sample data on the arrival time of their aircraft all over the world. Standardized clocks are used to collect the times to the nearest minute. The specifications are set for the arrival time (considered a data point of 0) plus and minus 15 minutes. The data points for subsamples of four ($n = 4$) are given in the following table. Solve for the $\overline{\overline{X}}$ and $\overline{R}$ and the control limit information, and construct the chart using all of the appropriate conventions. Round off all calculations to the nearest minute.

| *Sample* | *Observed Times* | | | | *Sample* | *Observed Times* | | | |
|---|---|---|---|---|---|---|---|---|---|
| 1 | 7 | −8 | −1 | 9 | 11 | 1 | 5 | 9 | 6 |
| 2 | −6 | 1 | 4 | −13 | 12 | −3 | −8 | −8 | −3 |
| 3 | −1 | 2 | 0 | 7 | 13 | −3 | −10 | −20 | −7 |
| 4 | −6 | −8 | −3 | 7 | 14 | 13 | −9 | −4 | −9 |
| 5 | −6 | 4 | 2 | 8 | 15 | −10 | −3 | 0 | −1 |
| 6 | 6 | 8 | −2 | 9 | 16 | 9 | 2 | 4 | −18 |
| 7 | 9 | −8 | −9 | 6 | 17 | −6 | −4 | 3 | 0 |
| 8 | −6 | 9 | 4 | −17 | 18 | 9 | 3 | −15 | −4 |
| 9 | 3 | −4 | 6 | 9 | 19 | −5 | −6 | −4 | 9 |
| 10 | 19 | −9 | −3 | 0 | 20 | 8 | 1 | −5 | −23 |

5. List all of the out-of-control points you find on the chart completed in problem 4. Be sure to consider the rule of sevens and the rule of thirds. What further studies would you recommend for the airlines?

6. Review Example 4.3. Discuss how an unsuitable gauge sensitivity could have been responsible for the conditions cited (review the text explanation of Figure 2.9f and Laboratory—Chapter 3). Could the training of the operator who recorded the data have affected the outcome? Briefly discuss how important appropriate training is to obtaining valid SPC studies.

7. Find the $C_p$ and the $C_{pk}$ for the following processes:
   a. $\overline{\overline{X}} = 90, \overline{R} = 5, n = 4$, LSL = 82, USL = 102
   b. $\overline{\overline{X}} = 7, \overline{R} = 0.4, n = 3$, LSL = 6, USL = 9

c. $\overline{\overline{X}} = 108, \overline{R} = 0.54, n = 5$, LSL = 106, USL = 109
d. $\overline{\overline{X}} = 0, \overline{R} = 0.07, n = 8$ (see Appendix C), LSL = −0.02, USL = 0.02

8. Perform the five-card stud experiment described in the "Averages and Individuals" subsection of Section 4.3. Draw the histograms of the individual distributions and the distribution of the averages. Use Figure 4.12 as a model. Discuss, in one paragraph, your findings concerning the variability of the individual data compared with the averages and the shapes of the two distributions.
9. A part has a nominal dimension of 2.010 inches. The specification limits are ± 0.003 inch. The part was produced on three machines of the same type and on three parallel lines. The following data were taken from the process, and are coded data. The readings are recorded in thousandths of an inch above 2.000 inches.

| *Sample Number* | *Machine A* | *Machine B* | *Machine C* |
|---|---|---|---|
| 1 | 15 | 10 | 10 |
| 2 | 7 | 8 | 8 |
| 3 | 13 | 6 | 9 |
| 4 | 14 | 8 | 9 |
| 5 | 14 | 8 | 11 |
| 6 | 10 | 9 | 12 |
| 7 | 9 | 8 | 11 |
| 8 | 12 | 9 | 10 |
| 9 | 9 | 11 | 11 |
| 10 | 4 | 9 | 10 |
| 11 | 14 | 11 | 9 |
| 12 | 11 | 9 | 12 |
| 13 | 8 | 9 | 10 |
| 14 | 10 | 10 | 10 |
| 15 | 4 | 8 | 9 |
| 16 | 11 | 5 | 13 |
| 17 | 9 | 7 | 9 |
| 18 | 12 | 4 | 11 |
| 19 | 10 | 6 | 11 |
| 20 | 10 | 8 | 10 |
| 21 | 4 | 5 | 10 |
| 22 | 12 | 8 | 12 |
| 23 | 9 | 5 | 12 |
| 24 | 12 | 6 | 13 |
| 25 | 11 | 7 | 11 |

Construct the chart for all three machines, and discuss any in-control and out-of-control conditions.

10. For problem 9, calculate the capability $C_p$. Keep in mind that you are working with coded data.
11. Draw a histogram of machine A's data (problem 9), showing the upper and lower specification limits. Make a definitive statement regarding machine A's contribution and what steps you might take to improve the overall process.
12. Calculate the $C_{pk}$ of the process in problem 9. Is centering a major problem with this process? If so, what should be done to remedy the situation?
13. Analyze the charts in Figure 4.20. Note that some are $\overline{X}$ charts and some are $R$ charts. Match the following descriptions to the applicable chart. Hint! There are no double matches.
    a. Only chance causes are acting on this process.
    b. A gross error has occurred in the process.
    c. A batch process is being charted, and there are mixture problems.
    d. There is a shift in the average of the process.
    e. This process is a die-cutting operation, and the die is wearing.
    f. The variability of the process becomes less.
14. Given $\overline{\overline{X}} = 2.8$ and $\overline{R} = 0.7$ for a particular process, solve for the control limits, $UCL_{\overline{x}}$ and $LCL_{\overline{x}}$, and $UCL_R$. Use sample sizes of 2, 4, and 6. What happens to control limits as sample size increases? See Example 4.4.
15. Can control limits ever be wider than the limits for the individuals? Explain your answer.
16. Can the specification limits be wider than the limits for individuals? Would this situation be desirable or undesirable? Explain your answer.
17. Which of the following are most likely to lead to assignable causes, as contrasted with those that would most likely lead to chance causes that arise from system conditions?
    a. A specific machine requires maintenance.
    b. A company has performed capability studies on all of its equipment, and all of the equipment is capable and well maintained.
    c. The maintenance department of a manufacturing plant is on strike.

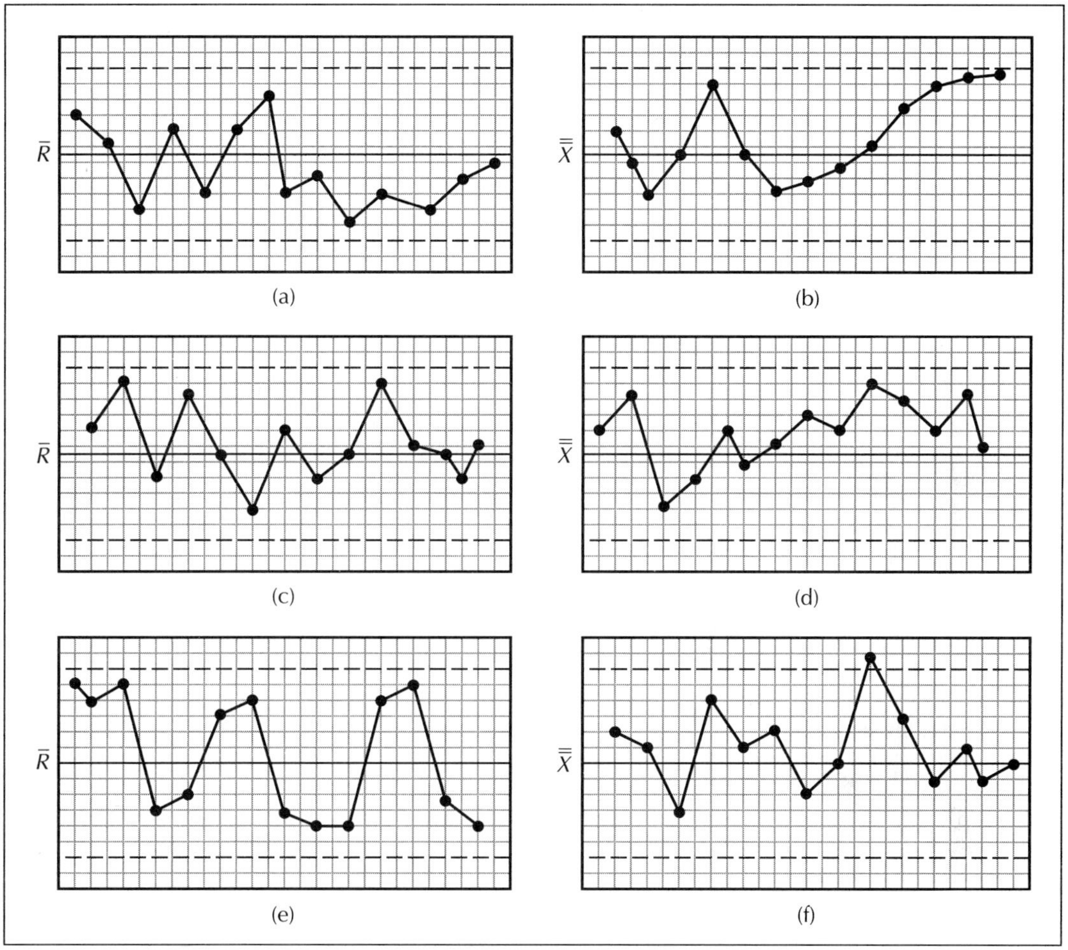


**FIGURE 4.20**
Six charts to be matched to the descriptions in problem 13.

d. Waiters at a restaurant are well trained and well paid, and morale is high.
e. A cashier at a grocery is untrained.
f. A hotel clerk has developed laryngitis.
g. A golfer has improper follow-through on some shots.
h. A well-trained golfer plays a well laid out course.

18. If the $C_p$ of a normally distributed and centered process is 0.75, what percent of product is being produced beyond the specification limits?

19. In a one-page paper, discuss the advantages of the control chart in eliminating overcontrol in manufacturing and service industries. Cite brief examples of overcontrol from both types of industries.
20. Many quality practitioners will use the statistics of the standardized normal curve (Chapter 3) to solve for the actual product beyond the specification limits, rather than using $C_{pk}$ and $C_p$. Solve for the percent of product beyond specification limits for the process used in Example 4.6. Use the statistics of the normal curve, as you studied in Chapter 3. Use the equation

$$Z = \frac{X - \overline{X}}{3s}$$

where the $X$ is the USL and LSL. Consider the $s$ equal to the $\hat{\sigma}$ of 1, as in the example.

## LABORATORY · CHAPTER 4

## Gauge Repeatability and Reproducibility (R&R) Study—Long Method

*Materials needed:* Measurement tools, 20 to 30 parts (depending on number of operators in study), R&R forms (attached).

Using the long method for studying the capability of a gauge results in the separation of the two variability factors, repeatability (gauge variation) and reproducibility (operator variation).

*Laboratory example:* Gary, Craig, and Andy have just completed collecting the data on the diameter of an axle's bearing. The data were collected after properly cleaning the axle bearings and performing two sets of measurements on each of the parts. The parts were randomly measured in each set of measurements. The data are shown in Figure 4.21. Complete the calculations to solve for the percent of R&R to specification.

*Solution:* Confirm the calculations shown in Figure 4.21. Which of the operators did the best measurement job? (Examine $\overline{R}_a$, $\overline{R}_b$, and $\overline{R}_c$ in Figure 4.21.)

After transferring the $\overline{R}$ and the $\overline{X}_{\text{diff}}$ to the associated calculation form (Figure 4.22), Gary performed the calculations as shown. Note

Data is coded from 2.2xxx

Target = 2.2490
Zero = 2.2

**Gauge Repeatability and Reproducibility Data Sheet (Long Method)**

| | 1 | 2 | 3 | 4 | 5 | 6 | 7 | 8 | 9 | 10 | 11 | 12 |
|---|---|---|---|---|---|---|---|---|---|---|---|---|
| Operator | A. Gary | | | | B. Craig | | | | C. Andy | | | |
| Sample # | 1st Trial | 2nd Trial | 3rd Trial | Range | 1st Trial | 2nd Trial | 3rd Trial | Range | 1st Trial | 2nd Trial | 3rd Trial | Range |
| 1 | 487 | 487 | | 0 | 488 | 489 | | 1 | 488 | 488 | | 0 |
| 2 | 489 | 489 | | 0 | 490 | 490 | | 0 | 491 | 490 | | 1 |
| 3 | 486 | 487 | | 1 | 487 | 487 | | 0 | 487 | 486 | | 1 |
| 4 | 489 | 489 | | 0 | 489 | 489 | | 0 | 489 | 488 | | 1 |
| 5 | 488 | 489 | | 1 | 489 | 489 | | 0 | 489 | 489 | | 0 |
| 6 | 487 | 488 | | 1 | 487 | 487 | | 0 | 488 | 489 | | 1 |
| 7 | 489 | 489 | | 0 | 489 | 489 | | 0 | 491 | 491 | | 0 |
| 8 | 490 | 490 | | 0 | 490 | 490 | | 0 | 493 | 491 | | 2 |
| 9 | 486 | 487 | | 1 | 487 | 487 | | 0 | 487 | 487 | | 0 |
| 10 | 489 | 489 | | 0 | 489 | 490 | | 1 | 491 | 490 | | 1 |
| Totals | 488.0 | 488.4 | | 4 | 488.5 | 488.7 | | 2 | 489.4 | 488.9 | | 7 |
| | → | 488.0 | | 0.4 | → | | | 0.2 | → | | | 0.7 |
| | | | ← $\bar{R}_A$ | | | | ← $\bar{R}_B$ | | | | ← $\bar{R}_C$ | |
| | SUM | | | | SUM | | | | SUM | | | |
| | $\bar{X}_A$ | 488.2 | | | $\bar{X}_B$ | 488.6 | | | $\bar{X}_C$ | 489.1 | | |

Divide final sum by twenty or thirty to compute $\bar{X}_A$, $\bar{X}_B$, $\bar{X}_C$,

| | |
|---|---|
| $\bar{R}_A$ | 0.4 |
| $\bar{R}_B$ | 0.2 |
| $\bar{R}_C$ | 0.7 |
| SUM | 1.3 |
| $\bar{R}$ | 0.433 |

| | |
|---|---|
| Max. $\bar{X}$ | 489.1 |
| Min. $\bar{X}$ | 488.2 |
| $\bar{X}$ Diff. | 0.9 |

Notes: ____________________

**FIGURE 4.21**
The data collected for the axle bearing diameters.

**Gauge Repeatability and Reproducibility Calculations (Long Method)**

Part Number / Name Axle Bearing    Gauge Number / Name G570    Date 2-23-94

Characteristic LARGE Bearing Surface    Performed by Team Best

Specification 2.2485 to 2.2495    Using .0001" accuracy (Master is 2.25075, use .00005 accuracy)

From Data Sheet: $\overline{R}$ = 0.433    $\overline{X}_{diff}$ = 0.9

*Measurement Unit Analysis*

Repeatability-Gauge Variation (G.V.)

$$G.V. = (\overline{R}) \times (K_1)$$

$$= (0.433) \times (4.56) = 1.97$$

| Trials | 2 | 3 |
|---|---|---|
| $K_1$ | 4.56 | 3.05 |

Reproducibility-Operator Variation (O.V.)

$$O.V. = (\overline{X}_{diff}) \times (K_2)$$

$$= (0.9) \times (2.70) = 2.43$$

| Operators | 2 | 3 |
|---|---|---|
| $K_2$ | 3.65 | 2.70 |

Repeatability and Reproducibility (R&R)

$$R\&R = \sqrt{(G.V.)^2 + (O.V.)^2}$$

$$= \sqrt{(1.97)^2 + (2.43)^2} = 3.13$$

*Percent Specification Analysis*

$$\%\ G.V. = 100\ [(G.V.) / (Specification)]$$

$$= 100\ [(1.97) / (10)]$$

$$= 19.7\ \%$$

$$\%\ O.V. = 100\ [(O.V.) / (Specification)]$$

$$= 100\ [(2.43) / (10)]$$

$$= 24.3\ \%$$

$$\%\ R\&R = \sqrt{(\%\ G.V.)^2 + (\%\ O.V.)^2}$$

$$= \sqrt{(19.7)^2 + (24.3)^2}$$

$$31.3\ \%$$

*Note:* All calculations are based upon predicting 5.15σ (99% of the area under normal curve).

**FIGURE 4.22**
The R&R calculations.

**Gauge Repeatability and Reproducibility Data Sheet (Long Method)**

Target = ______
Zero = ______

| | 1 | 2 | 3 | 4 | 5 | 6 | 7 | 8 | 9 | 10 | 11 | 12 |
|---|---|---|---|---|---|---|---|---|---|---|---|---|
| Operator | A. | | | | B. | | | | C. | | | |
| Sample # | 1st Trial | 2nd Trial | 3rd Trial | Range | 1st Trial | 2nd Trial | 3rd Trial | Range | 1st Trial | 2nd Trial | 3rd Trial | Range |
| 1 | | | | | | | | | | | | |
| 2 | | | | | | | | | | | | |
| 3 | | | | | | | | | | | | |
| 4 | | | | | | | | | | | | |
| 5 | | | | | | | | | | | | |
| 6 | | | | | | | | | | | | |
| 7 | | | | | | | | | | | | |
| 8 | | | | | | | | | | | | |
| 9 | | | | | | | | | | | | |
| 10 | | | | | | | | | | | | |
| Totals | | | | | | | | | | | | |

| | |
|---|---|
| | |
| | |
| SUM | |
| $\overline{X}_A$ | |

$\overline{R}_A$

| | |
|---|---|
| | |
| | |
| SUM | |
| $\overline{X}_B$ | |

$\overline{R}_B$

| | |
|---|---|
| | |
| | |
| SUM | |
| $\overline{X}_C$ | |

$\overline{R}_C$

Divide final sum by twenty or thirty to compute $\overline{X}_A$, $\overline{X}_B$, $\overline{X}_C$,

| | |
|---|---|
| $\overline{R}_A$ | |
| $\overline{R}_B$ | |
| $\overline{R}_C$ | |
| SUM | |
| $\overline{R}$ | |

| | |
|---|---|
| Max. $\overline{X}$ | |
| Min. $\overline{X}$ | |
| $\overline{X}$ Diff. | |

Notes: ______________________________

______________________________

______________________________

**FIGURE 4.23**
A blank R&R form, page 1.

**Gauge Repeatability and Reproducibility Calculations (Long Method)**

Part Number / Name ______ Gauge Number / Name ______ Date ______

Characteristic ______ Performed by ______

Specification ______

From Data Sheet: $\overline{R}$ = [____] $\overline{X}_{diff}$ = [____]

*Measurement Unit Analysis*

Repeatability-Gauge Variation (G.V.)

G.V. = $(\overline{R}) \times (K_1)$

= (____) × (____) = [____]

| Trials | 2 | 3 |
|---|---|---|
| $K_1$ | 4.56 | 3.05 |

Reproducibility-Operator Variation (O.V.)

O.V. = $(\overline{X}_{diff}) \times (K_2)$

= (____) × (____) = [____]

| Operators | 2 | 3 |
|---|---|---|
| $K_2$ | 3.65 | 2.70 |

Repeatability and Reproducibility (R&R)

R&R = $\sqrt{(G.V.)^2 + (O.V.)^2}$

= $\sqrt{(\_\_\_\_)^2 + (\_\_\_\_)^2}$ = [____]

*Percent Specification Analysis*

% G.V. = 100 [(G.V.) / (Specification)]

= 100 [(____) / (____)]

= [____ %]

% O.V. = 100 [(O.V.) / (Specification)]

= 100 [(____) / (____)]

= [____ %]

% R&R = $\sqrt{(\% G.V.)^2 + (\% O.V.)^2}$

= $\sqrt{(\_\_\_\_)^2 + (\_\_\_\_)^2}$

[____ %]

*Note:* All calculations are based upon predicting 5.15σ (99% of the area under normal curve).

**FIGURE 4.24**
A blank R&R form, page 2.

that the three operators performed two trials. These two facts led to the selection of the $k_1$ and $k_2$ factors as shown in the calculations. Please confirm the calculations shown in Figure 4.22.

To summarize the study, the gauge alone is using 20 percent of the specification. Most quality practitioners use 20 percent (some use 10 percent) as the maximum gauge R&R. This gauge is marginally acceptable for measuring the axle bearing diameters.

Using the blank forms provided (Figures 4.23 and 4.24), conduct an R&R study with the parts provided by the instructor. Review previous experiments to use appropriate data collection techniques.

# 5

# Optional Control Charts

**Charts for Individuals** 5.1
**Median Charts** 5.2
**Trend Charts** 5.3
**Precontrol Charts** 5.4

## 5.1 *Charts for Individuals*

Now that you have been exposed to the principal methods of control charting, other types of charts may be introduced. The *alternative charts* in this chapter are not an exhaustive collection, but are representative of the variety of alternative charting forms used in business and industry. Any one of the examples that follow may allow you someday to control a process that would be difficult or impossible to control with the more traditional charting techniques. The first alternative chart is the chart for individual data points.

An **individuals chart** is a variable control chart that requires only one piece of data per subgroup. A review of the $\overline{X}$ and $R$ chart (Section 4.2) will reveal that the subsample size $n$ is greater than one. Otherwise the range cannot be calculated, and the range is used to calculate the control limits for the $\overline{X}$s ($UCL_{\bar{\bar{X}}}$, $LCL_{\bar{\bar{X}}}$). So how can a subgroup of one permit a control chart to be developed? The answer is to use the concept of the moving range. A **moving range** uses the difference between individual subgroup values (for instance, $X_2 - X_1$) as the value of the range. The following example explains the method further.

## Chart Construction

Figure 5.1 shows a completed individuals chart for controlling haze in polymeric film. A polymer is the chief ingredient in plastic products. The film is used to attractively wrap foods, such as meats and cheeses, for easy viewing.

The clarity of the film is the opposite of what is called the haze of the film. So, the lower the haze the better. Some haze will always be present, however, and the specification range in the chart's header information is 3 to 10 percent. The percentage figure is based on the relative difference between a test specimen and the laboratory standard as measured by a haze meter.

The blown process for film is to extrude a continuous polymer tube through a round die. Air pressure is introduced underneath the tube. The air pressure maintains the thickness of the film. The haze factor is controlled by maintaining the quality of the polymer material and the consistency of the flow of the material through the die.

The data have been collected from the 2-mil test specimens pulled from finished rolls of the material. Each film specimen is sent to the lab for haze measurements as well as other tests. The individuals chart is ideal for such batch-processing operations.

Note that the range data have only 24 points, one subsample less than the individual data ($X$) information. Each moving range value is the absolute difference (never negative) between the last and present $X$ data point. Confirm the individual range values $R$ at this time.

EXAMPLE 5.1 Solve for both averages for the individuals chart example.

*Solution:* The average of the individuals chart is known as the *X*-bar ($\overline{X}$, not the *X*-double bar, $\overline{\overline{X}}$, as before). You may confirm with your

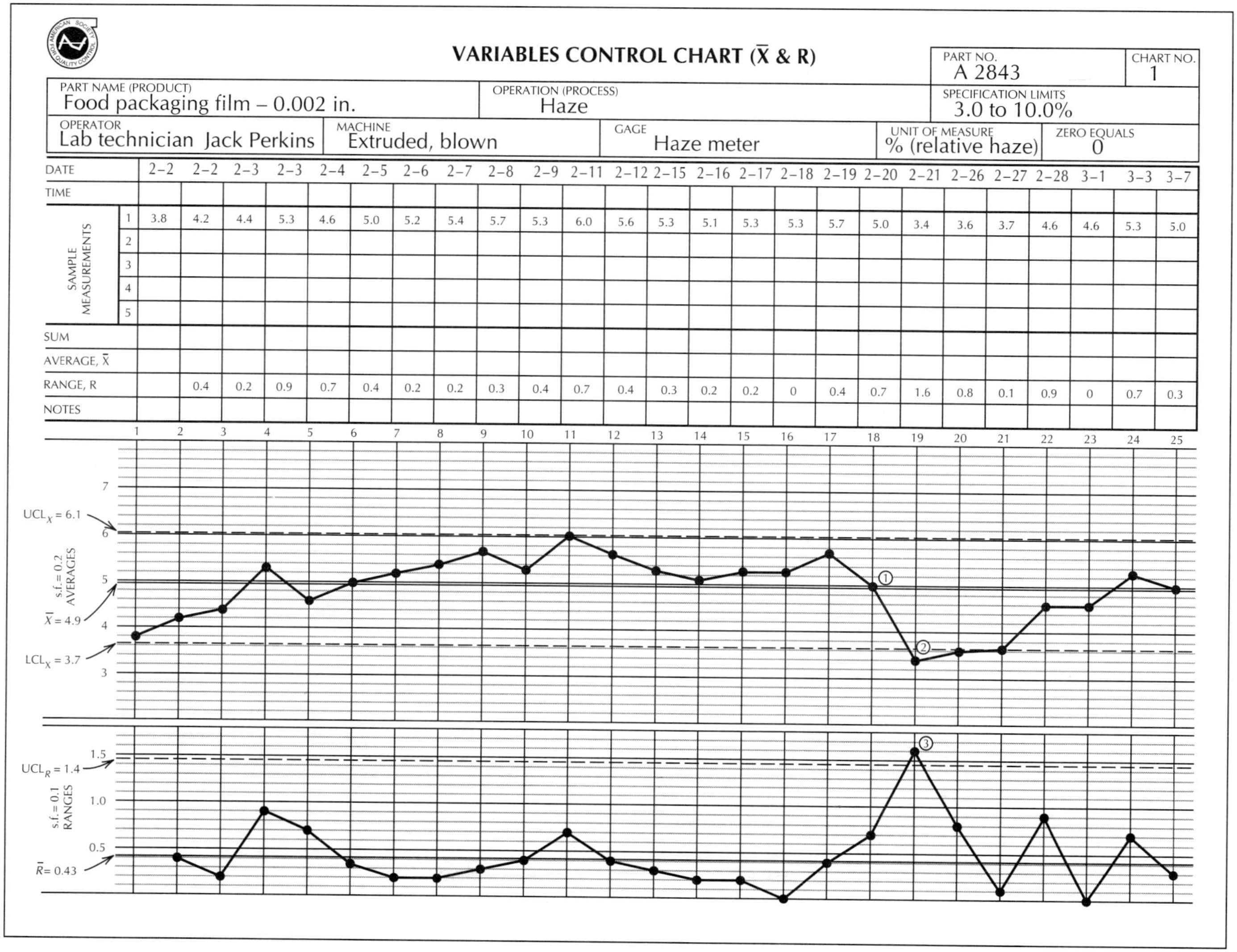


**FIGURE 5.1**
The individuals chart.

calculator that the average of the individual $X$ data is

$$\overline{X} = 4.9$$

Calculating the $\overline{R}$ we obtain (remembering that the $n$ is 24 for the $\overline{R}$ calculation)

$$\overline{R} = 0.43$$

The equations used to calculate the control limits for the chart are of the same general form as for the $\overline{X}$ and $R$ chart. The calculation for the upper control limit for $R$ ($UCL_R$) is exactly the same as before. The value of $D_4$ is 3.267 (inside-front cover, $n = 2$), and the calculation is

$$UCL_R = D_4\overline{R} = 3.267 \times 0.43 = 1.4$$

The value for $LCL_R$ is zero. This is true for the $\overline{X}$ and $R$ chart for subsample sizes $n$ less than seven.

The weighting factor for the individuals chart control limits ($UCL_x$, $LCL_x$) is 2.66. The calculations are as follows:

$$UCL_x = \overline{X} + (2.66)\overline{R} = 4.9 + (2.66)(0.46) = 6.1$$

$$LCL_x = \overline{X} - (2.66)\overline{R} = 4.9 - (2.66)(0.46) = 3.7$$

The factor 2.66 is not the factor $A_2$, as with $\overline{X}$ and $R$ charts, and is derived from three times the estimated standard deviation of the population ($\hat{\sigma}$) by using $3/d_2$ ($d_2$ for $n = 2$ is 1.128, inside-front cover).

## Chart Interpretation

The points out of control (OOC) on the chart are marked with the numbers 1, 2, and 3. The operator, from notes taken while the process was running, made the following observations concerning assignable causes:

*Note 1:* Twelve points in a row above the process average. The pressure in the bubble was lower than usual at the time, causing the film to be thicker than usual. Thickness, another variable tracked by control chart, was the reason for higher than normal haze levels.
*Note 2:* Two points (19 and 20) are below the lower control limit ($LCL_x$). Due to lack of homogeneity of the material feeding into the process (see Note 3).

*Note 3:* Point out (19). Due to a malfunction in the silo's mechanical feeder. All of the company's film machines had an OOC haze condition at this time.

One obvious advantage of the individuals chart is revealed by this actual case taken from the plastic film industry. It is often as much as a day later that samples from the process are collected from the rolls of material (batches) and analyzed in the lab. Before control charting was implemented, the operators had no feedback concerning the lab outcomes, and much scrap was generated because the laboratory was not discussing real-time problems with the operators.

By keeping thorough notes, the operators fixed assignable causes to OOC conditions indicated by the chart. Suddenly causes were found for generating scrap (Note 1). A conservatively estimated savings of \$60,000 to \$100,000 a month was the result of effective chart implementation.

An even more significant finding was that operators could now identify system causes that were acting over which they had no control (Notes 2 and 3). Now that operators know when a system cause is operating, they do not have to feel as though they may have been responsible for the bad product. Accurate knowledge of the process generates confidence and pride in the workforce.

## Advantages of Individuals Charts

- One form of a statistical process control chart (SPC chart), the individuals chart, quickly reveals what works and doesn't work.
- Individuals charts can be used for short runs or inspection requiring destructive testing.
- They can be used where batch processes that are relatively homogeneous do not require charts having greater sensitivity to process change.

## Disadvantages of Individuals Charts

- Individuals charts are not as sensitive to process change as are $\overline{X}$ and $R$ charts.
- They do not present as good a picture of piece-to-piece variability as does the range chart on an $\overline{X}$ and $R$ chart, even if the size of the subgroups is only two.

## *Section 5.1 Review Problems*

1. The individuals chart is used to judge the capability of ______________________ (short, long) runs.
2. The average of the individuals chart is known as the
   a. *X*
   b. *X*-bar
   c. *X*-double bar
   d. *R*-bar
3. Use the individuals chart when ______________________ (destructive, nondestructive) testing is necessary.
4. List, from memory, at least two advantages of an individuals chart.

# 5.2 *Median Charts*

**Median charts** are used mostly where management wants to keep analysis as simple as possible. The median is found by simply finding the middle value of a subsample.

A company using a significant number of temporary workers is an example of a company that might wish to control processes with median charts. Using a large number of temporaries is not recommended, as you will note later in our discussion.

One major reason the median chart is presented here is that it will help you to understand the $\overline{X}$ and *R* chart. In order to do so most effectively, we will use the data from the $\overline{X}$ and *R* chart in Figure 4.8. After we have completed the construction of the median chart, a comparison of the two charts will be instructive.

## *Chart Construction*

Figure 5.2 shows the completed median chart. It is from the fill-weight process discussed in Chapter 4. You should use a bookmark to refer to Figure 4.8 as we discuss the median chart. The header information is the same as for the $\overline{X}$ and *R* chart in Figure 4.8.

## VARIABLES CONTROL CHART ($\bar{X}$ & R)

| PART NAME (PRODUCT) | OPERATION (PROCESS) | PART NO. | CHART NO. |
|---|---|---|---|
| 170 g (6 oz.) yogurts | Fill | N/A | One |

| SPECIFICATION LIMITS |
|---|
| 182.0 to 190.0 |

| OPERATOR | MACHINE | GAGE | UNIT OF MEASURE | ZERO EQUALS |
|---|---|---|---|---|
| Quality team | Filler bowl and valve assy. | Top-loading scale | Grams | 0 |

| | 1 | 2 | 3 | 4 | 5 | 6 | 7 | 8 | 9 | 10 | 11 | 12 | 13 | 14 | 15 | 16 | 17 | 18 | 19 | 20 | 21 | 22 | 23 | 24 | 25 |
|---|---|---|---|---|---|---|---|---|---|---|---|---|---|---|---|---|---|---|---|---|---|---|---|---|---|
| DATE | | | | | | | | | | | | | | | | | | | | | | | | | |
| TIME | | | | | | | | | | | | | | | | | | | | | | | | | |
| SAMPLE MEASUREMENTS 1 | 187.0 | 176.2 | 187.0 | 187.0 | 176.3 | 184.0 | 181.3 | 183.1 | 184.3 | 176.9 | 181.0 | 183.3 | 178.2 | 181.5 | 183.6 | 184.5 | 184.9 | 175.0 | 182.1 | 185.2 | 184.9 | 187.6 | 187.7 | 186.0 | 185.7 |
| 2 | 179.6 | 179.5 | 182.9 | 179.6 | 180.0 | 184.2 | 183.0 | 184.2 | 177.0 | 181.0 | 183.3 | 184.5 | 181.3 | 183.5 | 184.8 | 178.5 | 177.9 | 182.2 | 183.9 | 179.1 | 185.1 | 185.6 | 185.8 | 186.0 | 185.7 |
| 3 | 181.6 | 181.7 | 182.9 | 183.7 | 182.9 | 183.0 | 184.2 | 176.9 | 181.2 | 183.2 | 184.4 | 178.4 | 183.5 | 184.7 | 178.0 | 180.3 | 182.3 | 183.7 | 185.1 | 181.6 | 179.9 | 185.6 | 185.8 | 185.6 | 187.6 |
| 4 | 184.0 | 184.0 | 184.0 | 184.1 | 184.1 | 184.2 | 177.0 | 180.1 | 183.2 | 184.3 | 177.6 | 180.1 | 184.6 | 178.6 | 182.4 | 183.4 | 183.7 | 189.8 | 179.3 | 183.9 | 182.0 | 176.7 | 176.9 | 185.6 | 187.6 |
| 5 | 185.3 | 185.3 | 185.4 | 185.4 | 185.5 | 183.9 | 180.6 | 183.1 | 184.4 | 177.3 | 180.6 | 183.3 | 178.7 | 182.5 | 183.6 | 184.6 | 185.1 | 179.6 | 182.0 | 185.2 | 183.7 | 185.7 | 185.9 | 185.5 | 188.2 |
| SUM | | | | | | | | | | | | | | | | | | | | | | | | | |
| AVERAGE, $\bar{X}$ | | | | | | | | | | | | | | | | | | | | | | | | | |
| RANGE, R | | | | | | | | | | | | | | | | | | | | | | | | | |
| NOTES | | | | | | | | | | | | | | | | | | | | | | | | | |

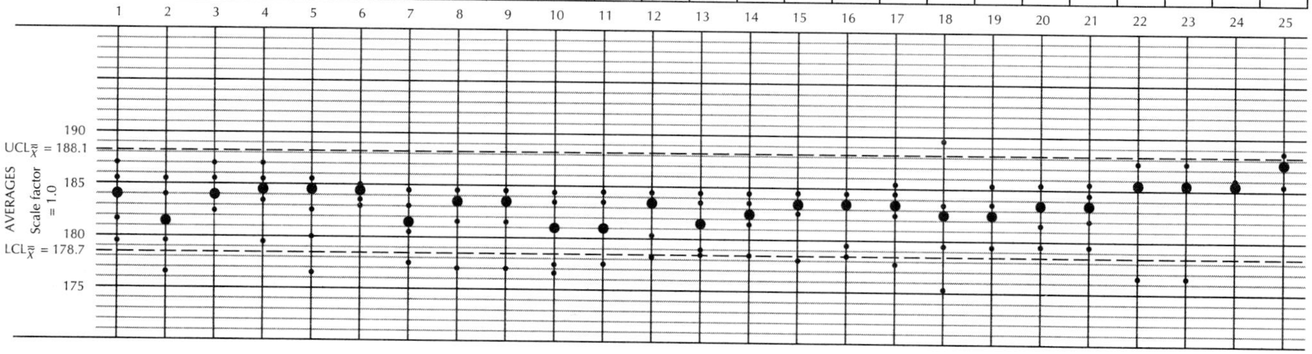


**FIGURE 5.2**
The median chart.

The chart's data are the same, but the data points circled on the median chart represent the median values of each subsample. You will recall from Chapter 3 (Section 3.1) that the median is the middle value of the ranked subsample data.

EXAMPLE 5.2 Scale the median chart from the data in Figure 5.2 (also by referring to Figure 4.8).

*Solution:* The highest and lowest data point (the full range of the process's distribution) determines the scaling of the chart and the plotting of the points. Confirm $X_{high}$ as 189.8 and $X_{low}$ as 175.0. This may be best determined by using Figure 4.8, where the highs and lows of each subsample are circled and boxed, respectively. From this maximum range of points the scale factor of 1 is determined. This procedure is intended to make the plotting of the points easier for the operators. In Figure 4.8, a scale factor of 0.5 made better use of the area of that chart.

Determining the scale factor of the chart is done by knowledgeable QC technologists or managers. Charts used at the process will be laid-out (scaled) prior to use.

Confirm the correct plotting of the individual data points on the chart. Some subsamples seem to have less than five points plotted. Where two or more points fall in the same place that point is simply made larger. The median point in each subsample is circled.

The average line for the median chart will be referred to as the **X-tilde-bar ($\bar{\tilde{X}}$)** or the mean of the medians. Confirm that the average of the medians is 183.4. The average range $\bar{R}$ is calculated as for the $\bar{X}$ and $R$ chart. First you must find the range of each subgroup. This step has already been done for the fill-weight distribution (see Figure 4.8). Confirm the $\bar{R}$ value to be 6.8.

To calculate the control limits for the medians, a special weighting factor must be used. The factor is **$\tilde{A}_2$ ($A_2$-tilde),** and its value is shown in the following table for various subsample sizes ($n$).

| $n$ | $\tilde{A}_2$ |
|---|---|
| 2 | 1.88 |
| 3 | 1.19 |
| 4 | 0.80 |
| 5 | 0.69 |
| 6 | 0.55 |
| 7 | 0.51 |

We will use the $\tilde{A}_2$ value for $n = 5$ of 0.69. Note that this value is larger than the value of $A_2$ for the $\overline{X}$ and $R$ chart ($A_2 = 0.58$). This larger value means that the control limits will be wider on the median chart. This is the reason for the median chart's lack of sensitivity in detecting a process shift, when compared to the $\overline{X}$ and $R$ chart's sensitivity.

EXAMPLE 5.3 Calculate the control limits for the median chart.

*Solution:* Use the same equations as used for calculating the control limits for the $\overline{X}$ and $R$:

$$UCL_{\tilde{X}} = \bar{\tilde{X}} + (\tilde{A}_2)(\overline{R}) = 183.4 + (0.69)(6.8) = 188.1$$

$$LCL_{\tilde{X}} = \bar{\tilde{X}} - (\tilde{A}_2)(\overline{R}) = 183.4 - (0.69)(6.8) = 178.7$$

The upper control limit for $R$ ($UCL_R$) is the one remaining calculation. (The lower control limit for $R$ is zero for $n < 7$). The calculation is, as for the $\overline{X}$ and $R$ chart,

$$UCL_R = D_4\overline{R} = (2.11)(6.8) = 14.3$$

Often the range is also charted on the median chart, but it need not be. If the chart is to be completed and interpreted by semiskilled or temporary operators, it will be an unused portion of the chart.

## Chart Interpretation

Chart interpretation involves the use of a card designed from historical data that has the maximum allowable range ($UCL_R$) marked. The limits on the card were set by using 14 lines on the control chart form. The rounded value of 14, rather than 14.3 (see the calculation for the $UCL_R$ in the preceding subsection), has been used in this example. The card is laid over the working chart in Figure 5.3 to check the spread of the individual points on the chart. It is located at point 18, where the range has a point out (refer to the $\overline{X}$ and $R$ chart, Figure 4.8). You can see that the card indicates the OOC condition. With untrained operators using the chart, they must stop the process or seek help whenever the individual plots lie beyond the maximum allowable range.

The maximum allowable range, then, is the value of the process's $UCL_R$. Completed charts may be sent to the QC department where new **maximum-allowable-range cards** may be constructed.

**VARIABLES CONTROL CHART ($\bar{X}$ & R)**

| PART NAME (PRODUCT) | OPERATION (PROCESS) | PART NO. | CHART NO. |
|---|---|---|---|
| 170 g (6 oz.) yogurts | Fill | N/A | One |
| | | SPECIFICATION LIMITS: 182.0 to 190.0 | |

| OPERATOR | MACHINE | GAGE | UNIT OF MEASURE | ZERO EQUALS |
|---|---|---|---|---|
| Quality team | Filler bowl and valve assy. | Top-loading scale | Grams | 0 |

| | | 1 | 2 | 3 | 4 | 5 | 6 | 7 | 8 | 9 | 10 | 11 | 12 | 13 |
|---|---|---|---|---|---|---|---|---|---|---|---|---|---|---|
| DATE | | | | | | | | | | | | | | |
| TIME | | | | | | | | | | | | | | |
| SAMPLE MEASUREMENTS | 1 | 187.0 | 176.2 | 187.0 | 187.0 | 176.3 | 184.0 | 181.3 | 183.1 | 184.3 | 176.9 | 181.0 | 183.3 | 178.2 |
| | 2 | 179.6 | 179.5 | 182.9 | 179.6 | 180.0 | 184.2 | 183.0 | 184.2 | 177.0 | 181.0 | 183.3 | 184.5 | 181.3 |
| | 3 | 181.6 | 181.7 | 182.9 | 183.7 | 182.9 | 183.0 | 184.2 | 176.9 | 181.2 | 183.2 | 184.4 | 178.4 | 183.5 |
| | 4 | 184.0 | 184.0 | 184.0 | 184.1 | 184.1 | 184.2 | 177.0 | 180.1 | 183.2 | 184.3 | 177.6 | 180.1 | 184.6 |
| | 5 | 185.3 | 185.3 | 185.4 | 185.4 | 185.5 | 183.9 | 180.6 | 183.1 | 184.4 | 177.3 | 180.6 | 183.3 | 178.7 |
| SUM | | | | | | | | | | | | | | |
| AVERAGE, $\bar{X}$ | | | | | | | | | | | | | | |
| RANGE, $R$ | | | | | | | | | | | | | | |
| NOTES | | | | | | | | | | | | | | |

| | | 14 | 15 | 16 | 17 | 18 | 19 | 20 | 21 | 22 | 23 | 24 | 25 |
|---|---|---|---|---|---|---|---|---|---|---|---|---|---|
| DATE | | | | | | | | | | | | | |
| TIME | | | | | | | | | | | | | |
| SAMPLE MEASUREMENTS | 1 | 181.5 | 183.6 | 184.5 | 184.9 | 175.0 | 182.1 | 185.2 | 184.9 | 187.6 | 187.7 | 186.0 | 185.7 |
| | 2 | 183.5 | 184.8 | 178.5 | 177.9 | 182.2 | 183.9 | 179.1 | 185.1 | 185.6 | 185.8 | 186.0 | 185.7 |
| | 3 | 184.7 | 178.0 | 180.3 | 182.3 | 183.7 | 185.1 | 181.6 | 179.9 | 185.6 | 185.8 | 185.6 | 187.6 |
| | 4 | 178.6 | 182.4 | 183.4 | 183.7 | 189.8 | 179.3 | 183.9 | 182.0 | 176.7 | 176.9 | 185.6 | 187.6 |
| | 5 | 182.5 | 183.6 | 184.6 | 185.1 | 179.6 | 182.0 | 185.2 | 183.7 | 185.7 | 185.9 | 185.5 | 188.2 |
| SUM | | | | | | | | | | | | | |
| AVERAGE, $\bar{X}$ | | | | | | | | | | | | | |
| RANGE, $R$ | | | | | | | | | | | | | |
| NOTES | | | | | | | | | | | | | |

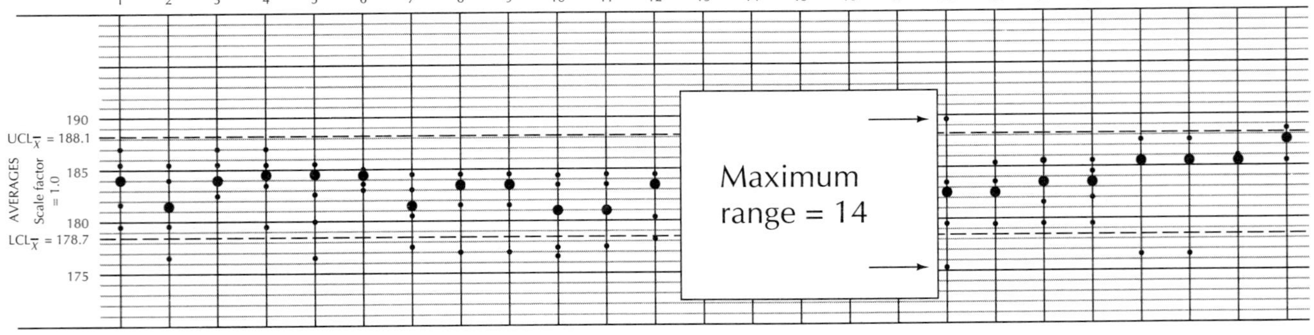


**FIGURE 5.3**
The maximum-allowable-range card.

## Advantages of Median Charts

The median chart displays the individual data values. The operators and others examining the chart might find it easier to understand than an $\overline{X}$ and $R$ chart.

It is not recommended, however, that untrained, semiskilled operators run processes. The heart of SPC is to continually train workers so they understand the processes they run or operate. Deming calls the content of such training **"profound knowledge,"** a vital component of a company's success.

(It is interesting to note that *some of the individual data points on the chart are beyond the control limits on the chart*. This phenomenon occurs because of the central limit theorem discussed in Section 4.3, "Averages and Individuals." Remember that the control limits on any of the control charts for variables are not for the individual readings, but for the central tendencies of the subsamples.)

## Disadvantages of Median Charts

The chart should only be used with odd-number subsample sizes ($n$ = 3, 5, 7), allowing easy determination of the median value. The median, when compared to the average, does not consider extreme values of data, making the median chart about 25 percent less sensitive to process change than is the $\overline{X}$ and $R$ chart. In some cases this lack of sensitivity to extreme data points is counted as an advantage. Note that the point out for the $\overline{X}$ and $R$ chart (Figure 4.8) is not a point out for the median chart. The point is close to out, however, and the trend (rule of sevens) can be as clearly seen as it was with the $\overline{X}$ and $R$ chart.

## Section 5.2 Review Problems

1. A median chart may be used in a company employing a large number of ______________________ workers.
2. The median is the ______________________ (average, middle) value of ranked data.

3. The scale factor of the median chart is determined by ______________ data point(s).
   a. one
   b. two
   c. all
   d. a representative subsample's
4. The average line of the median chart is referred to as the ______________ – ______________ – ______________, which is calculated from the mean of the medians.
5. The median chart should only be used with ______________ (even, odd)-number sample sizes.

## 5.3 *Trend Charts*

What would you do if in constructing an $\overline{X}$ and $R$ chart you noticed a trend that was an inherent or natural part of the process? Many times a manufacturing process involves tool or die wear. In this case there is a naturally occurring trend that must be accounted for in order to examine the process with the usual $\overline{X}$ and $R$ chart methods. The **trend chart** may be used to show a process with a trend, yet examine other assignable causes as well by establishing control limits that are parallel to the process average or $\overline{\overline{X}}$.

The example we will use for a trend chart application involves a die extrusion process, where the die wears as the parts are extruded. The trend chart for the example is illustrated in Figure 5.4

The header information in Figure 5.4 reveals the following: electrodes are being extruded through a die, the specification limits for the diameter of the electrode vary only 0.08 millimeter from target, and the subsamples are gathered and checked every hour.

In Figure 5.4, only the $\overline{X}$s and $R$s are shown as computed from the individual data. The subgroup size is only three ($n = 3$) because of the time needed to measure the small parts accurately. The averages and ranges of each subsample are accurate to 0.01 mm. In this case the $R$ data do not show a trend, so they will be charted in the normal fashion.

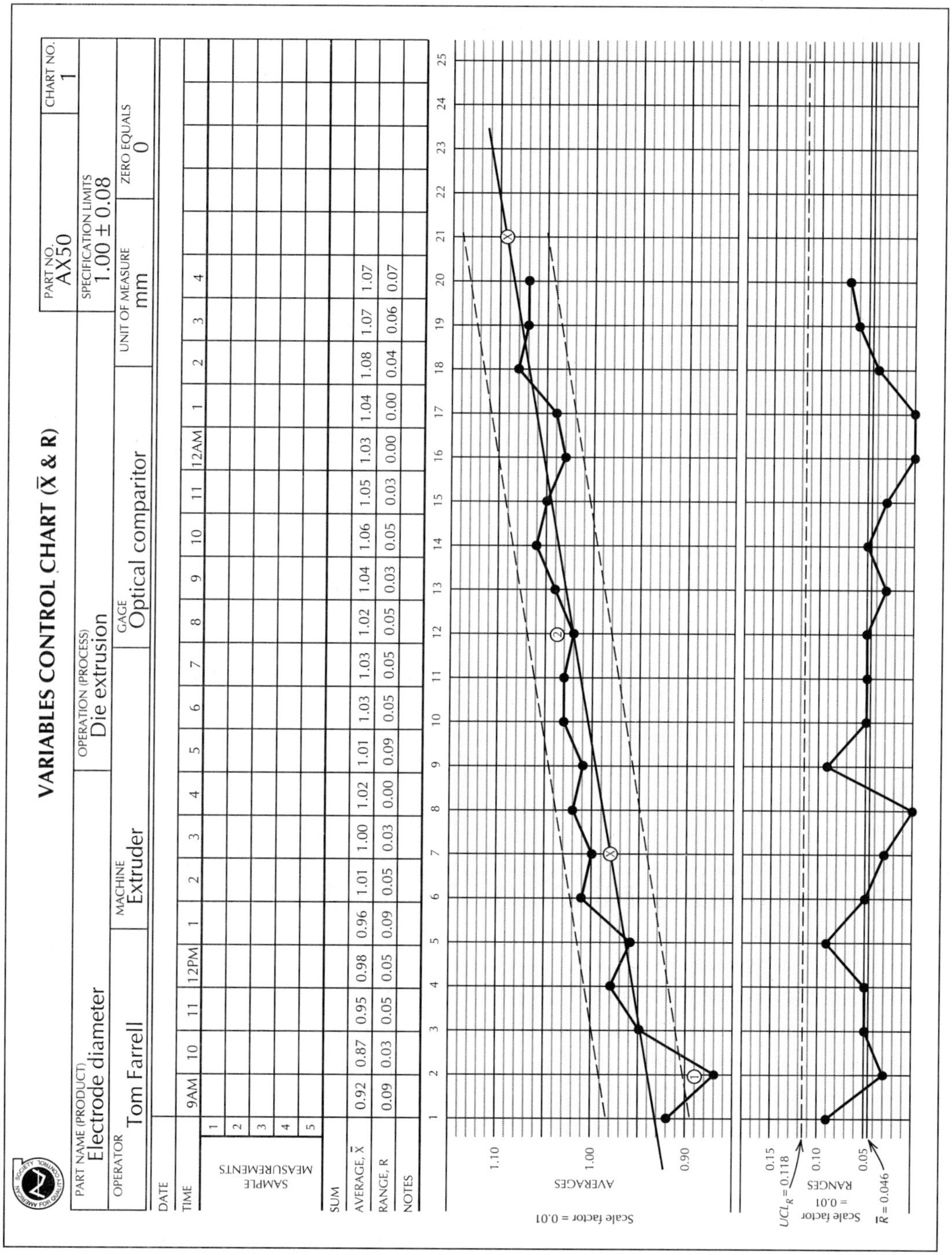


**FIGURE 5.4**
The trend chart.

The chart for the averages ($\overline{X}$) was scaled prior to determining the control limits. First, the highest and lowest measurements were determined, and then the chart was scaled. This process resulted in the scale factor of 0.01. Confirm the correct scaling of the average chart and the correct plotting of the $\overline{X}$ points at this time.

After the $\overline{X}$ points are plotted, a line of best fit is drawn through the points. We could use linear regression (refer to Section 2.2 and check your calculator manual's statistical section) to find the exact slope of the line, but the **line-of-best-fit method** is a lot easier and often more appropriate.

EXAMPLE 5.4

Calculate the slope of the line of best fit as drawn in Figure 5.4. Explain what the slope of the line represents.

*Solution:* To calculate the slope of the line, use the *X*- and *Y*-axes coordinates, in this case the circled *X*s marked on the average ($\overline{X}$) line. Using subgroup numbers for the *X* axis, the coordinates are [7, 0.98] and [21, 1.09]. The slope of the line is found to be

$$\text{Slope} = \frac{\Delta Y}{\Delta X} = \frac{\Delta \overline{X}}{\text{Subgroup no.}} = \frac{1.09 - 0.98}{21 - 7} = 0.0079$$

The slope calculation tells us that the die wears about eight one-thousandths of a millimeter (0.008 mm) between subgroups, or after every hour of operation. From the slope of the line you can predict when to change the die. This knowledge enables the operators to have the die ready to change at the right time, and the downtime on the extruder is minimized.

Please note that the die wear was greatest earlier in the process, with the $\overline{X}$ points tailing off later on the chart. This is the usual case when a new tool or die replaces the one in previous service. This is not a perfect linear regression, but the assumption of linearity will work for estimating die change.

Next, the *R* chart parameters are calculated. Confirm the correct calculation of the parameters shown on the chart (Chapter 4). Using the $\overline{R}$, the control limits are calculated by

$$\overline{R} = 0.046, \text{ and } UCL_R = (2.57)(0.046) = 0.118$$

$$A_2\overline{R} = (1.02)(0.046) = 0.047$$

$$UCL_{\overline{x}}, LCL_{\overline{x}} = \overline{X} \pm 0.047$$

The control limits are drawn parallel to the average line of best fit ($\bar{\bar{X}}$) at the distance of 4.7 hundredths of a millimeter.

## *Chart Interpretation*

The process has two OOC signals. The first is a point out (point 1), and the second is seven points above the line of best fit for the averages (point 2). The operators, with management's support, identified the assignable causes connected with these OOC conditions.

For OOC point 1 (the second subgroup), the point is out below the lower control limit. The assignable cause was identified as a material impurity problem.

For OOC point 2, the large number of points above the centerline could have been caused by the estimation of the line of best fit. Also, the trend is not perfectly linear. The operators decided that an assignable cause may not be present, but will be alert to a possible assignable cause operating between subsamples 6 and 15 in future runs.

The most important observations involve the control limits. The lower control limit shows that the process is capable of producing out-of-specification (OOS) parts at the present machine setting, as well as at the beginning of the run. The operators decide to center the machine a little higher on the next run. They use the distance from the lower control limit on the chart at the beginning of the run to the lower specification limit (0.92 mm). Based on this calculation, they set the machine higher by 0.03 millimeter.

The trend chart's upper control limit also reveals that the process is capable of making parts OOS by the end of the 15th subgroup (14 hours into the run). In the future, the die change will be accomplished after 14 hours rather than after 24.

## *Advantages of Trend Charts*

- The trend chart is a straightforward approach to statistically controlling a process with a trend.
- The trend chart is the only way to isolate assignable causes other than the factor causing the noticeable trend. Without a trend chart, these other factors could not be identified and controlled.
- The trend chart can aid in predicting tool or die changes.

- The historical control limits can help to define the machine setup (centering) and the length of a run before tooling change is necessary.

## Disadvantages of Trend Charts

- The construction of the trend chart and the slope calculation can only be accomplished satisfactorily by technologists trained in good charting techniques.
- Chart interpretation is complicated because of the constantly changing process average. (With some tooling, the range chart may also show a trend.)

## Section 5.3 Review Problems

1. To find the slope of the $\overline{X}$ points on a trend chart, __________ may be used.
    a. linear regression
    b. a line of best fit
    c. both a and b are appropriate
    d. none of the above are true
2. Draw a line of best fit between subgroup points 11 and 20 in Figure 5.4. Calculate the slope of the line. Compare your result with the slope of the line of best fit for all of the data (0.0079, Example 5.4).
3. How does the trend chart help in the initial centering of a process? See the text discussion.
4. List, from memory, at least two advantages of trend charts.

# 5.4 Precontrol Charts

The statistical charts presented so far (except possibly for the median chart) demand significant training for operators, and it is the operators who are the most effective in controlling processes. These charts also demand calculations and interpretations during production runs, and this requirement places further responsibility on the very people responsible for looking after the machinery, materials, and other aspects of the process.

The charts must be stored and maintained over a period of time in order to demonstrate statistical control of processes. Even the individuals chart has an inspection frequency that may be quite demanding, as well as expensive. Yet the advantages of such charts in detecting process shifts and in identifying assignable causes adversely affecting the process usually far outweigh the energy and expense consumed by the charting process.

Consider, however, a process that has been in control for a period of time. All assignable causes have been discovered and effectively dealt with, resulting in a process that yields only a few OOC parts out of thousands or even millions of parts produced. The capability of this hypothetical process ($C_p$—the ability of the process to produce parts within specification—Section 4.4) is 1 or more, which means that a six-standard-deviation spread of the process is within the upper and lower specification limits when the process is centered. Do we continue to use Shewhart control charts, or is there a less expensive and less time-consuming way to control the quality of the parts? There is another control chart available that offers such advantages, and it is known as the **precontrol chart.**

A typical form used in precontrol studies is shown in Figure 5.5. The different shaded columns on the middle-left side of the form are the colors red (outside columns), yellow, and green (center column). The colors represent a certain percentage of the width of the specification. The green area in the center column, for instance, represents one-half, and the yellow area columns on either side the remaining one-half (one-quarter per side) of the total specification. Because of the use of familiar traffic-light colors of green, yellow, and red, the precontrol chart is often known as the **stoplight chart.** Dial indicator overlays that have the green and yellow areas colored in may be purchased from suppliers of measuring instruments (Figure 5.6).

## *Precontrol Rules*

There are rules that must be followed in using the precontrol chart. The rules are often put on laminated pocket cards for the operators to carry with them. The rules may vary depending on the sensitivity

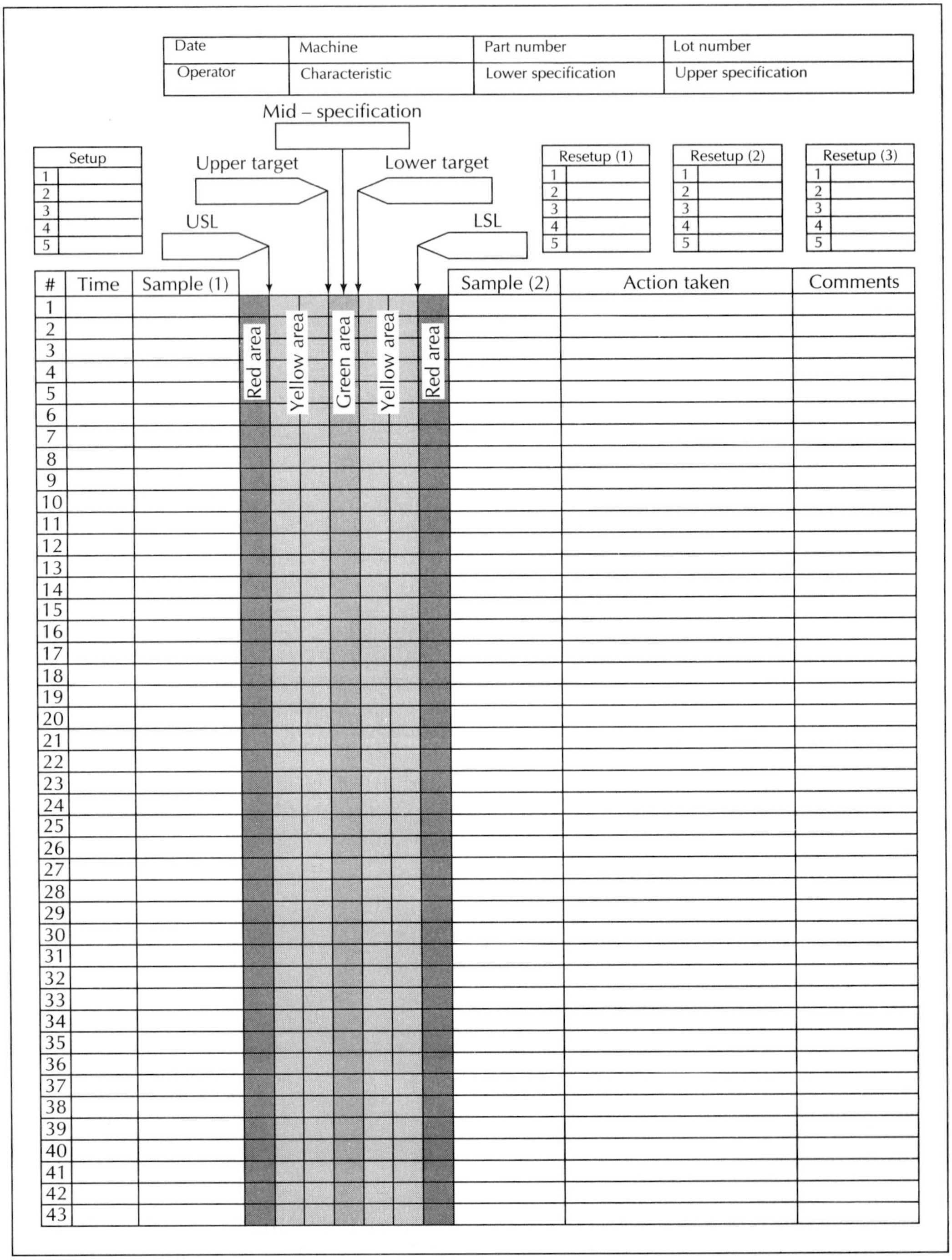


**FIGURE 5.5**
The precontrol chart form.

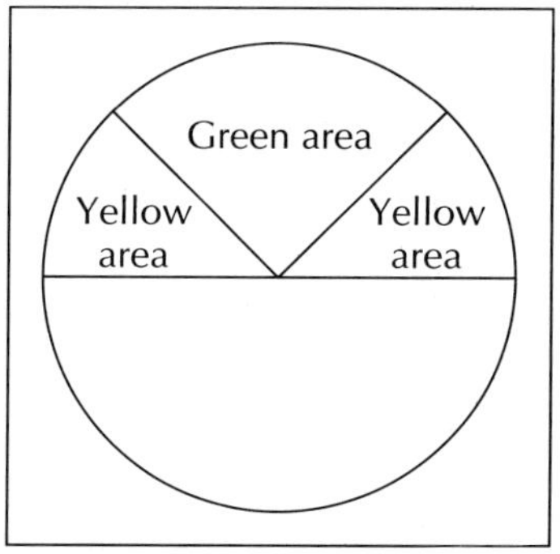


**FIGURE 5.6**
An overlay for a dial indicator gauge, used for precontrol.

desired in detecting process changes, but some typical rules are as follows:

**Precontrol Rules**

1. Setup: process is okay to run when five pieces in a row are inside the green area.
2. Running: sample two consecutive pieces, and
   - if the first piece is within the green area (go), continue to run (do not measure the second piece).
   - if the first piece is in the yellow area (caution), check the second piece.
   - if the second piece is within the green area, continue to run.
   - if both pieces are in the yellow area, go back to setup and follow rule 1.
   - if any of the pieces are in the red area (stop), go back to setup and follow rule 1.

Variations on these rules might involve sampling more pieces than two in a subsample; or if the second piece is in the yellow area, you must find three successive pieces inside the green area. Such variations increase the sensitivity of the chart.

From the rules you can see that a run is not initiated until the process yields five "good pieces" (measurements in the green area) in a row. You will also recognize the criterion as that for attribute data. Continuous scale measurements (variable data) may be taken and recorded, but this action is not necessary.

Sampling frequency is determined by how often, on average, the process needs adjustment. Most users recommend at least six samples between the average time of adjustment.

## Precontrol Chart Analysis

It has already been stated that the use of the precontrol chart is recommended only when the process is in control or stable and capable (capability implies a normally distributed process). After this criterion has been met, the probabilities for the process may be determined.

Figure 5.7 shows a normally distributed and centered process, with precontrol lines superimposed at half the specification distance from the nominal on each side. The upper and lower specification limits are designated as the USL and LSL, respectively. Since the distribution is just within the specification limits, the capability is one ($C_p = 1$).

We might be interested in the probability of acceptance $P(A)$ of a subsample of two during a run, while assuming the process has held steady. In this case a stop signal would be false.

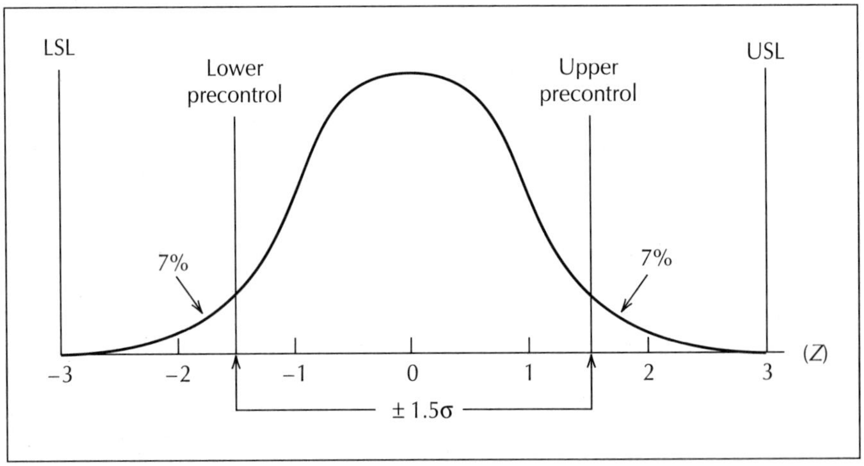


**FIGURE 5.7**
Calculating the probability of acceptance of a precontrol chart.

EXAMPLE 5.5 Find the probability of acceptance $P(A)$ of a subsample of two if the process remains steady.

*Solution:* To calculate $P(A)$ we must use the table "Probability Areas Under the Standardized Normal Curve" in Appendix A. In Figure 5.7 the six standard deviations have been transformed to the $Z$ parameters used for the chart. If each precontrol limit is set between the nominal and the specification limit on that side, then the $Z$ value for either side is 1.5. Using the table in Appendix A we find the area under the normal curve to be 0.0668 at $Z = -1.5$. The probability of a part being in the left tail and beyond 1.5 standard deviations, then, is 6.68 percent, or approximately 7 percent.

Since the normal curve is symmetrical about the mean and the process is assumed to be centered, there is a 7 percent probability of a part being in each yellow area of the chart. The joint probabilities of both parts falling in a yellow area are independent and must be multiplied (Section 6.1). The probability of both parts falling in a yellow area must be

$$0.07 \times 0.07 = 0.0049 \approx 0.5\%$$

The probability of acceptance then is equal to 99.5 percent. This means that there is only a slim chance (0.5%) of having a false signal of two yellows (resulting in a run being stopped when the process is stable).

Many people using precontrol charts recommend that the six-standard-deviation variability occupy only 75 percent of the specification width. This is a capability of

$$C_p = \frac{\text{USL} - \text{LSL}}{6\sigma} = \frac{1}{0.75} = 1.33$$

Even though the precontrol chart essentially uses attribute or go/no-go data, troubleshooting of the process can be accomplished. For instance, the parts in the subsamples may be measured and the variable data recorded. Then, $\overline{X}$ and $R$ charts may be constructed from the data, and process troubleshooting may be accomplished from these charts. Some customers of original equipment manufacturers, especially automotive assemblers, demand that control charts be maintained.

Process troubleshooting may also be accomplished simply by using the precontrol chart. Consider the following situations and their interpretations:

- Both subsample parts (two per subsample) fall in the yellow area on the same side—the process has shifted.

- The two parts fall on either side of the green zone and in the yellow area—the process spread has increased.
- During setup, one out of five parts is in the yellow every time the process setup is run—the six-standard-deviation process spread is slightly greater than specification.

By using these troubleshooting methods a process may be kept stable and produce high-quality parts.

## *Advantages of Precontrol*

- Easy to implement, precontrol requires only attribute (go/no-go) measurements on very few parts.
- The chart uses specification limits and not control limits. Control limits are more difficult for operators and others to understand. Using the chart does not require graphing skills or calculations.
- Operators require less training and are often more understanding of the basic rationale.
- Short runs may be controlled most effectively by precontrol techniques.

## *Disadvantages of Precontrol*

- Unless the distribution is relatively normal, the process well-behaved and centered, and the process capability ($C_p$) greater than 1, precontrol should not be used.
- Precontrol does not contribute as well toward reducing variation as do variable charts. One of the advantages listed, that of working to specification rather than the natural control limits of the process, represents a disadvantage in effectively working to target.
- Narrower limits than specification, referred to as **narrow limit gauging,** causes operators to believe that management does not trust their work. However, the control limits for variable control charts often are even narrower. *Narrow limit gauging, in all cases, must be explained to operators and other shop-floor personnel.*

## *Section 5.4 Review Problems*

1. A precontrol run is initiated when the process yields ________ good pieces.
   a. 3
   b. 4
   c. 5
   d. 6
2. The term "good pieces" implies the use of ________ (variable, attribute) data.
3. List, from memory, at least two advantages of precontrol.

# *Summary*

- The alternative charts discussed in the chapter do not represent all of the charts used, or even the most frequently used, in business and industry today. Instead, the charts presented are representative of the broad variety of charts you might be exposed to during your career.
- Individuals charts use a moving range. They are used for short runs, when it is not feasible to collect a lot of data. They are less sensitive to process changes than are the $\overline{X}$ and $R$ charts that have subsample sizes of two or more.
- Median charts have control limits calculated from the median values of the subgroups. Use median charts where control charting is necessary and operators are not trained on other statistical charts, such as $\overline{X}$ and $R$ charts.
- Trend charts are used for processes that involve tool or die wear—processes that display a constant trend. Construction of the chart requires the technologist to view the process average ($\overline{\overline{X}}$) as a line of best fit drawn through the plotted subgroup averages ($\overline{X}$). Control limits are constructed parallel to the average line.
- Precontrol charts use narrow limits, based on specification limits, to control a process. They may be especially useful for processes proven to be in control using control charts and having capabilities greater than 1.

- Of the variable charts discussed in this chapter, none are as all-around effective as are $\overline{X}$ and $R$ charts. You should exercise prudence in selecting something other than an $\overline{X}$ and $R$ chart to see cycles, shifts, and trends in processes. The trend chart, of course, is an $\overline{X}$ and $R$ chart.

## CASE STUDY · CHAPTER 5

You work for a company that manufactures bomb fuses. One particular quality characteristic is the electrical current ($I$) at which the fuse detonates. To perform the inspection the fuse must be destroyed. The company is considering how to perform the inspections.

When inspecting fuses, as with any inspection involving destructive testing, as few units as possible should be inspected. If actual current measurements are collected, the data will be variable. After completing this chapter you should have listed the chart for individuals as the control chart to be used to bring the process into control.

Once the process is in control and capable ($C_p > 1$), precontrol would definitely be a viable option. Precontrol does not require much sampling (destroying fuses in this case) and as long as the precontrol rules are followed, the process will make acceptable products.

During the individuals-charting phase, a technologist or manager familiar with control charting could lead the effort to bring the process into control and assure that its capability is greater than 1. With this accomplished, the training for the production workers may only involve their understanding of precontrol rules. Also, the necessity of using narrow-limit gauging to produce adequate quality should be explained to the operators.

After precontrol has been implemented, individuals charts could be constructed over time if the variable data are still measured and recorded. Warranty costs could be significant with this product, and the increased assurance a control chart gives may be worth the additional effort.

## *Chapter Review Problems*

1. Structural plastic foam used for bulk insulation has a critical quality characteristic of density. From the following data prepare a chart for individuals. The data are coded from 1 (for example, the first data point is read as 1.66 lb/ft$^3$).

| *Sample Number* | *X* | *Sample Number* | *X* |
|---|---|---|---|
| 1 | 66 | 11 | 65 |
| 2 | 64 | 12 | 66 |
| 3 | 62 | 13 | 66 |
| 4 | 69 | 14 | 65 |
| 5 | 67 | 15 | 68 |
| 6 | 64 | 16 | 68 |
| 7 | 63 | 17 | 69 |
| 8 | 67 | 18 | 68 |
| 9 | 65 | 19 | 65 |
| 10 | 67 | 20 | 67 |

Identify the point where the level of the blowing agent was changed by locating the OOC point on the range chart. Did the process adjustment seem to improve the process?

2. Calculate the capability of the process in problem 1. The density should be 1.65 lb/ft$^3$, with a bilateral tolerance of $\pm$0.05 lb/ft$^3$. The capability of an individuals chart may be calculated as it was presented in Section 4.4. First, the estimated sigma ($\hat{\sigma}$) must be calculated by

$$\hat{\sigma} = \frac{\overline{R}}{d_2}$$

$$C_p = \frac{\text{USL} - \text{LSL}}{6\hat{\sigma}} = \frac{1.7 - 1.6}{6\hat{\sigma}}$$

3. Calculate the $C_{pk}$ of the process in problem 1.

4. In many chemical process industries a **moving-average-and-moving-range chart** is used instead of an individuals chart. The moving average and moving range are calculated by using two or more individual data points. For instance, using the bulk insulation data in problem 1, we find the first moving average and moving range (using $n = 3$) to be 64 and 4, respectively. Construct a moving average and moving range chart for the data in problem 1. Discuss the smoother chart that results and whether the out-of-control point(s) are more numerous or more prominent than before. (The rule of sevens and the rule of thirds cannot be used, because one point is used in three different subsamples. Thus, the subsamples are no longer independent.)

5. See problem 9 in Chapter 4. Complete an individuals chart for machine A. Discuss machine A's contribution to the process.
6. Calculate the capability of the haze process (see Figure 5.1). Calculate both $C_p$ and $C_{pk}$. See Chapter 4, section 4.4. Use a $d_2$ of 1.128.
7. A length measurement results in the following data. Complete a median chart of the data as demonstrated in Section 5.2. Point out all OOC conditions except for the range (see problem 8). The measurements were by vernier caliper with zero equaling 6.00 inches. All data are coded in one-hundredth of an inch. (For example, in the first subsample the uncoded data are as follows: 6.20 in., 6.24 in., and 6.25 in.)

| *Subsample* | *Data* | *Subsample* | *Data* | *Subsample* | *Data* |
|---|---|---|---|---|---|
| 1 | 20–24–25 | 8 | 24–24–20 | 15 | 25–25–25 |
| 2 | 26–23–25 | 9 | 30–22–26 | 16 | 26–24–28 |
| 3 | 21–27–25 | 10 | 23–25–26 | 17 | 28–33–30 |
| 4 | 23–24–22 | 11 | 25–26–24 | 18 | 21–20–23 |
| 5 | 22–23–24 | 12 | 23–28–24 | 19 | 24–22–27 |
| 6 | 19–21–23 | 13 | 23–19–24 | 20 | 21–26–25 |
| 7 | 23–25–20 | 14 | 30–24–20 | | |

8. Make a scale drawing of the maximum range card for the chart in problem 7. Point out any OOC condition revealed by the card.
9. A die punch operation yields the following data on inner diameters of holes punched. Prepare a trend chart for the operation as demonstrated in Section 5.3. Samples were taken every hour and the units are in thousandths of an inch. Calculate the slope of the trend in inches per hour.

| *Subsample* | *Data* | *Subsample* | *Data* | *Subsample* | *Data* |
|---|---|---|---|---|---|
| 1 | 26–28 | 10 | 30–26 | 18 | 43–44 |
| 2 | 26–22 | 11 | 31–35 | 19 | 46–48 |
| 3 | 30–26 | 12 | 36–34 | 20 | 44–45 |
| 4 | 25–32 | 13 | 34–34 | 21 | 39–42 |
| 5 | 26–20 | 14 | 38–36 | 22 | 38–43 |
| 6 | 26–30 | 15 | 33–37 | 23 | 46–42 |
| 7 | 30–32 | 16 | 40–42 | 24 | 46–26 |
| 8 | 34–30 | 17 | 37–35 | 25 | 48–44 |
| 9 | 36–28 | | | | |

10. Use the linear regression function on your calculator to verify the line of best fit as drawn on the trend chart in Figure 5.4. Does your result compare favorably with the line-of-best-fit method used by the author?
11. During a run, and while using precontrol, subsamples of two were found to be as follows. For each of the conditions cited, explain whether you would continue to run the process or stop it and return to setup, and how the process may have changed, if it changed at all. In some cases only one piece needs to be checked. Mention this fact if it applies to any of the following conditions:
    a. One piece is in the yellow and the second in the green.
    b. Both pieces fall in the green.
    c. Two pieces fall in the red on one side.
    d. One piece is in the yellow area on one side and the other in the yellow area on the other side.
    e. Two pieces in a row fall into the yellow area on the same side.
12. See the probability calculations in Example 5.5. Assume that the distribution is flat rather than normal. What is the probability of the subsample of two parts being in the yellow area? What do you think the chances are that a flat distribution may pass setup?
13. List the advantages and disadvantages of each of the alternative charts discussed in this chapter or of those selected by the instructor.

# LABORATORY · CHAPTER 5
# Using the Individuals Chart

*Materials needed:* Measurement tools (a sine plate for measuring angles is a good choice for an individuals chart), 20 to 30 parts, $\overline{X}$ and $R$ chart forms.

After making the measurements, prepare an individuals chart of the data. If the order of manufacture of the parts is known, record the data in the proper order. If the order is not known, assume the order of measurement to be the order of manufacture. Complete the individuals chart. Calculate the capability. Prepare a histogram of the data and check the capability using the individual data.

The individuals chart is very useful in the service industries. An alternative laboratory could be to collect data from the following service processes:

- time of service;
- number of customers served per hour, day, or week (useful in predicting need for part-time staffing during peak periods);
- number of potential customers passing a business location per hour or per day.

# 6

# Probability of Attributes

**Fundamentals of Probability** 6.1
**Probability Flowchart** 6.2
**Permutations and Combinations** 6.3
**Selected Discrete Probability Distributions** 6.4

## 6.1 *Fundamentals of Probability*

Statistics offers the foundation for most quality control methods. In Chapter 3 the probability of the normal curve was studied. Probability distributions other than the normal curve were mentioned, and a representative sample of these distributions appears in Appendix B. In this chapter, the nature of probability will become more familiar to you. The examples, most of which were taken from actual quality control applications, will supplement any previous instruction you may have had in statistics. If you have had no previous instruction, the chapter's material should be understandable.

Statistics is a science that deals with the collecting, simplifying, and describing of data (descriptive statistics) and using the data to make predictions (statistical inference). **Probability** involves the use of data in making predictions of future events. We use probability every day because there is variability in all things. The weather from day to day varies. Measurements on a process quality characteristic may vary from measurement to measurement. And, of course, the quality characteristics of products and services vary. Most of what you have been introduced to in other courses has been deterministic. Probability is the opposite of determinism. The following deterministic approach to solving a problem may make the idea of probability clearer.

A force of 160 pounds exists over an area of 2 square feet. *Providing the data are correct,* find the pressure on the area in pounds per square foot (lb/ft$^2$). You determine the pressure on that area to be

$$P = \frac{F}{A} = \frac{160 \text{ lb}}{2 \text{ ft}^2} = 80 \frac{\text{lb}}{\text{ft}^2} .$$

This calculation is deterministic, and accepted as fact.

Contrast the pressure example (determinism) with the following: "The type of weather front that is entering your area has caused it to rain nine out of ten times in recorded history." The meteorologist may then predict that there is a 90 percent probability of rain. The prediction of a 90 percent probability of rain does not mean it *will* rain, but only that there is a 90 percent *probability* of rain. A given probability does not mean that something will happen, but only that there is a given chance (probability) of its happening.

Since variability is a very real part of experience, it is important to be able to predict the probability that an individual event or a group of selected events will occur. What is the probability of a manufacturing process causing one or more scratches (nonconformities) on a camera's case? What is the probability that you will be served within two minutes after you arrive in a restaurant?

Only groups of events can be predictable. You may predict that 50 percent of the time a fair coin will produce heads when tossed. But we all know not to bet much money that the coil will yield one head and one tail on the first two tosses. Turning to an example from business, how do insurance companies decide the rates for life insurance policies? The actuaries who develop the probabilities for setting the rates know it is impossible to say how long any one of us will live, but they realize that it is possible to collect longevity data from the human population. With this information, actuaries calculate the probability of anyone in the designated population living to a certain age. To make reasonable predictions we must use a fair number of data.

When collecting data it is also critical to be clear about the difference between a sample and a population (a population is also referred to as a universe). A **sample** is a limited number of units taken from a larger source. A **population** is that larger source of units the sample is drawn from. The sample is used to make predictions, or calculate probabilities, involving the defined population.

The probability of individual events may be calculated directly if the events can be counted without experimentation. This approach is known as **classical probability.** For example, you know that there are 13 spades in a deck of 52 cards, so you can calculate the probability of drawing a spade from the deck without experimentation.

You can also predict the occurrence of individual events by **experimental probability.** Process control studies incorporate experimental probability. For example, in a production run of 1,000 plastic fans, 100 of the fans are likely to be scratched. This result has been determined by averaging the number of scratched fans from 60 production lots of 1,000 over a period of one year (a relatively large sample from the population of plastic fans).

Whether calculating classical or experimental probability the relationship is

$$P(E) = \frac{E}{N} = \frac{\text{Selected events to occur}}{\text{All total events that may occur}}$$

where $P(E)$ symbolizes the probability of an event occurring. For the classical probability example, the probability of picking a spade from a well-shuffled deck of cards is

$$P(E) = \frac{13}{52} = 0.25 = 25\%$$

You may wish to confirm this probability by experiment. You need a fairly large number of trials (at least 50) to approach the value predicted by using the classical, or nonexperimental, method.

For the experimental probability example, the probability of selecting a scratched part from the production lot when all parts have an equal chance of being selected (random selection), is

$$P(E) = \frac{100}{1{,}000} = 10\%$$

In both situations, classical and experimental, the number in the denominator of the fraction is known as the **sample space.** The sample space for the deck of cards is 52, and the sample space for the production lot is 1,000.

EXAMPLE 6.1 In any one flip of a fair coin, what is the probability of a head?

*Solution:* A fair coin is one that has an equal chance of landing on either side—heads or tails. The selected event, a head, represents a possible event of one. The sample space is two—a head or tail can occur. Therefore, the probability of a head occurring is

$$P(H) = \frac{1}{2} = 0.5 = 50\%$$

EXAMPLE 6.2 When two fair dice are thrown, what is the probability of throwing a seven?

*Solution:* In this problem you can visualize the sample space and the number of selected events by constructing a six-by-six matrix of all possible combinations. The six-by-six matrix reveals that 36 (6 × 6) combinations exist in the entire sample space. The *x*s designate the selected events, all combinations adding to seven.

| | 1 | 2 | 3 | 4 | 5 | 6 |
|---|---|---|---|---|---|---|
| 1 | | | | | | *x* |
| 2 | | | | | *x* | |
| 3 | | | | *x* | | |
| 4 | | | *x* | | | |
| 5 | | *x* | | | | |
| 6 | *x* | | | | | |

The probability of a seven is

$$P(7) = \frac{6}{36} = 0.167 = 16.7\%$$

Using the matrix, find the probability of a obtaining a four (answer: 3/36 = 0.0833 = 8.33%). What is the number with the greatest probability of occurrence when throwing two dice?

Another important relationship is the probability of an event not happening:

$$P(E[\text{not}]) = 1 - P(E)$$

What is the probability of *not* throwing a seven? The easiest way to determine this probability is first to solve for the probability of a seven, which has already been done, and then to use

$$P(\text{not a seven}) = 1 - P(7) = 1 - 0.167 = 0.833 = 83.3\%$$

## Section 6.1 Review Problems

1. What is the difference between classical and experimental probability (one or two sentences)? Give an example of each.
2. Find the probability of obtaining a five when throwing two dice. See the matrix in Example 6.2.
3. Also from Example 6.2, determine the number with the greatest probability of occurrence.
4. What is the probability of drawing a nine from a well-shuffled deck of 52 cards?
5. A box contains two nonconforming parts and two conforming parts. Given that a nonconforming part has been drawn, what is the probability of drawing two consecutive conforming parts in the next three draws?
   a. 1/6
   b. 2/3
   c. 1/3
   d. 1/4

# 6.2 Probability Flowchart

In Section 6.1 we calculated probability without the need to consider a combination of probabilities. There are some important probability equations that permit combining probabilities. Consider these various combinations by referring to the probability flowchart presented in

Figure 6.1. The flowchart is continued later (Figure 6.3), and a complete probability flowchart is offered at the end of the chapter (Figure 6.7).

## Mutually Exclusive Events

First, we will calculate the probability of only one event. Moving horizontally across the flowchart we find that for **mutually exclusive events**—which means that only one selected event ($A$ or $B$) can happen at a time—the equation, known as the addition equation, is

$$P(A \text{ or } B) = P(A) + P(B)$$

You read the equation as "The probability of $A$ or $B$ is equal to the probability of $A$ plus the probability of $B$."

EXAMPLE 6.3 Find the probability of drawing a heart or a spade in one draw from a well-shuffled, complete deck of cards.

*Solution:* The two events are mutually exclusive; that is to say, only a heart *or* a spade may be drawn at one time. The probability calculation is

$$P(H \text{ or } S) = P(H) + P(S) = \frac{13}{52} + \frac{13}{52} = 0.50 = 50\%$$

The equation can be expanded to handle more than two cases. Example 6.4 will illustrate.

EXAMPLE 6.4 Find the probability of drawing a heart, spade, or club from the same well-shuffled deck.

$$P(H \text{ or } S \text{ or } C) = P(H) + P(S) + P(C)$$

$$= \frac{13}{52} + \frac{13}{52} + \frac{13}{52} = 0.75 = 75\%$$

The results of Examples 6.3 and 6.4 reveal a 50 percent and 75 percent probability, respectively. Does this finding seem to make sense? What is the probability of drawing a diamond from the deck? You already know the probability is 13/52, or 25 percent. Plug the informa-

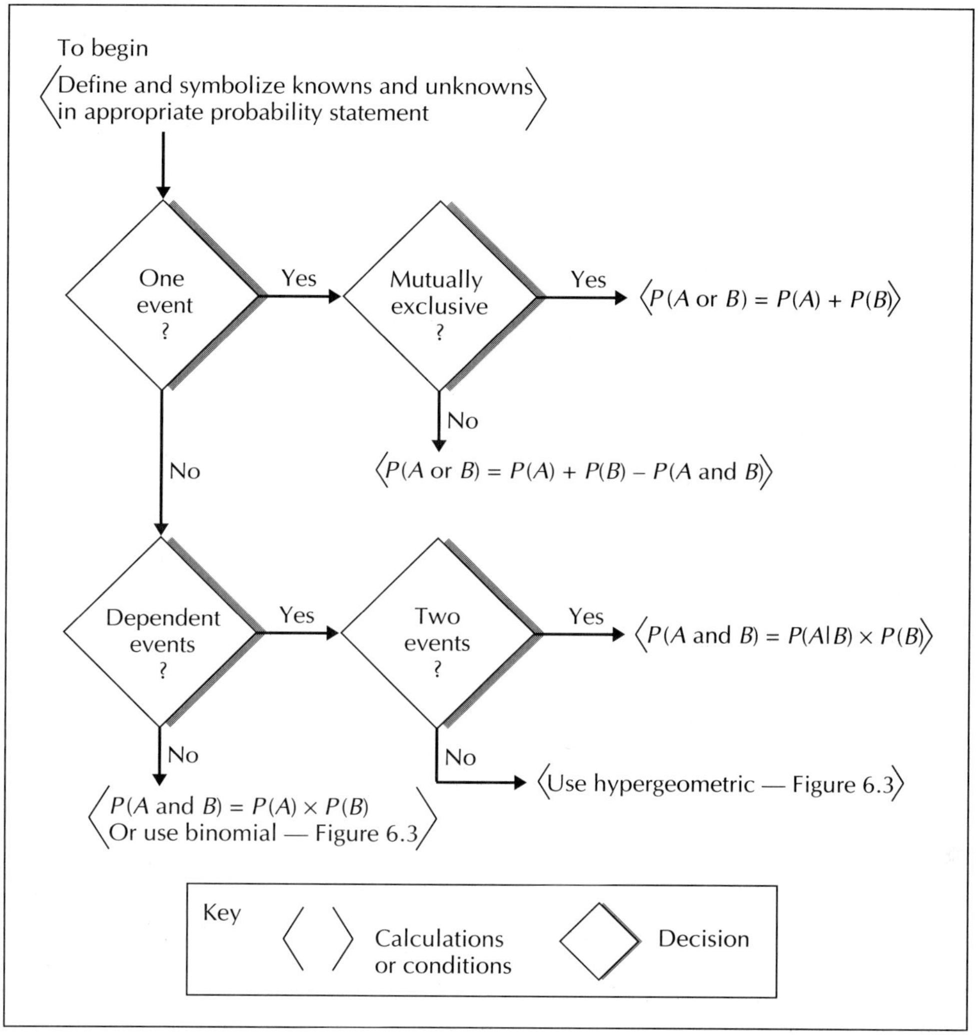


**FIGURE 6.1**
The probability flowchart helps determine what probability equation to use.

tion in Example 6.4 into the following equation:

$$P(D) = 1 - P(H, S, \text{ or } C) = 1 - 0.75 = 0.25 = 25\%$$

The equation yields the expected result.

# *Non–Mutually Exclusive Events*

**Non–mutually exclusive** events are events that may occur simultaneously. Returning to the flowchart, if the event is not mutually exclusive, the equation must be further expanded. To see why this is so, consider Example 6.5.

EXAMPLE 6.5 Calculate the probability of drawing a king *or* a spade from the well-shuffled deck of cards. The events, "a king or a spade," are not mutually exclusive, for a draw could result in the king of spades.

*Solution:* The probability of drawing a spade is ¼ or 25 percent. The probability of drawing a king is 1⁄13 or 7.69 percent. Simply adding the two results in a probability of 32.69 percent. But this answer is incorrect. The king of spades was counted twice: as one of the four kings in the fraction 1⁄13 and as one of the 13 spades in the fraction ¼. You can visualize the non–mutually exclusive event by using a *Venn diagram* (Figure 6.2), a useful tool in understanding logic systems.

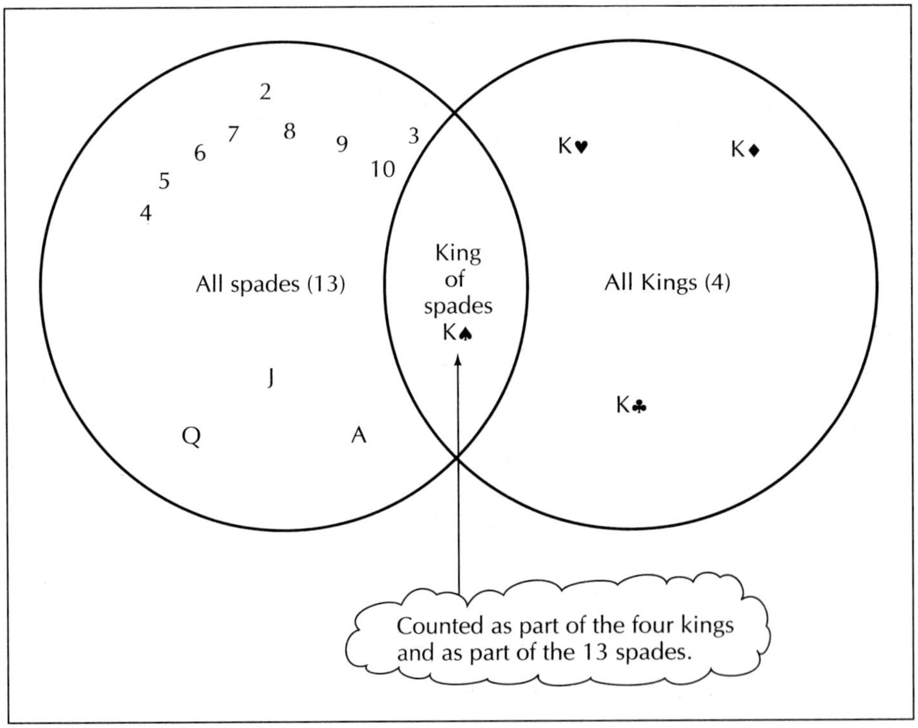


**FIGURE 6.2**
The king of spades is counted twice.

Using the fully developed addition equation (see Figure 6.1, the probability flowchart), the probability of drawing a king or a spade is

$$P(K \text{ or } S) = P(K) + P(S) - P(K \text{ and } S)$$

$$= \frac{4}{52} + \frac{13}{52} - \frac{1}{52} = 0.308$$

The probability is 30.8 percent.

In summary, use the addition equation whenever the probability of the outcome of one event is desired—the "or" condition. Expand the addition equation to include the situation of overlapping (non–mutually exclusive) events, because such events are counted twice. If the idea of "or" is present in the application, remember to always use addition.

## *Joint Probability—Dependent Events*

If the outcome of more than one event, often referred to as a **joint probability,** is desired, the flowchart shows that an "and" relationship is used. For instance, $P(A \text{ and } B)$ means that both $A$ and $B$ occur. To calculate the probability of multiple events occurring, it is important first to determine whether the events are dependent on each other. **Dependent events** require one event to happen before the other can happen. If the events are dependent, the equation is

$$P(A \text{ and } B) = P(B|A) \times P(A)$$

$$P(A \text{ and } B) = P(A|B) \times P(B)$$

where the vertical-bar symbol (|) means "given that." A useful rearrangement of this equation is used to solve for the probability of $B$ "given" $A$ (the probability of $B$ occurring after $A$ has occurred).

$$P(B|A) = \frac{P(A \text{ and } B)}{P(A)}, \text{ or for } A \text{ ``given'' } B \text{ use}$$

$$P(A|B) = \frac{P(A \text{ and } B)}{P(B)}$$

EXAMPLE 6.6 A manufacturer of large electric motors finds that three percent of all of the motors having shorted windings have supplier X's wire. Four percent of the motors have shorted windings. What is the probability that a motor with a shorted winding has wire provided by supplier X?

*Solution:* Let the motors having supplier X's wire = $A$
Let the motors having shorted windings = $B$
$P(A \text{ and } B) = 0.03 \qquad P(B) = 0.04$

$$P(A|B) = \frac{P(A \text{ and } B)}{P(B)} = \frac{0.03}{0.04} = 0.75 = 75\%$$

A motor having a shorted winding has a rather large (75%) chance of the wire being from supplier X. Supplier X's wire-insulation process should be reviewed for improved process control.

## Joint Probability—Independent Events

Often, the events are not dependent, but independent. For instance, could paint peeling from the outside case of the motors in Example 6.6 cause shorted windings? Probably not. Events are considered independent if the probability of $A$ given $B$ equals the probability of $A$. The next example will help to clarify dependence versus independence.

**EXAMPLE 6.7** A small lot of 25 parts has three nonconforming parts. What is the probability of obtaining two nonconforming parts in a random sample of two parts

1. when sampling without replacement;
2. when sampling with replacement?

*Solution, sampling without replacement:* Sampling without replacement means the probability of the second part being nonconforming is dependent on the condition of the first part. Since these are dependent events we must use the following equation. (*Note:* In the equations, nonconforming is abbreviated as *n.c.*)

$$P(A \text{ and } B) = P(A|B) \times P(B)$$

$$P(\text{both parts } n.c.) = P(\text{2nd part } n.c.|\text{1st part } n.c.) \times (\text{1st part } n.c.)$$

$$= \frac{2}{24} \times \frac{3}{25} = 0.01 = 1\%$$

*Solution, sampling with replacement:* Selecting the second part after the first part has been replaced (and the parts thoroughly mixed) means the second part has as likely a chance of being nonconforming or acceptable as it would have had if the first part had not been drawn. Therefore, the probabilities are independent. From the probability flowchart in Figure 6.1, and expressed rigorously, **independent events** means

$$P(A|B) = P(A)$$

$$P(\text{2nd part } n.c.|\text{1st part } n.c.) = P(\text{2nd part } n.c.) = \frac{3}{25}$$

Twenty-five parts remain after the first part drawn has been replaced. With the independent event satisfied,

$$P(A \text{ and } B) = P(A) \times P(B)$$

$$= \frac{3}{25} \times \frac{3}{25} = \frac{9}{625} = 0.0144 = 1.44\%$$

If the probability of independent events is desired, use the multiplication equation.

$$P(A \text{ and } B) = P(A) \times P(B)$$

which may be expanded to

$$P(A \text{ and } B \text{ and} \ldots \text{ and } Z) = P(A) \times P(B) \times \ldots \times P(Z)$$

EXAMPLE 6.8

In a state lottery, six numbers are drawn from a population of 50. What are the *odds* that the numbers drawn will satisfy (match) your set of numbers? In other words, what are the chances you'll be able to forget all about learning QC statistics and spend the rest of your life as an incredibly wealthy person?

*Solution:* Using the multiplication equation, the probability is calculated as follows:

$$P(\text{match all six numbers}) = P(\text{1st}) \times P(\text{2nd}) \times \ldots \times P(\text{6th})$$

$$P = \frac{6}{50} \times \frac{5}{49} \times \frac{4}{48} \times \frac{3}{47} \times \frac{2}{46} \times \frac{1}{45}$$

$$= 6.29 \times 10^{-8}$$

Each draw assumes that one of the six numbers was satisfied. After each draw there is one less number to satisfy, and the numerator decreases by one each time. Also, after each draw there is one less of the total number $n$ of 50. Thus the denominator or sample space decreases. This example of the multiplication rule is *a special case of dependency, in that each draw is assumed to be succesful.* To find the odds, we invert the probability $[1/(6.29 \times 10^{-8})]$ to see that the odds are 15.9 million to 1, of hitting all six numbers. Well, maybe it's best to continue to study QC statistics.

As may be seen from the example, the multiplication equation may be expanded to include as many independent events as needed. The multiplication equation is used whenever you need to determine the probability of multiple events. It will help you to remember to use multiplication when an "and" is verbalized.

**EXAMPLE 6.9** Given the data in Table 6.1, taken from the historical data of 1,000 parts in an industrial operation, answer the following questions by using the probability flowchart.

1. What is the probability of finding a part produced by machine A or machine B?
2. What is the probability of finding a part from machine A or a nonconforming unit?
3. What is the probability of finding a nonconforming unit that is from machine B?
4. What is the probability of finding a nonconforming unit from machine A and machine C when drawing two parts and replacing the first part before the second is drawn?
5. What is the probability of finding a nonconforming unit from machines A and C when drawing two parts without replacement?

*Solution 1:* The row and column totals are known as *marginal probabilities* (because they are found at the edges [margins] of the table). Using the addition equation (mutually exclusive events, see Figure 6.1) with the marginal probabilities

$$P(\text{mach. A or mach. B}) = P(\text{mach. A}) + P(\text{mach. B})$$

$$= \frac{560}{1{,}000} + \frac{280}{1{,}000} = \frac{840}{1{,}000} = 0.84 = 84\%$$

TABLE 6.1

| | *Machine A* | *Machine B* | *Machine C* | *Row Total* |
|---|---|---|---|---|
| Conforming unit | 400 | 200 | 130 | 730 |
| Nonconforming unit | 160 | 80 | 30 | 270 |
| Column total | 560 | 280 | 160 | 1,000 |

Machines A and B make 84 percent of the parts, yielding an 84 percent probability of picking a part from one of the machines.

*Solution 2:* These two events, a part from machine A and a nonconforming part, are not mutually exclusive. Use the fully developed addition equation (Figure 6.1).

$P$(part from mach. A or *n.c.* part) = $P$(part from mach. A) +

$$P(n.c.\text{ part}) - P(\text{both events}) = \frac{560}{1{,}000} + \frac{270}{1{,}000} - \frac{160}{560} = 0.544$$

$= 54.4\%$

*Solution 3:* Inspecting Table 6.1, get the answer directly from the table by dividing 80 by 1,000, and arriving at an answer of 8 percent. Using the multiplication equation from Figure 6.1 and assuming dependency

$$P(n.c.\text{ and from mach. B}) =$$

$$P(n.c.|\text{mach. B}) \times P(\text{mach. B}) = \frac{80}{280} \times \frac{280}{1{,}000} = 0.08 = 8.0\%$$

*Solution 4:* "With replacement" means that the two events are independent (see Example 6.7, "Solution, sampling with replacement").

$$P(n.c.\text{ mach. A and } n.c.\text{ mach. C}) =$$

$$P(n.c.\text{ A}) \times P(n.c.\text{ C}) = \frac{160}{1{,}000} \times \frac{30}{1{,}000} = 0.0048 = 0.48\%$$

*Solution 5:* "Without replacement" means that the two events are dependent. Assume that a nonconforming part from machine A ($P$ = 160/1,000) has already occurred. The probability of drawing the second part and having it be nonconforming from machine C is

$$P(n.c.\text{ mach. A and mach. C}) =$$

$$P(n.c.\text{ C}|n.c.\text{ A}) \times P(n.c.\text{ A}) = \frac{30}{999} \times \frac{160}{1{,}000} = 0.004805 = 0.4805\%$$

Comparing the results in Solutions 4 and 5 (0.4800% with 0.4805%) points out a very important statistical fact: *With a large number of parts, assuming independence does not introduce significant error.* This finding—small sample sizes from a large number of parts introduce little error when independency is assumed—will be assumed in some of the probability distributions developed in Section 6.4.

## Section 6.2 Review Problems

1. From a box of 50 parts, five have critical nonconformances, and 20 others have minor nonconformances. What is the probability of drawing a nonconforming part from the box?
2. Find the probability of drawing a club or a diamond from a well-shuffled deck of cards.
3. Find the probability of drawing an ace or a club from the well-shuffled deck.
4. Using the data in Table 6.1, calculate the probability of finding a part produced by machine A or machine C.
5. Using the data in Table 6.1, calculate the probability of finding a part produced by machine B or a nonconforming unit.

# 6.3 Permutations and Combinations

This section discusses the number of ways objects may be arranged. It is often important, for example, to determine the number of groups of items in determining the sample space in a probability calculation. We will find the use of counting techniques to be indispensable in Section 6.4, "Selected Discrete Probability Distributions," as well.

## Permutations

A **permutation** involves the orderly arrangement of a number of items. For example, the letters SQC may be arranged six different ways (forming six different permutations): SQC, SCQ, QSC, QCS, CSQ, and CQS. A **combination** only notes the number of elements and disregards any arrangement. Therefore, there is only one combination of the letters SQC.

The equation used to determine a permutation of $n$ items taken $r$ at a time is

$${}_nP_r = \frac{n!}{(n-r)!}$$

where the literal factor $_nP_r$ means "$n$ items taken $r$ at a time." We're using the mathematical symbol "!" for the first time. In statistics it is known as a *factorial*. The factorial signifies multiplication of the number by all discrete or whole numbers less than that number. For example,

$$3! = 3 \times 2 \times 1$$

$$6! = 6 \times 5 \times 4 \times 3 \times 2 \times 1$$

By definition 0! is equal to one.

EXAMPLE 6.10 Using the permutation equation, solve for the number of permutations of the three letters SQC ($n = 3$) taken three at a time ($r = 3$).

*Solution:*

$$_3P_3 = \frac{3!}{(3-3)!} = \frac{3!}{0!} = \frac{3 \times 2 \times 1}{1} = 6$$

Also, find on your statistical calculator the $_nP_r$ symbol. The statistical calculator will solve the permutation equation directly. Consult your owner's manual to learn how to enter the $n$ and the $r$ correctly.

## Combinations

The equation to determine the number of combinations of $n$ objects taken $r$ at a time is

$$_nC_r = \frac{n!}{r!(n-r)!}$$

See Example 6.11.

EXAMPLE 6.11 How many combinations of six numbers can be found in a total of 50 numbers?

*Solution:*

$$_nC_r = \frac{50!}{6!(44!)} = 15.9 \times 10^6$$

Find on your calculator the $_nC_r$ symbol. The statistical calculator will solve the combination equation directly. Consult your owner's manual for how to enter the $n$ and the $r$ correctly.

Compare the result in Example 6.11 with the odds calculated for a lottery game in Example 6.8. Can you explain why the combination equation yields the odds directly and the multiplication equation yields the probability?

## *Section 6.3 Review Problems*

1. Using the permutation equation, solve for the number of permutations of 30 parts taken four at a time.
   a. 27,400
   b. 657,720
   c. 986,450
   d. 1,235,640
2. A set of 81 gage blocks taken three at a time offers how many measurement combinations:
   a. 512
   b. 4,280
   c. 79,930
   d. 85,320
3. Explain why a combination lock might be more appropriately named a permutation lock.

# 6.4 *Selected Discrete Probability Distributions*

In this section we'll use discrete probability distributions to determine the probability of finding a certain number of nonconforming units in a given sample or population. **Discrete distributions** are those used to count the occurrence of nonconforming parts or nonconformities on a part. Such counts of nonconformities are known as attribute data. When measurements are made, the data are known as variable data, and **variable distributions** such as the normal distribution are used. See Appendix B for a representative group of probability distributions. Remember that this chapter's material has discussed only the probability of the occurrence of discrete events.

To consider the various probability distributions, refer to Figure 6.3, a continuation of the probability flowchart in Figure 6.1. The equa-

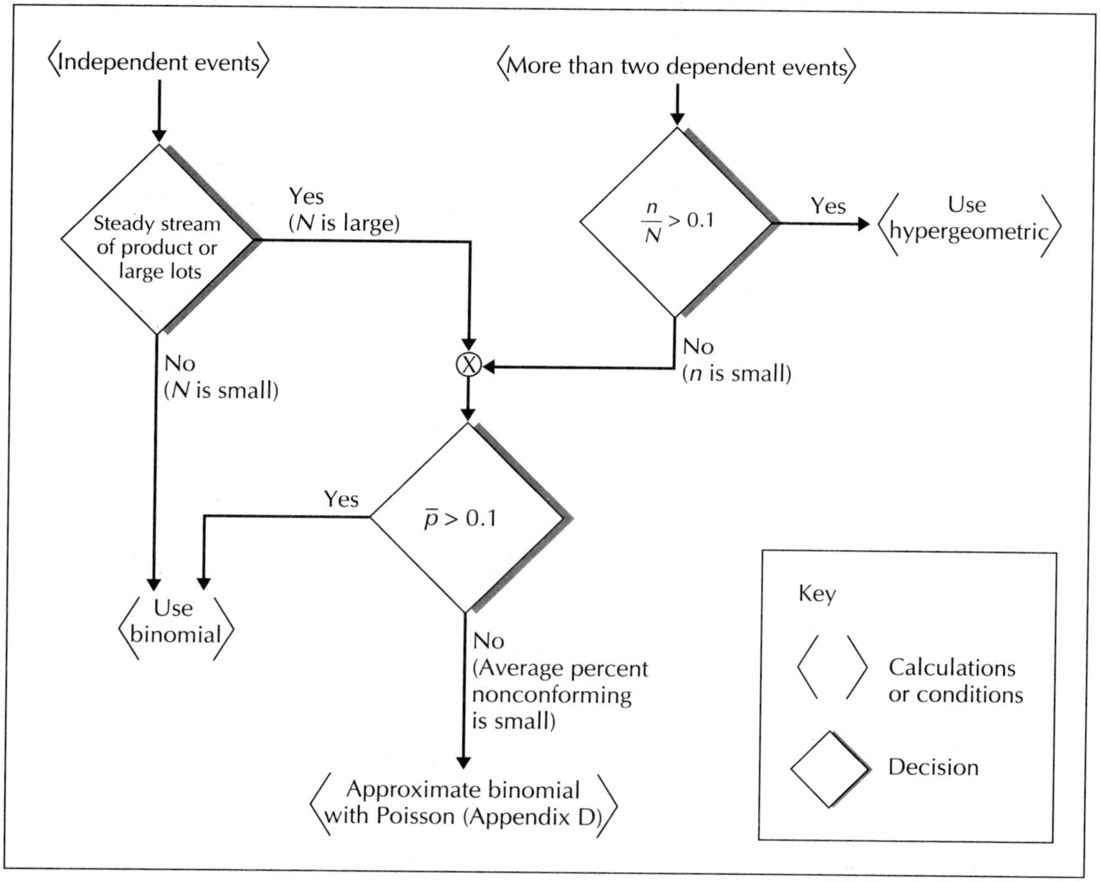


**FIGURE 6.3**
This flowchart is a continuation of Figure 6.1.

tions for the distributions are not given on the flowchart because they need explanation. The equations that define the hypergeometric, binomial, and Poisson distributions are frequently used in QC sampling plans (Chapter 9).

## *Hypergeometric Probability Distribution*

The hypergeometric probability distribution is the most basic of the discrete probability distributions. The hypergeometric equation is derived from simple counting techniques—combinations—and may be written as follows:

$$P(n.c.) = \frac{{}_{N.C.}C_{n.c.} \times {}_{N-N.C.}C_{n-n.c.}}{{}_{N}C_{n}}$$

where $n.c.$ = nonconforming units in sample
$N.C.$ = nonconforming units in population
$N$ = number of units in population
$n$ = number of units in sample

It is helpful to view the first combination (numerator) as all combinations of nonconforming items, the second combination (numerator) as all combinations of conforming items, and the third combination (denominator) as the number of samples of size $n$ that can be drawn from the population of size $N$. In setting up the equation, first determine the first numerator combination, next determine the denominator combination, and finally determine the second numerator combination by observing that the differences are for data already determined in the previously constructed combinations.

EXAMPLE 6.12 A government regulating organization suspects a new housing development of 50 homes to have excessively high levels of radon gas concentrations. If 30 of the 50 homes actually have high concentrations, the area is considered to have a high level of radon. Warnings from government regulators will follow, meaning the real-estate agency will have to provide adequate ventilation systems to prevent radon accumulation. What is the probability of finding no homes with high levels of radon [$P(0)$] out of a sample of 10? Of five homes P[(5)] out of a sample of 10?

*Solution:* The following values are taken to answer the first question [$P(0)$]: nonconforming homes in sample $n.c. = 0$, nonconforming homes in population $N.C. = 30$ (assumed), the number in the population $N = 50$, and the number actually sampled $n = 10$. Using the hypergeometric equation

$$P(0) = \frac{{}_{30}C_{0} \times {}_{20}C_{10}}{{}_{50}C_{10}} = 1.8 \times 10^{-5}$$

If the technologist performing the inspection for the regulating agency finds zero homes in the sample of 10, she should reject the conclusion (hypothesis) that 30 out of fifty homes (60% of the homes) have high radon levels. No government mandates will follow.

However, if she finds that five out of 10 homes are high in radon concentrations, the equation yields the following probability:

$$P(5) = \frac{{}_{30}C_5 \times {}_{20}C_5}{{}_{50}C_{10}} = 0.215$$

A 21.5 percent probability might lead the government regulating authority to conclude that there is a reasonable probability that the area has high levels of radon.

Referring to Figure 6.3, the hypergeometric distribution is the distribution used when the sample size is greater than 10 percent of the size of the population ($n/N > 0.1$), and for lots for which the total number of nonconforming parts is known. The advantage of the distribution is that replacement is not necessary to achieve valid results. In fact, if sampling *with* replacement is used, the hypergeometric will not produce valid results (see problem 16 at the end of the chapter).

The disadvantage of using the hypergeometric in manufacturing is that populations are often large and the exact number of nonconforming units is not known. Another distinct disadvantage of the hypergeometric is that calculations are laborious, even when using a statistical calculator that can automatically calculate combinations.

## Binomial Probability Distribution

Sir Isaac Newton, the great physicist, developed the binomial theorem. The prefix *bi* means two. Of course, when studying nonconforming units—*discrete* or *attribute data*—there are only two conditions: nonconforming units or conforming units. One form of equation for this important distribution is

$$P(c) = {}_{n}C_{c}\bar{p}^{c}(1 - \bar{p})^{n-c}$$

where $c$ = nonconformities in sample (The letters *n.c.* were previously used. The letter $c$ is commonly used to designate the number of nonconformities. Also refer to the Glossary of Symbols inside the back cover.)

$n$ = number in sample

$\bar{p}$ = average fraction nonconforming in population

The binomial distribution is used whenever independency is the case or may be assumed. Take, for example, the independent events of obtaining three sixes on six throws of a single die (or of throwing six dice at once). The events are independent because each throw of a die does not depend on any other throw of a die. The calculation using the binomial becomes

$$P(\text{three sixes in six throws}) = {}_6C_3\left(\frac{1}{6}\right)^3\left(\frac{5}{6}\right)^3 = 20 \times 0.00463 \times 0.579$$

$$= 0.0536$$

The binomial equation may also be used when independency may not be assumed, and, in fact, this is the way the equation is used in most quality control work. Please review Example 6.9, which showed how large enough numbers in a population can make an assumption valid—in this case the assumption of independency. Also, refer to Figure 6.3 and follow the flowchart from "More than two dependent events" through "$n/N > 0.1$" and "$\overline{p} > 0.1$." Independency will be assumed in Examples 6.13 and 6.14.

EXAMPLE 6.13

As an assembly line runs, producing a large number of units, a stratified-random sample of five units is selected during a one-hour period. If the average percent nonconforming in the process $\overline{p}$ is 0.10 or 10 percent, how many nonconforming units $c$ might be expected in the sample? What does the probability distribution look like?

*Solution:*

$$P(c) = {}_nC_c\overline{p}^c(1-\overline{p})^{n-c}$$

$$P(0) = {}_5C_0 0.1^0 0.9^5 = 0.590$$
$$P(1) = {}_5C_1 0.1^1 0.9^4 = 0.328$$
$$P(2) = {}_5C_2 0.1^2 0.9^3 = 0.0729$$
$$P(3) = {}_5C_3 0.1^3 0.9^2 = 0.00810$$
$$P(4) = {}_5C_4 0.1^4 0.9^1 = 0.00045$$
$$P(5) = {}_5C_5 0.1^5 0.9^0 = 0.00001$$

The probability distribution is shown in Figure 6.4. Confirm that the probabilities on the $Y$ axis [$P(c)$] correspond with the number of nonconforming units on the $X$ axis ($c$).

You can see from the binomial probability distribution in Figure 6.4 that the distribution of the probabilities calculated from the binomial equation is skewed positively, with zero nonconforming units being most probable. Finding more than two nonconforming units is very improbable. (If more than two are drawn, the population has probably changed or has more nonconforming product.) Adding the probabilities

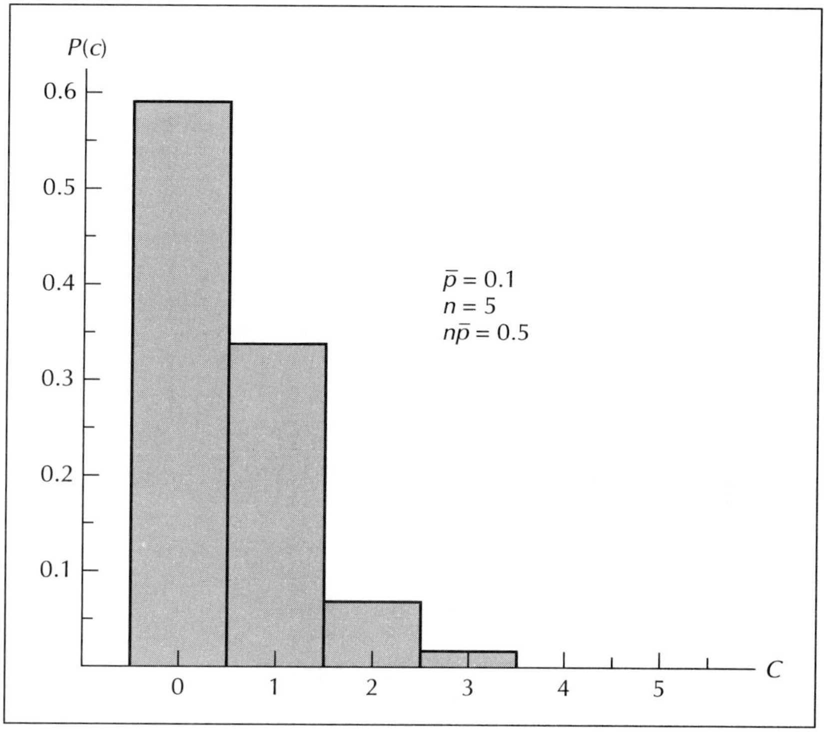


**FIGURE 6.4**
The binomial probability distribution for Example 6.13.

of finding 0 through 2 nonconforming units yields 0.999 or 99.9 percent. That result means there is only a 0.1 percent chance of obtaining more than two nonconforming units in the sample if the population has not changed.

It is interesting to view the probability distribution for a larger $n$ when using the binomial equation, as Example 6.14 demonstrates.

EXAMPLE 6.14 Use the data in Example 6.13 and change the sample size to $n = 50$. Calculate the probability distribution and construct the histogram.

*Solution:*

$$P(c) = {}_nC_c\bar{p}^c(1 - \bar{p})^{n-c}$$

$$P(0) = {}_{50}C_0 0.1^0 0.9^{50} = 0.00515$$
$$P(1) = {}_{50}C_1 0.1^1 0.9^{49} = 0.0286$$
$$P(2) = {}_{50}C_2 0.1^2 0.9^{48} = 0.0779$$
$$P(3) = {}_{50}C_3 0.1^3 0.9^{47} = 0.139$$
$$P(4) = {}_{50}C_4 0.1^4 0.9^{46} = 0.181$$

$$P(5) = {}_{50}C_5 0.1^5 0.9^{45} = 0.185$$
$$P(6) = {}_{50}C_6 0.1^6 0.9^{44} = 0.154$$
$$P(7) = {}_{50}C_7 0.1^7 0.9^{43} = 0.108$$
$$P(8) = {}_{50}C_8 0.1^8 0.9^{42} = 0.0634$$
$$P(9) = {}_{50}C_9 0.1^9 0.9^{41} = 0.0333$$
$$P(10) = {}_{50}C_{10} 0.1^{10} 0.9^{40} = 0.0152$$

The probability distribution is shown in Figure 6.5. Please confirm that the probabilities on the *Y* axis [$P(c)$] correspond with the number of nonconforming units on the *X* axis ($c$).

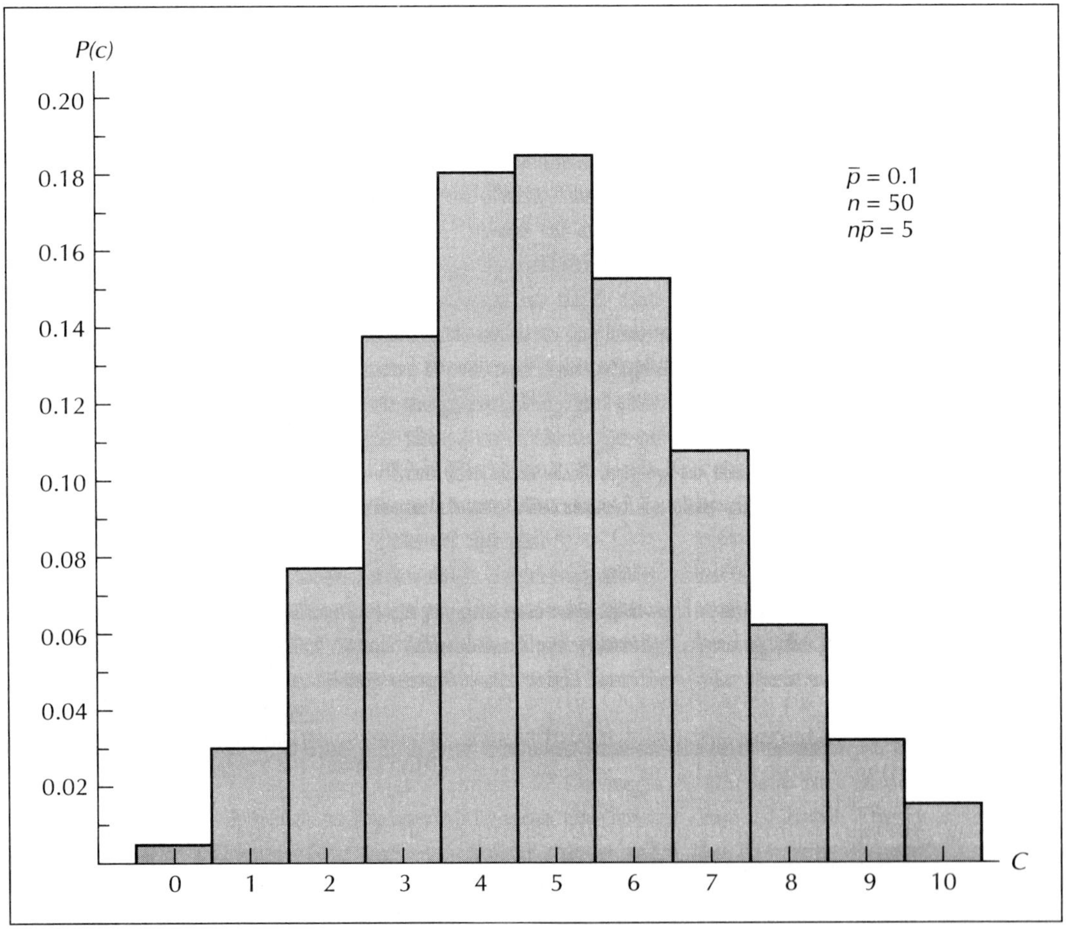


**FIGURE 6.5**
Compare this binomial distribution with the distribution in Figure 6.4.

You can see in Figure 6.5 that the distribution is tending toward the normal distribution. It is also similar to the Poisson distribution, which appears in Figure 6.6 and is discussed in the following subsection.

In summary (see Figure 6.3), if the sample size is less than 10 percent of the population ($n/N < 0.1$), the hypergeometric approximates the binomial. You can see in the defining equation that one advantage of the binomial over the hypergeometric distribution is that the total *lot* size and the total nonconforming units in the population do not have to be known. Indeed, they are not even used.

The binomial distribution has its own set of disadvantages however, including the following:

- A major assumption is that $\overline{p}$, the average percent nonconforming, remains constant.
- Independence is assumed only with large population sizes, usually a steady stream of product (see Example 6.9 and the conclusions that follow the example).
- Calculations are simpler than the hypergeometric, but are still laborious.

Fortunately, many industrial processes produce large numbers of products. Therefore, the approximation of the hypergeometric by the binomial, or by the Poisson which follows, is valid.

## Poisson Probability Distribution

The Poisson probability distribution is named for the French mathematician Siméon Denis Poisson, who published its derivation in 1837. The Poisson is often referred to as the law of small numbers. The distribution is most useful for average percent nonconformities ($\overline{p}$) of less than 10 percent. *The Poisson is the most frequently used probability distribution in quality control sampling plans and control charts.*

The equation defining the Poisson may be derived from the binomial. The result is

$$P(c) = \frac{(n\overline{p})^c}{c!} e^{-n\overline{p}}$$

where $c$ = number of nonconformities in sample
$n\overline{p}$ = average number of expected nonconformities
$e$ = the natural logarithm (2.718 . . . ; $e^x$ may be found on your calculator)

The Poisson may be used to approximate the binomial if the population is large (i.e., a large lot or steady stream of product) and if the average fraction nonconforming $\overline{p}$ is less than 0.1 (see Figure 6.3).

In Appendix D you will find a table generated from the equation defining the Poisson. The Poisson is preferred and is used to approximate the binomial because of the existence of this table. To use the Poisson table, first determine the average number of nonconformities or nonconforming units $n\overline{p}$ and the number of nonconformities or nonconforming units expected in the sample $c$. The table gives the probability of finding the expected number of nonconformities. For instance, if the average number of expected nonconformities is 1.0 ($n\overline{p} = 1.0$), the 1.0 (see the bold numbers across the top of the chart) column yields the following probabilities: the probability of 0 nonconformities is 0.368 or 36.8 percent, with the cumulative probability (in parentheses) being the same (less than 0 nonconformities is impossible); the probability of 1 nonconformity ($c = 1$) is also 0.368 or 36.8 percent, with the cumulative probability of having *0 or 1* nonconformity being the sum of the probabilities or 0.736; the probability of exactly 2 nonconformities is 0.184 or 18.4 percent, with a cumulative probability—having 0, 1, or 2 nonconformities—of 92 percent.

EXAMPLE 6.15

Using the Poisson table, Appendix D, look up the following probabilities. Your sample size is 3 and the average number of nonconformities per unit is 3.

- 0 nonconformities
- 2 nonconformities
- 4 nonconformities or fewer
- 10 or more nonconformities
- more than 10 nonconformities

*Solution:* The sample size $n$ is 3 and the average number of nonconformities per unit $\overline{p}$ is also 3. Therefore, $n\overline{p} = 3 \times 3 = 9$. We first find the $n\overline{p}$

of 9.0 across the top of the table. Using the various values of $c$ (left side of table), we find the following probabilities:

- $c = 0$, $P(c) = \underline{0.000}$
- $c = 2$, $P(c) = \underline{0.005}$
- $c = 4$ or less, $P(c) = \underline{0.055}$ (use cumulative values in parentheses)
- $c = 10$ or more, $P(9 \text{ or less}) = 0.588$ and $P(10 \text{ or more}) = 1 - 0.588 = \underline{0.412}$
- $c =$ more than ten, $P(\text{more than } 10) = 1 - 0.707 = \underline{0.293}$

EXAMPLE 6.16 Use the calculations in Examples 6.13 and 6.14 to see how close the Poisson approximates the binomial. Note that the examples use the maximum $\overline{p}$ ($\overline{p}$ should not be more than 10%) recommended in approximating the binomial by the Poisson.

*Solution:* The results are shown in the following tables, and the probability distributions in Figure 6.6 show how well the Poisson approximates the binomial.

*Solution (using the Poisson table, Appendix D, for Example 6.13):* $c = 0$ through 5, $n = 5$, $p = 0.1$ (or 10%), $n\overline{p} = 0.5$.

| *Probability of c* | *Cumulative Probability* |
|---|---|
| $P(0) = 0.607$ | 0.607 |
| $P(1) = 0.303$ | 0.910 |
| $P(2) = 0.076$ | 0.986 |
| $P(3) = 0.013$ | 0.999 |
| $P(4) = 0.002$ | 1.000 |
| $P(5) =$ — | — |

*Solution (using the Poisson table for Example 6.14):* $c = 0$ through 9, $n = 50$, $\overline{p} = 0.1$, $n\overline{p} = 5$.

| *Probability of c* | *Cumulative Probability* |
|---|---|
| $P(0) = 0.007$ | 0.007 |
| $P(1) = 0.034$ | 0.041 |
| $P(2) = 0.084$ | 0.125 |
| $P(3) = 0.140$ | 0.265 |
| $P(4) = 0.176$ | 0.441 |
| $P(5) = 0.176$ | 0.617 |
| $P(6) = 0.146$ | 0.763 |
| $P(7) = 0.105$ | 0.868 |
| $P(8) = 0.065$ | 0.933 |
| $P(9) = 0.036$ | 0.969 |

Figure 6.6 clearly shows that the approximation of the binomial by the Poisson is close enough for most industrial studies. Also notice that, as $n$ grows larger, the approximation improves—at least up to $n\overline{p} = 5$. Figure 6.3 shows that the Poisson is a good approximation for small values of $n$ and $\overline{p} < 0.1$.

One of the most important applications of the Poisson in industrial studies (and it just happens to involve the Case Study—Chapter 6) is to project the amount of rework or scrap resulting from setting a limit for the number of nonconformities before products are scrapped or reworked. Example 6.17 illustrates this point.

**EXAMPLE 6.17** In producing cameras, the average number of nonconformities per camera is two ($\overline{p} = 2$). What is the percent of scrap or rework involved if all the cameras with more than four nonconformities are scrapped or reworked?

*Solution:* From the Poisson table (Appendix D), using $n = 1$, $\overline{p} = 2$, and $c = 4$

$$P(4 \text{ or less}) = 0.947$$

$$P(\text{more than } 4) = 1 - 0.947 = 0.053$$

This means that the probability of seeing four or more nonconformities on any one camera is only 5.3 percent. Therefore, a little over 5 percent of the cameras would have to be scrapped or reworked.

## The Normal Distribution

The normal distribution (discussed in Chapter 3) is also a good approximation of the binomial. Conditions of $\overline{p} > 0.5$ and $n > 10$ are necessary, but it is most useful for high values of $n$. In most industrial operations, the normal distribution is not used to approximate the binomial, because it is a continuous probability distribution. Extra steps are necessary in order to use a continuous probability distribution when solving for discrete probabilities. Refer at this time to Appendix B.

## Section 6.4 Review Problems

1. Can the hypergeometric distribution be used when it is impossible to replace the parts after inspection?

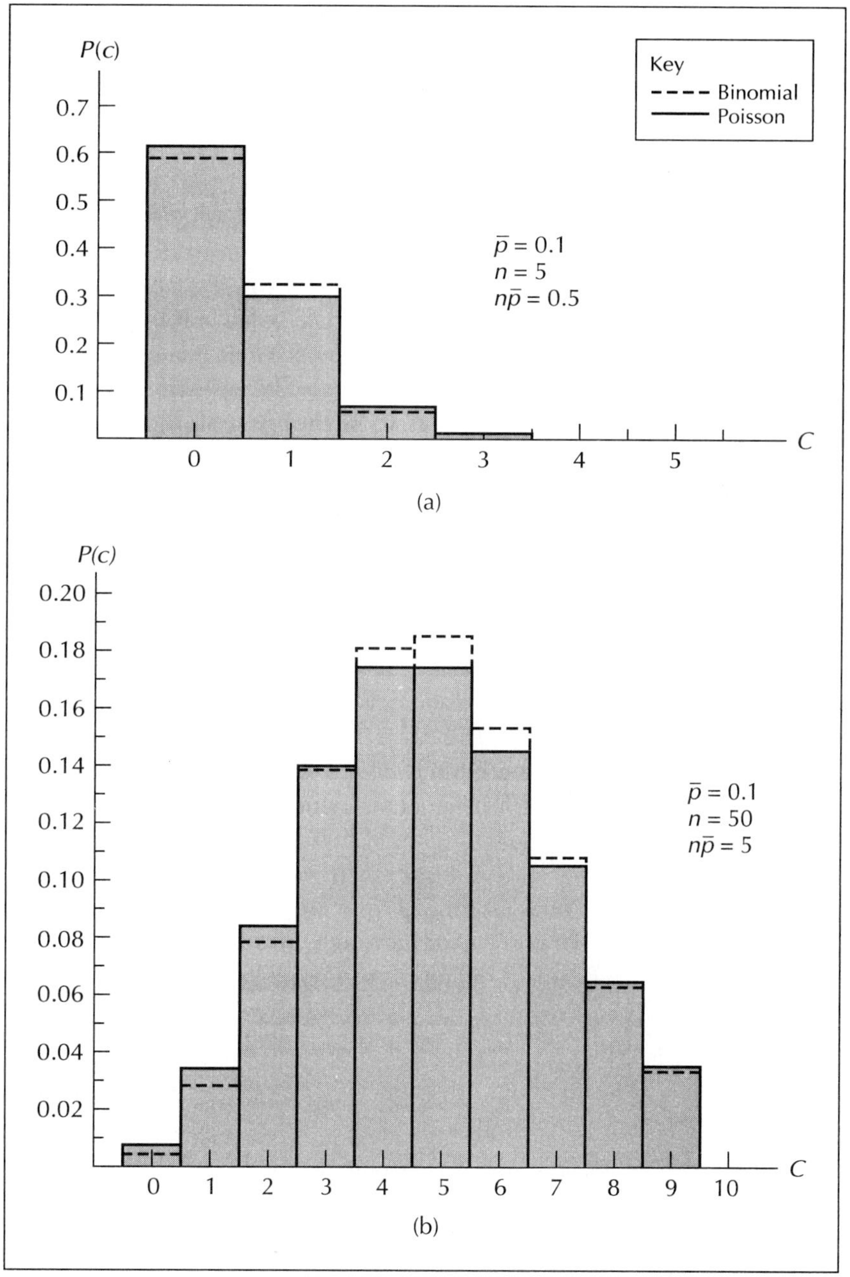


**FIGURE 6.6**
A comparison of the binomial and Poisson distributions.

2. A small lot of 30 parts has two nonconforming parts. What is the probability of obtaining one nonconforming part in a random sample of two parts when sampling without replacement?
   a. 6.3%
   b. 8.7%
   c. 10.2%
   d. 12.9%
3. Select the best distribution to use for a process producing a steady stream of goods with a $\overline{p}$ of 0.08 and the lot size not known.
   a. hypergeometric
   b. binomial
   c. Poisson
   d. normal
4. A plastic extrusion process has an average of three mars on every piece it produces. The chance of finding a piece with five mars is approximately (okay to use the Poisson table)
   a. 10%
   b. 30%
   c. 60%
   d. 90%
5. In problem 4, the chance of finding a piece with more than five mars is approximately
   a. 10%
   b. 30%
   c. 60%
   d. 90%

# *Summary*

Some fundamental concepts provided by the field of probability include the following:

- A probability does not mean something will happen, but only specifies the chance of it happening.
- Samples are drawn from populations to make predictions about the population.
- Classical probability is used when events and sample spaces are known, and experimental probability is used when events and sample

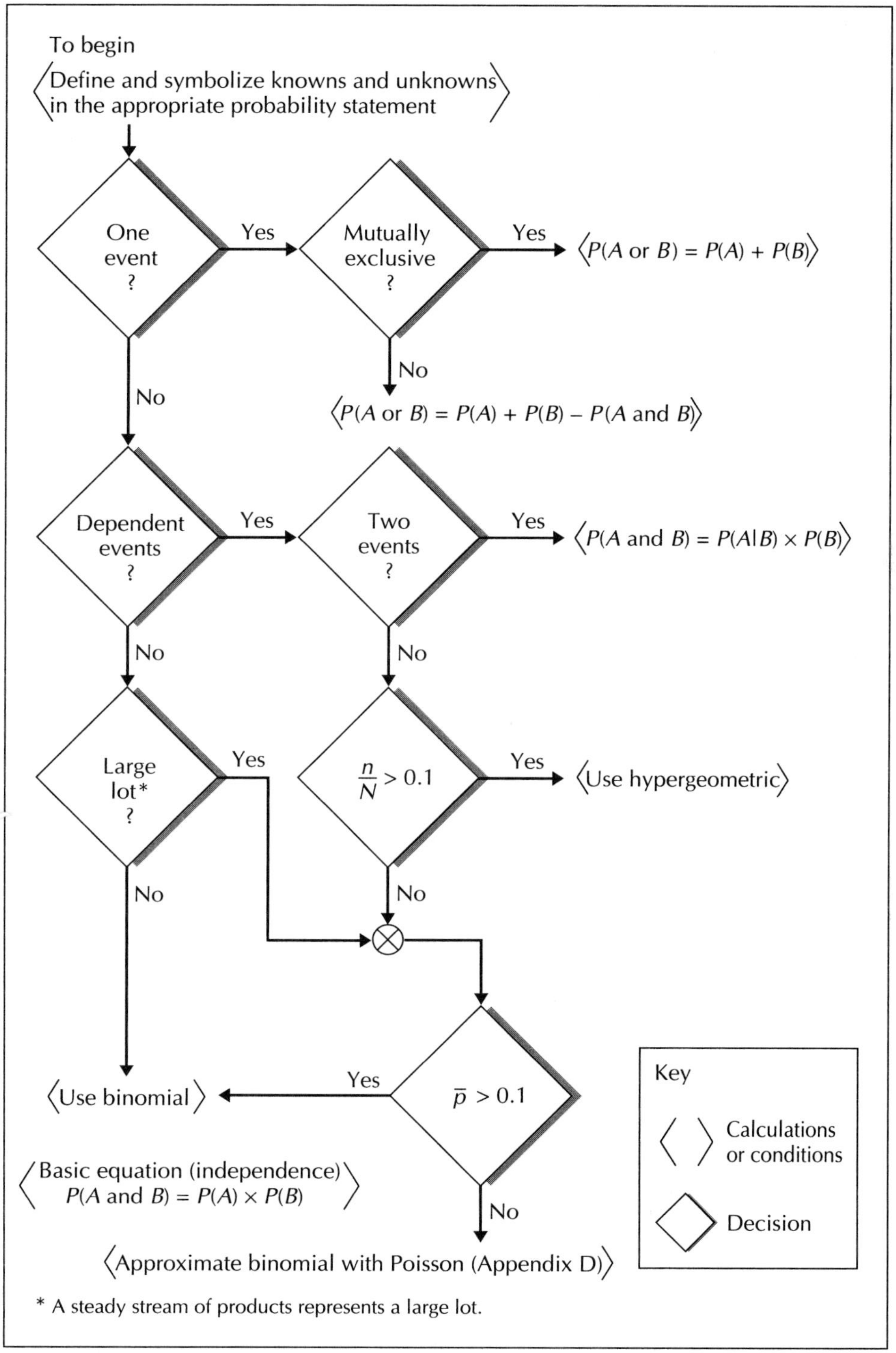


**FIGURE 6.7**
The complete flowchart from Figures 6.1 and 6.3.

spaces are determined by experiment. QC technologists and managers usually use experimental probabilities.

- The fundamental probability equation is expressed as

$$P(E) = \frac{E}{N} = \frac{\text{Selected events to occur}}{\text{All total events that may occur}}$$

- It was also pointed out that $P(E[\text{not}]) = 1 - P(E)$.
- The probability flowchart (see Figure 6.7) was used to aid in deciding what probability equations, theorems, or distributions to use.

Important definitions needed in using the flowchart include these:

- mutually exclusive events: only one selected event can happen at one time;
- dependent events: one event must happen before the next can occur;
- independent events: $P(A|B) = P(A)$.

- Permutations and combinations—counting techniques—were discussed in Section 6.3. The combination equation

$$_nC_r = \frac{n!}{r!(n - r)!}$$

was found to be especially useful in discussing the hypergeometric probability distribution.

- Probability distributions (Appendix B) are central to understanding the sampling plans for attributes and the control charts used for attribute studies in the chapters that follow.
- The importance of probability and of the binomial and Poisson distributions will become most apparent after further study in QC. After confirming their importance, you may wish to return to this chapter's material.

## CASE STUDY · CHAPTER 6

You are a quality technologist in a manufacturing company that produces cameras. One of the quality characteristics inspected is blemishes or imperfections on the external camera surface. The blemishes or nonconformities are considered to be minor, not impairing the intended function of the camera. The average number of minor nonconformities is one. This "fact" has been established through unspecified and undocumented spot inspections by manufacturing.

Your manager has just informed you that the plant manager wishes to scrap or rework all cameras that have more than one minor nonconformity. Your job is to predict the percent of cameras that will have to be reworked or scrapped, based on the average nonconformity count of one, because of this new edict. You are also to recommend the elements of a documented inspection plan.

## Your Solution

Example 6.17 showed the use of the Poisson to solve quickly and easily for the probability of finding a certain number of nonconformities in a random sample of product—in this case, cameras. From your assignment the average nonconformities per camera is one, or $n\bar{p} = 1$. The plant manager wishes to rework or scrap all cameras that have more than one nonconformity. Recalling that a nonconformity can be as minor as a scratch, major **internal failure costs** due to scrap and rework (see discussion of quality costs in Chapter 10) could result.

You use the Poisson distribution to predict the probability of one camera having more than one nonconformity. The data used to enter the Poisson table are $n\bar{p} = 1$ and $c = 2$ or greater. From the Poisson table, Appendix D, the probability of $c = 1$ or less will be determined.

$$P(1 \text{ or less}) = 0.736$$

The probability of finding two or more nonconformities then becomes

$$P(\text{more than } 1) = 1 - P(1 \text{ or less}) = 1 - 0.736 = 0.264 = 26.4\%$$

This means that more than 26 percent of the cameras will be reworked or scrapped! A frightening and expensive prospect.

You report your findings to your manager. You point out that the inspection frequency has not been determined or recommended. Up to this point, inspection frequency has been the decision of the operators. More importantly, there is really no limiting criterion for a nonconformity, so inspectors could vary in their interpretation of what is a nonconformity.

Your recommendations are first to define a nonconformity, second to review the average nonconformities per unit, and third to decide on an inspection frequency. You also offer a check sheet to aid in inspection. Figure 6.8 shows two views of the camera, and another check sheet will have

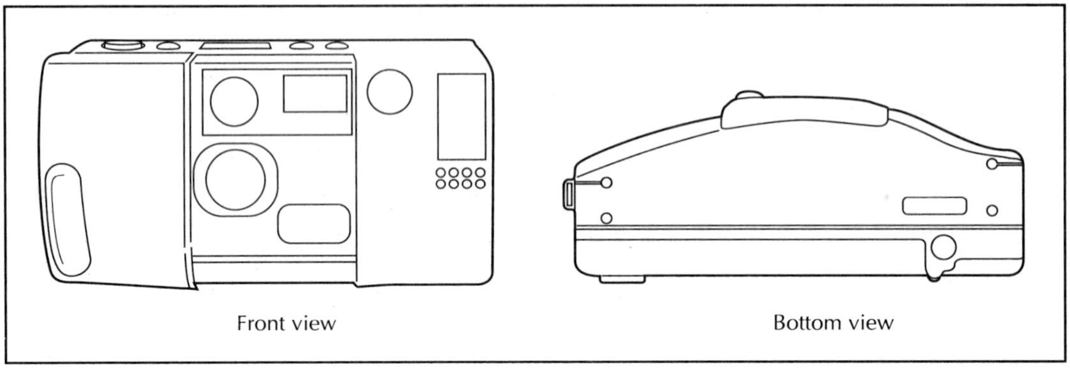


**Figure 6.8**
The check sheet for the front and bottom of the camera.

the remaining views. The inspector can mark where the nonconformities appear on the check sheet.

After new inspection procedures have been implemented, a new $n\bar{p}$ can be calculated. The revised $n\bar{p}$ will then be used, along with probabilities from the Poisson table, to predict how many nonconformities should be allowed before scrap or rework occurs.

# *Chapter Review Problems*

1. Describe in no more than two paragraphs how you would determine the gas mileage for all cars of a particular make and model. Use the words "sample" and "population" in your explanation.
2. What is your estimate of the probability of a human birth in the world during the next hour?
3. Based on your estimate in problem 2, what is the probability of not having a birth in the world during the next hour?
4. What is the probability of drawing a nine or a spade from a well-shuffled deck of cards?
5. In order to win a state lottery, six winning numbers must be selected from a total of 54. What is the probability of winning? What are the odds of winning?
6. A flashlight is assembled using four different components. The percentage of nonconforming parts in each of the four groups of

components is 0.3 percent. What is the probability of a nonconforming assembly?

7. An electronics product that has an intermittent problem must pass through four different inspections to be shipped. If each inspection has a 5 percent probability of catching the nonconformity, what is the probability that it will be shipped?
8. What is the probability that the faulty product will be sorted during the inspections in problem 7?
9. Using the data in Table 6.1, calculate the probability of finding a nonconforming unit from machine C.
10. A process has a 50 percent chance of producing a quality characteristic above or below the process average (Chapter 4, "Variable Control Charts"). What is the probability of obtaining seven points in a row above the process average?
11. A process has a 50 percent chance of producing nonconformities above or below the number of nonconformities counted in the last inspection. What is the probability of finding a trend where the number of nonconformities goes down seven times in a row? Does this probability mean the process is probably changing in the number of nonconformities produced?
12. How does the result found in Example 6.11 compare with that in Example 6.8? What is the difference between odds and probabilities?
13. See Example 6.7—sampling without replacement. Can the hypergeometric be used to determine the probability? Perform the calculation.
14. What are the odds of getting a pair of deuces in five-card stud or in the beginning hand of five-card draw?
15. What is the probability of obtaining two fours on four throws of the same dice. The binomial equation may be used because independency must be the case!
16. Explain why the hypergeometric cannot be used in solving problem 15. Do the calculation and note the difference between the two answers.
17. From a continuous production of integrated circuit chips (ICs), 1 percent ($\overline{p} = 0.01$) are nonconforming units. If six chips are selected from the process, what is the probability of 0, 1, or 2 nonconforming units? Use the binomial.

18. Use the Poisson table (Appendix D) to find the probabilities in problem 17. Is the Poisson a good approximation to the binomial in this case? (Refer to the probability flowchart in Figure 6.7.)
19. A carpet manufacturer finds that each 100-meter piece of carpet averages fifteen nonconformities. Use the Poisson table, Appendix D, to find the probabilities of a 100 m piece having exactly nine nonconformities, fewer than nine nonconformities, and nine or more nonconformities.
20. The carpet manufacturer in problem 19 reduces average nonconformities to ten. Solve for the same probabilities—exactly nine nonconformities, fewer than nine nonconformities, and nine or more nonconformities. Compare the results with those in problem 19.

# LABORATORY · CHAPTER 6
## Statistical Software for the Computer

*Materials needed:* Computer with SQC software, descriptions of SQC software packages obtained through vendors, or a manufacturing or service industry using SQC software.

After you have become oriented to the computer system, enter variable data into the system by using the keyboard. Use any of the data sets presented in previous chapters. Then practice using the SQC software to construct $\overline{X}$ and $R$ charts, median charts, individuals charts, and so on. After the various charts have been examined on the screen, they should be printed out. The data sets may then be used to construct histograms. The software tools offer many options for the construction of histograms. Determine the distribution's normality, the percent of product above and below the specification limits, and the capability ($C_p$ and $C_{pk}$) of the process under consideration. After printing out the histogram and associated information, affirm by calculation at least one of the parameters calculated by the computer.

The instructor may also offer a tour of an industry using SQC software. The tour should involve a description of the network used to collect data, as well as the actual use of the computer in collecting and preparing presentations of data.

# 7

# Attribute Control Charts

**Preparation for Charting** 7.1
**Constructing Charts—Nonconforming Units** 7.2
**Constructing Charts—Nonconformities** 7.3

## 7.1 *Preparation for Charting*

When a product or service is not fulfilling customers' needs or "things do not seem to be going just right" in a business or industry, solutions are implemented. Some of the solutions may appear to work at first, but later seem not to have the magic they once did. Other solutions will not seem to work at first, but with persistence and some fine tuning they can improve the situation.

One way to assure success in quickly separating good solutions from bad is to track the number of nonconformities or nonconforming units. This approach is quite different from using an $\overline{X}$ and $R$ chart to track variable data as discussed in Chapter 4. The major difference between charting attribute data and charting variable data is that attribute data reveal "big picture" information. *Attribute data often allow you to "see the forest rather than the trees."* As new ideas are implemented, the attribute chart will reflect how well the idea contributed to improved products, show that it had no effect, or indicate that the change actually had a negative effect over time.

First, we should carefully define the terms "nonconformity" and "nonconforming unit." A **nonconformity** (formerly known as a defect) is any undesirable element of a product or service. On a product, a nonconformity usually takes the form of a ding, scratch, tear, or soiled spot. With a service, the nonconformity could be a dirty seat on an aircraft or missing or incorrect data on a form. The nonconformity is undesirable, but it does not necessarily lead to a nonconforming unit.

The **nonconforming unit** (formerly known as a defective) is a unit of a product or service that is unacceptable. One or more nonconformities may exist to make the whole unit not fit for use. In a manufacturing process, a nonconforming unit is, for example, a radio without an output to the speakers—a condition rendering the product useless. In service industries, a nonconforming unit could be an incorrect keypunch input, when each input is considered a unit and all inputs are tabulated as conforming or nonconforming. This chapter will use the word **nonconformance** to mean either a nonconformity or a nonconforming unit.

There are four preliminary steps to attribute charting:

- The product attribute to be studied is selected.
- The type of attribute chart is determined.
- How to judge whether or not a nonconformance has occurred is clearly communicated.
- Data collection techniques (check sheets, chart forms, and sampling frequency) are decided.

## Selection of Attribute Study

Techniques used to determine that an attribute study is necessary have already been discussed at some length, and are much the same as those methods used to determine variable chart studies. These methods include the flowcharting of the process and Pareto analysis to determine the vital few nonconformances that cause the greatest cost in scrap or rework.

Once a process control study area has been determined, one attribute chart may be more economical to implement than several variable charts. Variable chart studies are very specific, and many are often necessary to study one facet of the product. Variable measuring equipment is expensive. So, instead of attempting to implement 20 or 30 variable chart studies with precision measurement equipment, an overall attribute study may be just as effective in improving the process. The attribute study will also show where variable chart studies are needed.

## *Type of Attribute Chart*

There are four distinct types of attribute charts—*p*, *np*, *c*, and *u* types. The type of attribute chart depends on

- whether or not subsample size is changing;
- whether a nonconformity or a nonconforming unit is being studied.

In many attribute studies it is possible to have a constant subsample size. If so, the constant subsample chart is the easiest to construct and interpret. But it is often impractical or impossible to hold the subsample size constant. The second factor that determines the type of chart used is whether we wish to track nonconformities or nonconforming units. A four-cell diagram is often used to clarify the explanation of the four types of attribute charts (Figure 7.1).

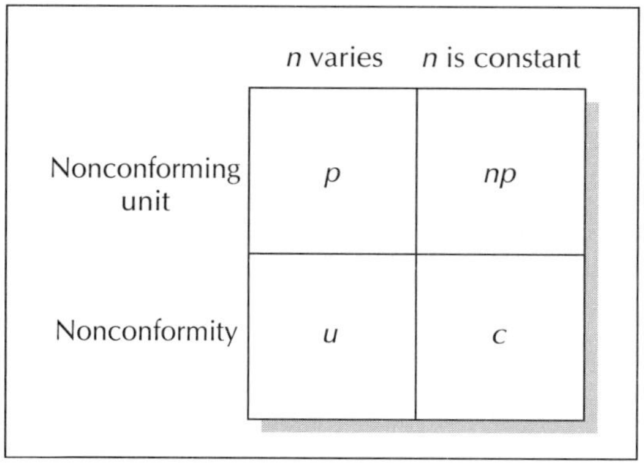


**FIGURE 7.1**
Attribute chart selection diagram.

Also, refer to the header information on the form used for attribute charting in Figure 7.4. In the center of the form, the four charts are listed, and the inspector is to check the appropriate block.

## Defining the Nonconformance

One of the surest ways to judge a nonconforming unit is to use the go/no-go principle. A snap gauge will be used as an example (Figure 7.2) A **snap gauge** uses a C-shaped frame to support two measuring jaws on one side of the open end that are referenced to a flat plate on the other end (Figure 7.2a). The piece to be tested must fit between the outside plate and the reference plate and not slip through the inside plate located closer to the reference plate. Other types of go/no-go gauges are plug gauges for measuring IDs and ring gauges for checking ODs. The **dial snap gauge** (Figure 7.2b) is a variable gauge, and it should not be confused with the attribute gauge in Figure 7.2a.

Most nonconformances are not so easily defined. How do you judge whether the colors of an advertising brochure are correct? When does a scratch on an automobile become a nonconformance? How many minor nonconformities does it take to make a camera a nonconforming unit? In order to inspect for attributes, specific guidelines must be decided. Guidelines are usually written by quality teams in brainstorming sessions.

Once the nonconformance is well defined, it must be communicated to all employees responsible for finding it. A **check sheet,** one of Ishikawa's seven tools of quality, is a good example of how to communicate. Figure 7.3 shows a check sheet used to aid inspectors or operators in listing nonconformances. In this case the product is a foundry casting through which pours molten glass. The molten glass is then formed into fibers to make the insulation used in homes and businesses. The basic "spinner" unit is cast in a foundry operation. The check sheet shows a cross section of the part, and critical points are identified by number. The instructions at the bottom of the check sheet clearly identify how to report nonconformities.

## Data Collection Techniques

The best way to collect attribute data is to use an attribute form such as the one shown in Figure 7.4. The inspector, many times the operator, will select a subsample $n$ and inspect all items within the subsample.

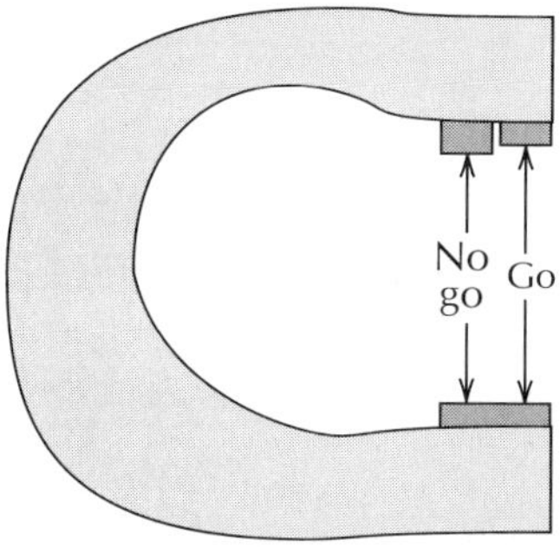


(a)

(b)

**Figure 7.2**
(a) The snap gauge is an attribute gauge. (b) The dial snap gauge is a variable gauge.

The subsample size (which may change) and the number of nonconformances are recorded on the form (lines identified by asterisks).

The subgroup size for the attribute chart is much larger than that for the variable chart. The subgroup must be large enough to have few or no groups with zero nonconformities. Consequently, to detect only five nonconformities in 1,000, a subgroup size of less than 1,000 would

**Check Sheet – Casting Nonconformities**

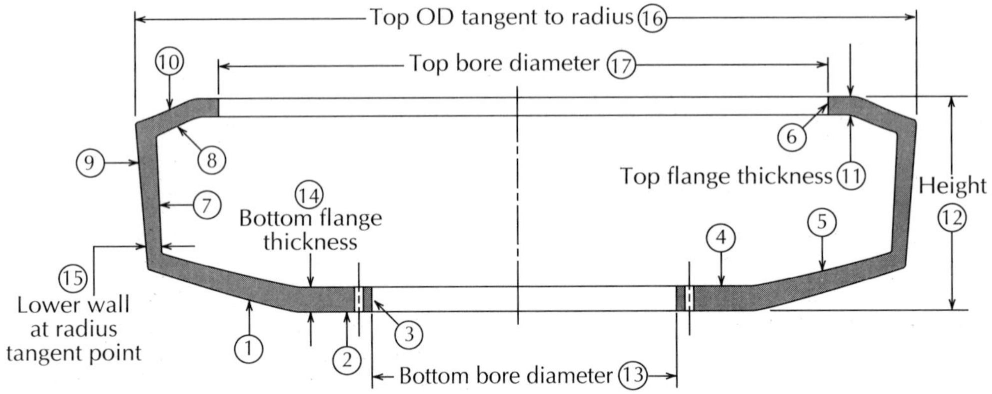


| | | | |
|---|---|---|---|
| A — Good | H — Short pour | O — Tool break | U — Pattern deviation |
| B — Slag | I — Glue | P — Dimensions | V — Missing holes |
| C — Gas | J — Core break | Q — Machine error | W — Operator error |
| D — Machining crack | K — Shell break | R — Mounting holes | X — Refractory |
| E — Shrink | L — Foundry crack | S — Not flat | Y — Heat treat |
| F — Inclusion | M — Pinholes | T — Hole diameters | Z — Sand hole |
| G — Cold run | N — Cold shot | | |

Instructions: Use alphabet reject code first; then follow with numerical location code. As an example, if you were to find slag in the top flange, the reject code would be B10.

**Figure 7.3**
A check sheet is useful in tracking nonconformities.

not detect enough nonconformities per subgroup to be meaningful as a detector of process shifts. The minimum size $n$ of a subgroup for most processes is 50. Naturally, this *need to have a larger sample size is a disadvantage of the attribute chart compared to the variable chart.* (The other disadvantage is that an attribute chart does not track a specific process or product quality characteristic.)

Attribute studies are often used to monitor incoming materials, especially from multiple-source suppliers. (Products from single-source suppliers that have implemented SQC principles often do not have to be monitored.) As the lots come in, a subsample is pulled from each supplier's lot and checked for nonconforming units. The manufacturer of radios, for instance, can check subsamples of capacitors from lots of incoming capacitors. The subsample units are measured and com-

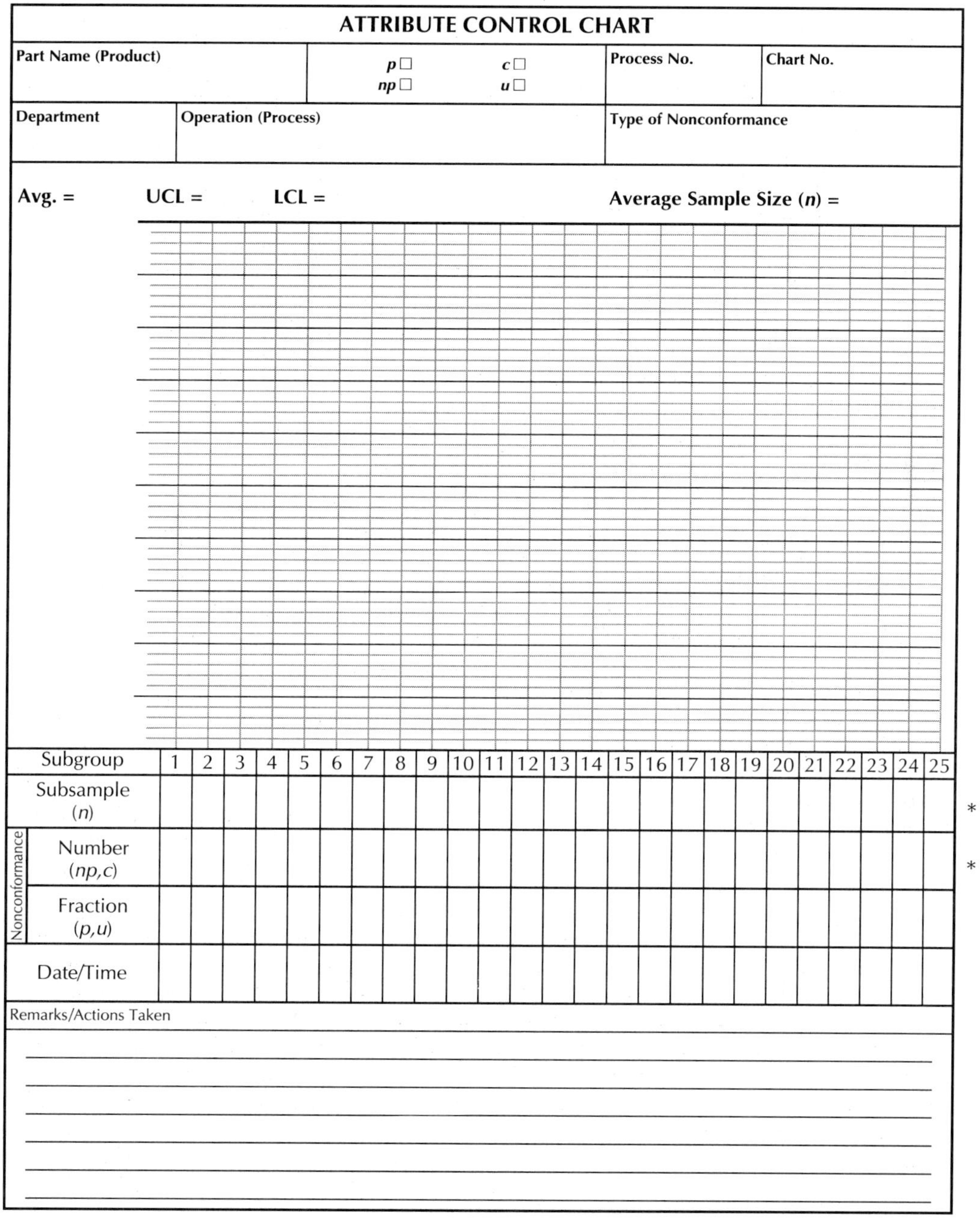

**ATTRIBUTE CONTROL CHART**

| Part Name (Product) | *p* ☐ *np* ☐ *c* ☐ *u* ☐ | Process No. | Chart No. |
|---|---|---|---|

| Department | Operation (Process) | Type of Nonconformance |
|---|---|---|

**Avg. =** **UCL =** **LCL =** **Average Sample Size (*n*) =**

| | | 1 | 2 | 3 | 4 | 5 | 6 | 7 | 8 | 9 | 10 | 11 | 12 | 13 | 14 | 15 | 16 | 17 | 18 | 19 | 20 | 21 | 22 | 23 | 24 | 25 | |
|---|---|---|---|---|---|---|---|---|---|---|---|---|---|---|---|---|---|---|---|---|---|---|---|---|---|---|---|
| Subgroup | | 1 | 2 | 3 | 4 | 5 | 6 | 7 | 8 | 9 | 10 | 11 | 12 | 13 | 14 | 15 | 16 | 17 | 18 | 19 | 20 | 21 | 22 | 23 | 24 | 25 | |
| Subsample (*n*) | | | | | | | | | | | | | | | | | | | | | | | | | | | * |
| Nonconformance | Number (*np*,*c*) | | | | | | | | | | | | | | | | | | | | | | | | | | * |
| | Fraction (*p*,*u*) | | | | | | | | | | | | | | | | | | | | | | | | | | |
| Date/Time | | | | | | | | | | | | | | | | | | | | | | | | | | | |

Remarks/Actions Taken

**Figure 7.4**
A form used to construct attribute control charts.

pared to specifications. All capacitors above or below specification are tallied as nonconforming units.

Sometimes historical data are used to construct attribute charts. The use of past data can often help to construct charts to be used as the process is running. Some caution must be advised when constructing charts from historical data, because the data may not be as well understood as they are when charting processes "while they are running." One important thing to remember about data collection is that the chart is not valid unless each item in a particular subgroup was produced under similar use of machinery, materials, measurement, methods, environment, and people (MMMMEP, Chapter 1).

## *Section 7.1 Review Problems*

1. List three examples of nonconforming units and another three of nonconformities. For each set of three, list at least one example from a service industry.
2. List at least one advantage and one disadvantage of an attribute chart, when compared with a variable chart.
3. Draw Figure 7.1 from memory.

# *7.2 Constructing Charts—Nonconforming Units*

Now that we have considered the four preliminary steps to chart implementation, a chart may be constructed. The first chart will track a computer keypunch (data-entry) process. The process takes place in the supply terminal for a large government repair facility. After the quality team uses brainstorming techniques and Pareto analysis, they conclude that one of the most serious problems occurs when keypunch operators type in a receipt for an item and the original order cannot be matched on the database. The team gave the problem a name—**process receipt without order.**

How could a receipt be issued unless the department the receipt came from had already submitted an order? The answer is that the order was placed and either not entered into the computer (keypunched) at the time, or the computer (a new system only recently programmed) could not find the original data on the order. *The nonconforming unit, then, is the missing or misplaced order when the receipt is entered into the computer.* The computer is programmed to total the number of receipts processed and to keep track of the number of missing orders by the date they should have been issued.

The header information in Figure 7.5a reflects the preceding description of the problem. As was stated before, it is important to complete the header information as accurately and completely as possible. Note that the *p* chart block has been checked. Since this is a study involving nonconforming units and the sample size is changing, the *p* **chart** is the appropriate attribute chart (refer to Figure 7.1).

**ATTRIBUTE CONTROL CHART**

| Part Name | | *p* ☒ *np* ☐ | *c* ☐ *u* ☐ | Process Number | Chart No. |
|---|---|---|---|---|---|
| Receipt | | | | 1070 | 1 |
| **Department** | **Operation (Process)** | | | **Type of Nonconformance** | |
| Supply | Keypunch | | | No order | |

(a)

| Subgroup | 1 | 2 | 3 | 4 | 5 | 6 | 7 | 8 | 9 | 10 | 11 | 12 | 13 | 14 | 15 | 16 | 17 | 18 | 19 | 20 | 21 | 22 | 23 | 24 | 25 |
|---|---|---|---|---|---|---|---|---|---|---|---|---|---|---|---|---|---|---|---|---|---|---|---|---|---|
| Subsample ($n$) | 171 | 129 | 188 | 106 | 151 | 133 | 170 | 166 | 164 | 421 | 242 | 164 | 185 | 788 | 120 | 249 | 78 | 71 | 235 | 397 | 202 | 367 | 212 | 208 | 183 |
| Nonconforming: Number (($np$), $c$) | 4 | 6 | 16 | 8 | 8 | 5 | 7 | 6 | 4 | 94 | 35 | 10 | 14 | 25 | 3 | 15 | 24 | 13 | 20 | 10 | 5 | 6 | 35 | 4 | 18 |
| Nonconforming: Fraction (($p$), $u$) % | 2.3 | 4.7 | 8.5 | 7.5 | 5.3 | 3.8 | 4.1 | 3.6 | 2.4 | 22.3 | 14.5 | 6.1 | 7.6 | 3.2 | 2.5 | 6.0 | 30.8 | 18.3 | 8.5 | 2.5 | 2.5 | 1.6 | 16.5 | 1.9 | 9.8 |
| Date/Time: June | 1 | 2 | 3 | 4 | 5 | 6 | 7 | 8 | 9 | 10 | 11 | 12 | 13 | 14 | 15 | 16 | 17 | 18 | 19 | 20 | 21 | 22 | 23 | 24 | 25 |

Remarks/Actions Taken

(b)

**FIGURE 7.5**
(a) Completed header information; (b) data for the *p* chart study.

The data are recorded as shown in Figure 7.5b. Before continuing you should examine the data. For subgroup 1, for instance, 171 receipts were entered and 4 were found not to have an order. Note that the sample size $n$ varies a great deal. The quality team realized that subsamples—collected per day—could change considerably. Later they will consider how to keep subsample size constant, yet keep conditions approximately the same in each individual subsample.

## *Calculating the p Chart Parameters*

Each individual subsample's percent (or decimal fraction) of nonconformities or nonconforming units is recorded on the line for $p$ or $u$ charts. The percent nonconforming ($p$) calculation for the first subgroup is

$$p = \frac{\text{Number nonconforming in subsample}}{\text{Number inspected in subsample}} = \frac{4}{171} = 0.023 = 2.3\%$$

The row on the form labeled "Fraction ($p$, $u$)" is only used for charts having varying subsample size. The $p$ is circled in this case indicating it is a chart for nonconforming units. (The $u$ chart is used for the study of nonconformities.)

The average for the $p$ chart ($\overline{p}$, or $p$-bar) is calculated by the equation

$$\overline{p} = \frac{\text{Total nonconforming}}{\text{Total inspected}} = \frac{\sum np^*}{\sum n}$$

EXAMPLE 7.1

Calculate the average nonconforming units $\overline{p}$ for the study of process receipts without orders.

*Solution:* The sum of the total nonconforming ($np$s) is first found to be 395. The total items inspected $n$ is 5,500. The final calculation is

$$\overline{p} = \frac{\sum np}{\sum n} = \frac{395}{5{,}500} = 0.072 = 7.2\%$$

* The $np$ symbol means the number in a subsample $n$ multiplied by the percent nonconforming $p$. This calculation yields the actual number of nonconforming units in the subgroup. So, the sum of the nonconforming units is simply divided by the total inspected.

It is important to realize that, in this calculation, had you merely averaged the *p*s already calculated, you would have a different value for $\bar{p}$ (7.9% in this case). This difference is due not only to rounding error, but also to differing subsample sizes.

The **capability** of any attribute chart is the average for the chart. In this study the process capability is the $\bar{p}$ value of 0.072 or 7.2 percent.

The control limits are calculated by using the standard deviation of the binomial distribution, the distribution used to derive the Poisson distribution (Section 6.4). The binomial distribution is used because attribute data are in either one of two states—"bad or good," "go or no-go." The standard deviation of the binomial distribution (as applied to the *p* chart) is

$$\sqrt{\frac{\bar{p}(1-\bar{p})}{n}}$$

Since three standard deviations are usually considered best for being sure a process has changed when a point lies outside them, the control limits are calculated by

$$UCL_p = \bar{p} + 3\sqrt{\frac{\bar{p}(1-\bar{p})}{n}}$$

$$LCL_p = \bar{p} - 3\sqrt{\frac{\bar{p}(1-\bar{p})}{n}}$$

## Constructing the p Chart

One way to calculate and construct control limits based on varying sample size is to calculate each individual subsample's control limits. This is decidedly a lot of work and makes for a difficult chart to interpret when control limits are constantly varying. Therefore, the most popular method of dealing with varying sample size will be the only method used in this chapter.

The method is to use the same control limits for all subsample sizes $n$ within plus or minus 25 percent of the average subsample size $\overline{n}$. In this case $\overline{n}$ is

$$\overline{n} = \frac{\text{Total number inspected}}{\text{Number of subgroups}} = \frac{\Sigma n}{k} = \frac{5{,}500}{25} = 220$$

Please confirm this calculation with your calculator.

Next, calculate the range of subsample sizes within plus or minus 25 percent of the average $n$. The range is calculated as follows:

$$220 \pm (220 \times 0.25) = 165 \text{ to } 275$$

The subgroups within the 25 percent range calculated are 1, 3, 7, 8, 11, 13, 16, 19, 21, 23, 24, and 25.

EXAMPLE 7.2

Calculate the control limits for all of the points within 25 percent of the average $n$.

*Solution:* Using the equations developed from the binomial, with an $\overline{n}$ of 220, and a $\overline{p}$ of 0.072,

$$UCL_p = 0.072 + 3\sqrt{\frac{0.072\,(1 - 0.072)}{220}}$$

$$LCL_p = 0.072 - 3\sqrt{\frac{0.072\,(1 - 0.072)}{220}}$$

This calculation is easiest to accomplish if you calculate the fraction under the radical sign first, and press the equals key before taking the square root. A sample calculation is as follows:

0.072 [×] [(] 1 [−] 0.072 [)] [÷] 220 [=] [√] [×] 3 [=] 0.0523

*UCL* = 0.072 + 0.052 = 0.0124 or 12.4%

*LCL* = 0.072 − 0.052 = 0.02 or 2%

At this point the scale factor for the chart is set up to accommodate one-half to two-thirds of the total space (see Figure 7.6). In this particular chart the bold lines must equal 2 percent in order to use an even-number scale factor for these lines, yet have enough space to accommodate the $UCL_p$ of 12 percent. The scale factor for each line on the chart must be 0.4 percent. The 0.4 percent is noted on the left-hand margin of the chart. The chart for all of the subgroups within 25 percent of

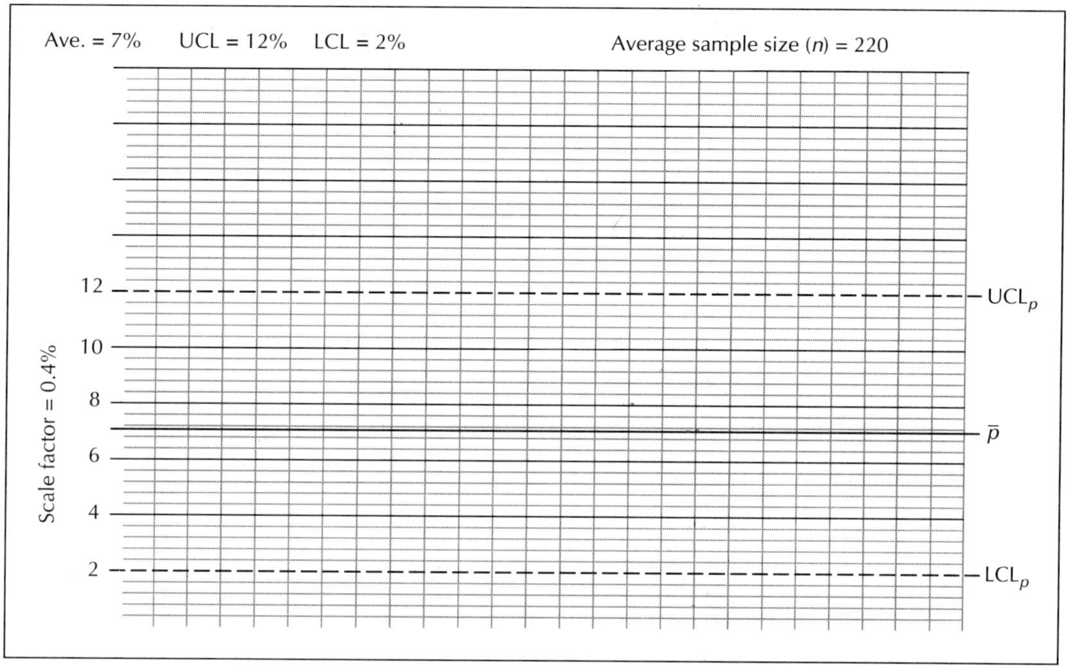


**FIGURE 7.6**
The $p$ chart's scaling has been verified.

the average subgroup size is shown in Figure 7.6. The dashed control limit lines are drawn completely across the chart. Now that the layout and chart scaling have been verified, the remaining control limits may be calculated.

EXAMPLE 7.3 Calculate the new control limits for points 18 (having the lowest $n$, 71) and 14 (the highest $n$, 788).

*Solution:* To do the four calculations a shortcut will be introduced. From your algebra course you learned that the square root of a fraction is equal to the square root of the numerator divided by the square root of the denominator. Therefore,

$$3\sqrt{\frac{\bar{p}(1-\bar{p})}{n}} = \frac{3\sqrt{\bar{p}(1-\bar{p})}}{\sqrt{n}}$$

By calculating the numerator term, a constant, we will not need to recalculate the same term over and over again. The use of this shortcut is demonstrated in the following calculations:

$$\mathrm{UCL}_p\,(\text{point }18) = 0.072 + \frac{3\sqrt{0.072\,(1-0.072)}}{\sqrt{71}}$$

$$= 0.072 + \frac{0.775}{8.43} = 0.164 \approx 16\%$$

$$\mathrm{LCL}_p\,(\text{point }18) = 0.072 - \frac{0.775}{8.43} = -0.020 = -2\%$$

Continuing on to calculate the control limits for point 14:

$$\mathrm{UCL}_p\,(\text{point }14) = 0.072 + \frac{0.775}{\sqrt{788}} = 0.10 \approx 10\%$$

$$\mathrm{LCL}_p\,(\text{point }14) = 0.072 - \frac{0.775}{\sqrt{788}} = 0.044 \approx 4.4\%$$

Figure 7.7 shows the control limits for points 18 and 14 calculated in Example 7.3, as well as those for the other 11 points yet to be calculated. The dashed control limit lines drawn completely across the form in Figure 7.6 have been erased for the points where subsample size $n$ varies more than 25 percent, and the actual control limit lines for these points have been drawn in.

Note the minus sign on the lower control limit for the percent nonconforming ($\mathrm{LCL}_p$, point 18). The minus sign indicates that we could have less than zero percent nonconforming. Of course, this is impossible; therefore, *the zero line on the chart will be the lower limit for all negative values of the lower control limit.*

Calculate the control limits for the remaining 11 subsamples and verify the correct location of their control limits on the chart. Finally, you should verify the correct location of the percent nonconforming $p$ points on the chart.

## Interpreting the p Chart

The out-of-control points are marked on the chart in Figure 7.7 with a number. Also in Figure 7.7 are the notes written while the process was running. The notes are keyed to the numbers used to mark the out-of-control points indicating assignable causes operating. Note that some of the points are off the chart. This situation often happens when a new study is implemented and before the employees and management can begin to attack the assignable causes.

**ATTRIBUTE CONTROL CHART**

| Part Name (Product) | | Process No. | Chart No. |
|---|---|---|---|
| Receipt | p ☒ np ☐ c ☐ u ☐ | 1070 | 1 |

| Department | Operation (Process) | Type of Nonconformance |
|---|---|---|
| Supply | Keypunch | No order |

Avg. = 7% UCL = 12% LCL = 2% Average sample size (*n*) = 220

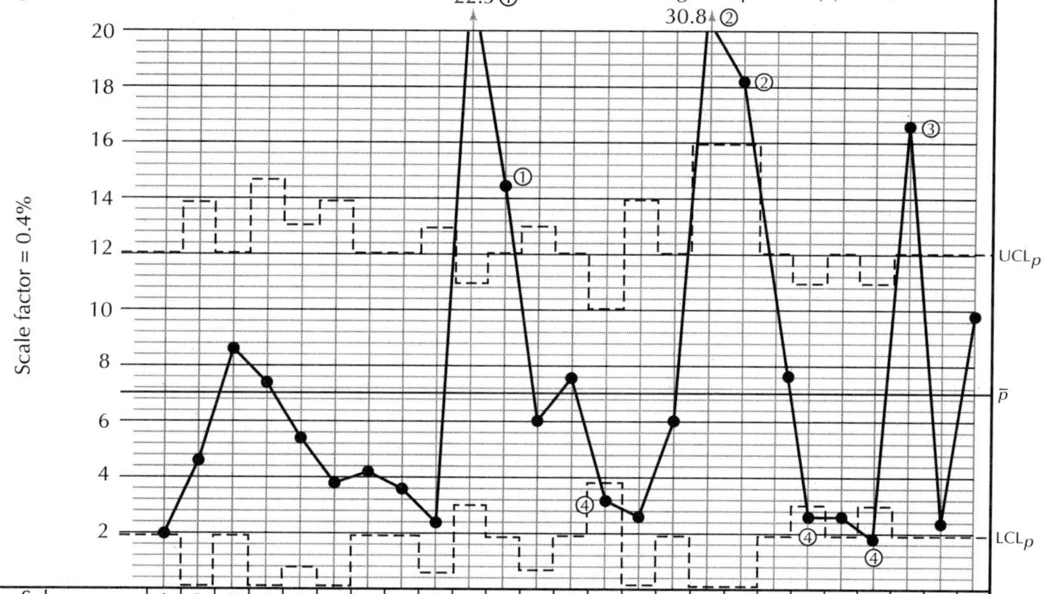


| Subgroup | 1 | 2 | 3 | 4 | 5 | 6 | 7 | 8 | 9 | 10 | 11 | 12 | 13 | 14 | 15 | 16 | 17 | 18 | 19 | 20 | 21 | 22 | 23 | 24 | 25 |
|---|---|---|---|---|---|---|---|---|---|---|---|---|---|---|---|---|---|---|---|---|---|---|---|---|---|
| Subsample (*n*) | 171 | 129 | 188 | 106 | 151 | 133 | 170 | 166 | 164 | 421 | 242 | 164 | 185 | 788 | 120 | 249 | 78 | 71 | 235 | 397 | 202 | 367 | 212 | 208 | 183 |
| Nonconformance: Number (*np*,*c*) | 4 | 6 | 16 | 8 | 8 | 5 | 7 | 6 | 4 | 94 | 35 | 10 | 14 | 25 | 3 | 15 | 24 | 13 | 20 | 10 | 5 | 6 | 35 | 4 | 18 |
| Nonconformance: Fraction (Ⓟ,*u*) | 2.3 | 4.7 | 8.5 | 7.5 | 5.3 | 3.8 | 4.1 | 3.6 | 2.4 | 22.3 | 14.5 | 6.1 | 7.6 | 3.2 | 2.5 | 6.0 | 30.8 | 18.3 | 8.5 | 2.5 | 2.5 | 1.6 | 16.5 | 1.9 | 9.8 |
| Date: June | 1 | 2 | 3 | 4 | 5 | 6 | 7 | 8 | 9 | 10 | 11 | 12 | 13 | 14 | 15 | 16 | 17 | 18 | 19 | 20 | 21 | 22 | 23 | 24 | 25 |

Notes

1. Order cards incorrectly completed by one department.
2. Receiving area under construction (drawing) [Figure 7.8].
3. Computer programming error.
4. Higher volumes.

**FIGURE 7.7**
The completed attribute chart for nonconforming units with variable subsample size (*p* chart).

Employee involvement is evident in this study. The notes explain that all of the out-of-control points to the high side have a distinct cause. One keypunch operator was so motivated by the study that she drew a cartoon on her own time (Figure 7.8) of construction workers cutting through a block wall in the receiving area where the keypunching takes place (note 2). Obviously, it must have been a traumatic experience for all of the employees on those two days.

The computer programming error (note 3) led to the discovery that some orders were misplaced in the computer database. This problem on the new computer was discovered because of the single out-of-control point, but it also was affecting the average percent nonconforming, as the next study will show.

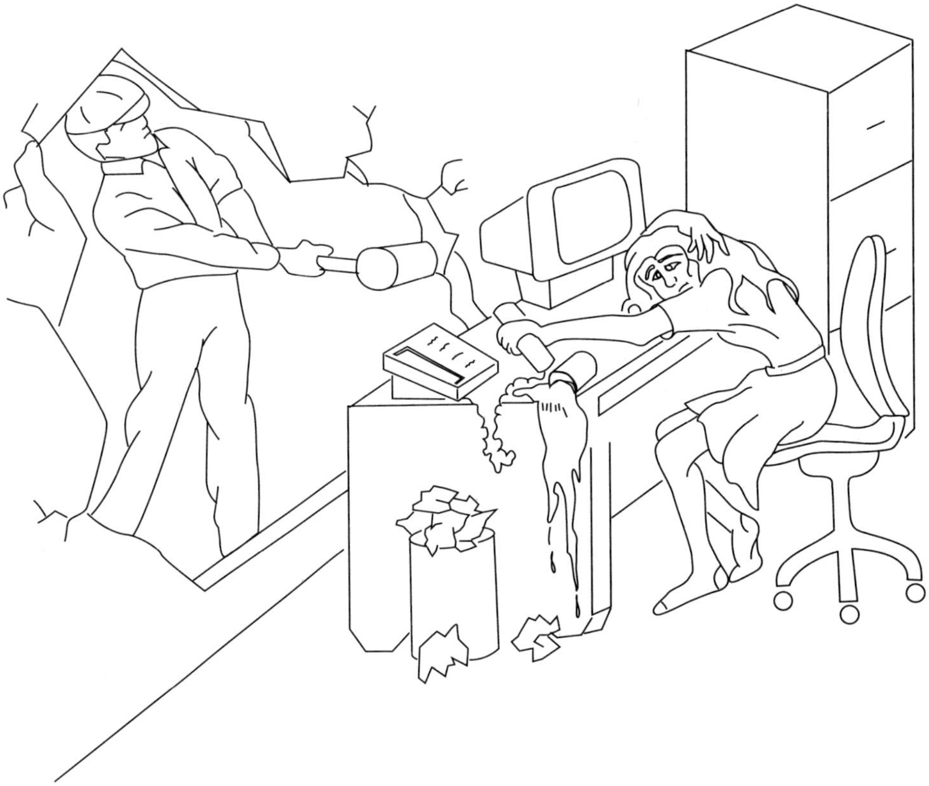

**FIGURE 7.8**
The cartoon is one employee's way of expressing her consternation when construction was taking place in her area.

The out-of-control points to the low side were only marked "high volumes." The employees did not identify any other assignable causes during these high-volume times, but they thought that high volume leads to less interruption in work flow and less chance for boredom to set in. *Experienced managers recognize how damaging to morale is a reduced workload situation.*

In analyzing the chart, the quality team decided to initiate a new study. They noted how difficult it was to construct the $p$ chart with varying subsample size, and decided to program the computer tracking the process to use subgroup sizes of 200, regardless of what time of day each subsample of 200 was completed. The relatively low constant sample size of 200 should still keep subsample conditions approximately the same during the study. The study will be continued in the next section.

## Constructing the np Chart

The chart used for the study of nonconforming units having a constant subsample size $n$ is the ***np* chart** (Figure 7.9). When the subsample size is constant, you no longer have to convert the number of nonconforming units $np$ to a percent nonconforming $p$. The row labeled "($np$, $c$)" will be the row of choice, shown with the $np$ circled, and the charting will reflect the product of $n$ times $p$ ($n \times p$) or the number of units nonconforming. You will see that the $np$ chart is much easier to construct and understand.

The header on the chart in Figure 7.9 is the same as that on the $p$ chart (Figure 7.7) with the following exceptions: the block for $np$ is checked, rather than the block for $p$, and the chart number has been changed to 2. The label 2 is simply to remind the quality team that there is a previous chart, the $p$ chart, used to track the process.

Referring to the data collected for the new study and recorded on the $np$ chart in Figure 7.9, we will first calculate the $p$-bar ($\overline{p}$). The equation is the same as that used before for the $p$ chart.

$$\overline{p} = \frac{\text{Total nonconforming}}{\text{Total inspected}} = \frac{\sum np}{\sum n} = \frac{135}{200 \times 25} = 0.027$$

The result for $\overline{p}$, 0.027, means that, on average, 2.7 percent of the units are nonconforming. This average percent nonconforming represents

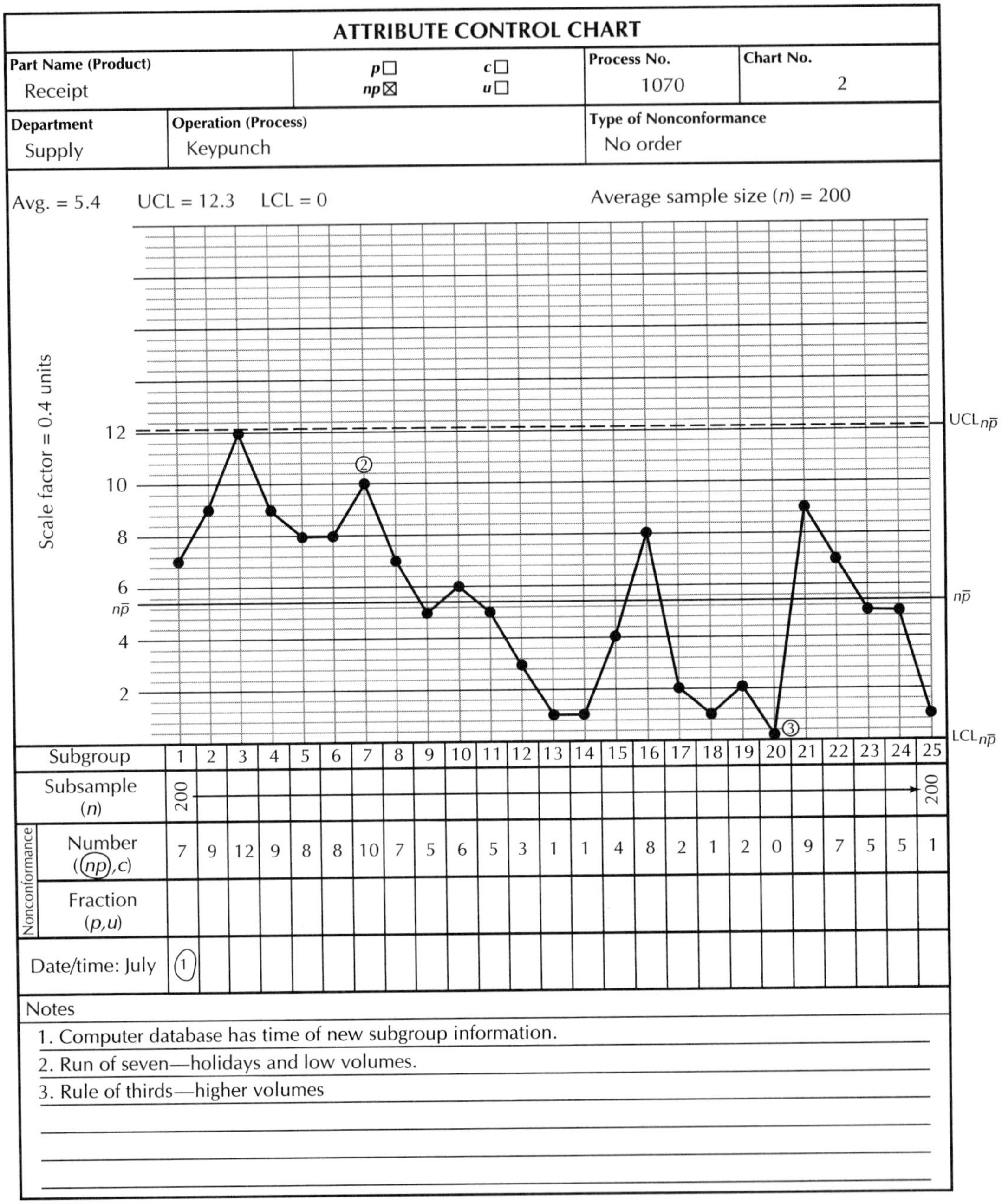

**ATTRIBUTE CONTROL CHART**

| Part Name (Product) | | Process No. | Chart No. |
|---|---|---|---|
| Receipt | $p$ ☐ $c$ ☐ $np$ ☒ $u$ ☐ | 1070 | 2 |

| Department | Operation (Process) | Type of Nonconformance |
|---|---|---|
| Supply | Keypunch | No order |

Avg. = 5.4 UCL = 12.3 LCL = 0 Average sample size ($n$) = 200



| Subgroup | 1 | 2 | 3 | 4 | 5 | 6 | 7 | 8 | 9 | 10 | 11 | 12 | 13 | 14 | 15 | 16 | 17 | 18 | 19 | 20 | 21 | 22 | 23 | 24 | 25 |
|---|---|---|---|---|---|---|---|---|---|---|---|---|---|---|---|---|---|---|---|---|---|---|---|---|---|
| Subsample ($n$) | 200 → | | | | | | | | | | | | | | | | | | | | | | | | 200 |
| Nonconformance: Number (($np$),$c$) | 7 | 9 | 12 | 9 | 8 | 8 | 10 | 7 | 5 | 6 | 5 | 3 | 1 | 1 | 4 | 8 | 2 | 1 | 2 | 0 | 9 | 7 | 5 | 5 | 1 |
| Nonconformance: Fraction ($p$,$u$) | | | | | | | | | | | | | | | | | | | | | | | | | |
| Date/time: July | (1) | | | | | | | | | | | | | | | | | | | | | | | | |

Notes
1. Computer database has time of new subgroup information.
2. Run of seven—holidays and low volumes.
3. Rule of thirds—higher volumes

**Figure 7.9**
The completed attribute chart for nonconforming units with constant subsample size ($np$ chart).

the new capability for the process. The new capability compares favorably with the 7 percent experienced before.

The constant subsample size $n$ of 200 is multiplied by the average units nonconforming $\bar{p}$, to calculate $n\bar{p}$.

$$n\bar{p} = 200 \times 0.027 = 5.4$$

The result of the $n\bar{p}$ calculation means that on the average 5.4 units (not percent of units) are nonconforming.

Now that we have the $n\bar{p}$ calculation, the control limits may be calculated. The equation used to calculate the control limits for the $p$ chart will be simplified by recognizing that the $n$ is constant, and by multiplying each term by the constant $n$ we complete the following derivation:

$$\text{UCL}_{np} = n\bar{p} + \frac{3\,\overset{\sqrt{n}}{\cancel{n}}\sqrt{\bar{p}(1-\bar{p})}}{\cancel{\sqrt{n}}} \quad (\sqrt{n} \text{ divided into } n \text{ is } \sqrt{n})$$

$$\text{UCL}_{np} = n\bar{p} + 3\sqrt{n\bar{p}(1-\bar{p})}$$

Of course, $\text{LCL}_{np}$ would be similar, as

$$\text{LCL}_{np} = n\bar{p} - 3\sqrt{n\bar{p}(1-\bar{p})}$$

EXAMPLE 7.4

Calculate the control limits for the $np$ chart.

*Solution:* The subsample size $n$ is 200, $\bar{p}$ has been established as 0.027, and $n\bar{p}$ has been calculated as 5.4. Using these values we calculate

$$\text{UCL}_{np} = 5.4 + 3\sqrt{5.4\,(0.973)} = 12.3$$

$$\text{LCL}_{np} = 5.4 - 3\sqrt{5.4\,(0.973)} = -1.5 \text{ (or zero)}$$

Since we cannot have a minus number of nonconforming units, $LCL_{np}$ will be zero. The values for the average ($n\bar{p}$) and the control limits are noted at the top of the form in Figure 7.9.

Chart scaling can now be determined, and, as before with the $p$ chart, we aim for the control limit lines to occupy no more than two-thirds of the space available. If you involve yourself in the decision, you would conclude that the easiest chart scaling is as shown, or two units per major division (bold line) and 0.4 units per minor division. The scale factor is noted on the left-hand side of the form.

The $n\bar{p}$ or average line and the control limits are now drawn in. It is important that you confirm the correct position of the lines (dashed for control limit, and solid for average) on the chart. The number of nonconforming units in each subsample *np* is plotted on the chart (you should confirm the correct plotting of these points), and the completed chart is as shown in Figure 7.9. You will agree that, compared to the construction of the *p* chart, the construction of the *np* chart has been much simpler.

## Interpreting the np Chart

Comparing chart number 2 in Figure 7.9 with chart number 1 in Figure 7.7, you can see that the process has improved remarkably. Most of the improvement is due to the elimination of the **assignable causes:** the incorrectly completed order cards, the receiving area under construction (Figure 7.8), and the computer programming error.

We will use the chart section "Notes" in Figure 7.9 to further interpret the control chart. Note 1 reminds anyone interpreting the chart that the exact time of completing a previous subsample and beginning a new one is available on the computer database. This information is often needed to find out if conditions radically changed during the collection of data for one subsample. (During high-volume workdays, more than one subsample may be completed.)

There are no points beyond the control limits in this study. However, the same three rules for out-of-control conditions discussed for the variable chart (Section 4.3) apply to the interpretation of each of the four attribute charts discussed in this chapter—**point out, rule of sevens,** and **rule of thirds.**

Note 2 identifies the assignable cause operating at that point. The more than seven points in a row above the average are attributed to low levels of work (discussed previously) during the July Fourth holiday season. Many employees with families take their vacations during this period.

Note 3 is a less obvious out-of-control condition. Points 13 and 14 and then again points 17 through 20 show a run of points in Zone 3 (refer to Figure 4.11) near the lower control limit. The quality team discussing the assignable cause with the keypunch operators again decided (as they had previously when interpreting the *p* chart) that the work volume was higher during the periods when the out-of-control conditions occurred. The out-of-control condition, in this case, signals an improved process (fewer nonconforming units).

# Completing the Shewhart Cycle

The quality team completed the second phase of the Shewhart cycle, project implementation, during the month of August. After one more chart was constructed, the team decided that it was time to complete the project evaluation phase (Figure 7.10).

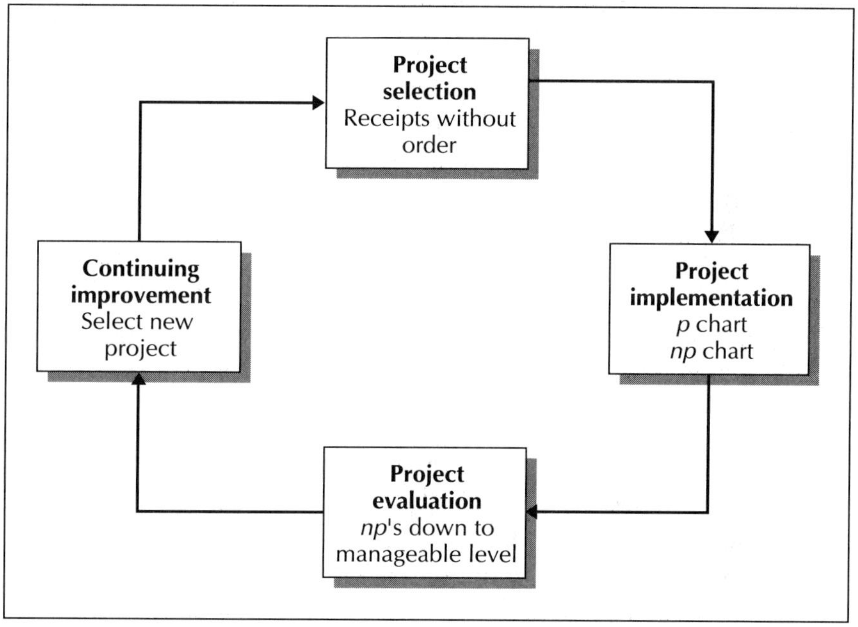


**FIGURE 7.10**
The Shewhart cycle revisited.

Chart 3 showed an average number of nonconforming units $n\bar{p}$ of 2.5. The team was satisfied with the much-improved process and documented the steps leading to their successful project.

The team recommended that charting continue using the historical average and control limits of the third chart ($n\bar{p} = 2.5$, $UCL_{np} = 5.2$, and $LCL_{np} = 0$). Also recommended was that the employees meet at least once each week to interpret the *np* charts. The meetings can take place during low-workload periods. This action should help to balance the workload more effectively.

After the report was generated, the team concluded that no further improvement was necessary (see the last step of the Shewhart cycle, Figure 7.10). The team decided to begin the cycle once again and select a new project to work on.

Their project evaluation was submitted to the office of the military commander (the CEO of the repair facility) for consideration of the yearly quality award. The cost saving was calculated and verified at $1,200.00 per week. The committee evaluating the reports submitted for the year cited the team for their excellent use of statistical process control (SPC) principles and awarded them first prize, which included a monetary award to each participant and special parking privileges.

## Section 7.2 Review Problems

1. If the number inspected in a subsample is 312 and 15 nonconformities are found, what is the percent nonconforming $p$ in the subsample?
   a. 3.7
   b. 4.8
   c. 6.3
   d. 9.2
2. If the total nonconforming from 25 subsamples on a $p$ chart is 455 and the total inspected is 6,500, what is the average ($\overline{p}$) of the chart?
   a. 3%
   b. 5%
   c. 7%
   d. 9%
3. What is the upper control limit for $p$ (proportion of nonconforming units) when the average daily production is 600 units with an established fraction nonconforming of 0.02?
   a. 0.0371
   b. 0.0485
   c. 0.0493
   d. 0.0562
4. An $np$ chart has a constant subsample number of 125. An average of 3.5 percent of the units are nonconforming in each of the 25 subsamples. Solve for the lower control limit for the chart.

a. 3.71
b. 0.849
c. −0.0493
d. −1.789

5. In problem 4, what value would you give the upper control limit on the chart?

# 7.3 Constructing Charts—Nonconformities

## The u Chart

The ***u* chart** is used to track nonconformities, rather than nonconforming units, when the subsample size changes. Typical uses for *u* charts include the tracking of

- carpet weave imperfections where sample area varies;
- engine defects where a low number of defects per unit forces the use of large subsamples of engines;
- housekeeping services at a motel where the subsample size (number of rooms cleaned in a sample in this case) varies.

An example taken from industry involves the assembly of two-cycle engines. The engines have a low number of nonconformities per engine, so more than one or even several engines are needed in a subsample to produce a meaningful chart. The engine-building department decided to use the daily count of engines assembled as the subsample number $n$. Since the number of engines assembled varies from day to day, the subsample number will change from day to day. Therefore, the *u* chart is the chart of choice. The chart for the assembly operation is shown in Figure 7.11.

First examine the data and find that the sample size varies, so both the ($np$, $c$) row (number of nonconformities, $c$) and the ($p$, $u$) row (percent nonconforming, $u$) must be completed. Since the nonconformities are being tracked rather than the nonconforming units, the $u$ is circled in the "Fraction" row. The header information also shows the

## ATTRIBUTE CONTROL CHART

| Part Name | | Process Number | Chart No. |
|---|---|---|---|
| 2-cycle engine | p □ np □ c □ u ☒ | 118 | 1 |

| Department | Operation (Process) | Type of Nonconformance |
|---|---|---|
| Engine building | Assemble 2-cycle engine | Engine part |

Avg. = 1.4 UCL = 2.1 LCL = 0.72 Average sample size ($n$) = 27

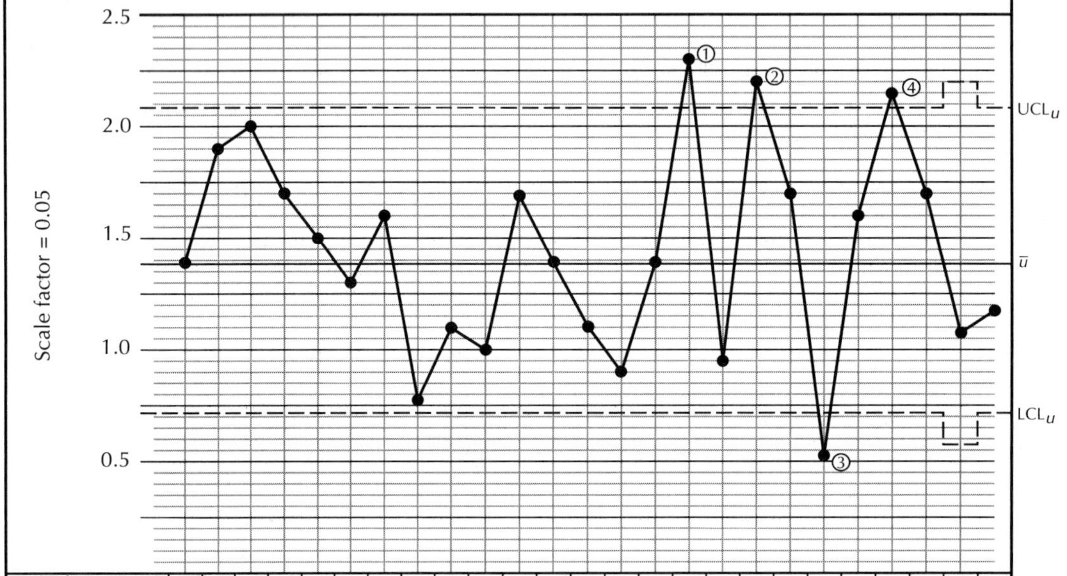


| Subgroup | 1 | 2 | 3 | 4 | 5 | 6 | 7 | 8 | 9 | 10 | 11 | 12 | 13 | 14 | 15 | 16 | 17 | 18 | 19 | 20 | 21 | 22 | 23 | 24 | 25 |
|---|---|---|---|---|---|---|---|---|---|---|---|---|---|---|---|---|---|---|---|---|---|---|---|---|---|
| Subsample ($n$) | 27 | 26 | 24 | 28 | 27 | 27 | 30 | 27 | 32 | 27 | 30 | 29 | 27 | 27 | 27 | 29 | 31 | 26 | 26 | 27 | 27 | 24 | 24 | 19 | 27 |
| Nonconforming: Number ($np$,$c$) | 38 | 50 | 48 | 47 | 40 | 34 | 49 | 21 | 34 | 26 | 50 | 40 | 29 | 25 | 37 | 68 | 30 | 56 | 43 | 14 | 44 | 52 | 41 | 21 | 33 |
| Nonconforming: Fraction ($p$, Ⓤ) | 1.4 | 1.9 | 2.0 | 1.7 | 1.5 | 1.3 | 1.6 | 0.78 | 1.1 | 1.0 | 1.7 | 1.4 | 1.1 | 0.93 | 1.4 | 2.3 | 0.97 | 2.2 | 1.7 | 0.52 | 1.6 | 2.2 | 1.7 | 1.1 | 1.2 |
| Date | 9/4 | 5 | 6 | 7 | 8 | 11 | 12 | 13 | 14 | 15 | 18 | 19 | 20 | 21 | 22 | 25 | 26 | 27 | 28 | 29 | 10/2 | 3 | 4 | 5 | 6 |

Notes

1. Faulty seals (cracks).
2. Faulty seals (wrong OD).
3. CEO visits supplier of seals on 9-27.
4. Short on belts.

**FIGURE 7.11**
A $u$ chart is used to control two-cycle engine nonconformities.

selection of the $u$ chart. Each of the subsample $u$ values is calculated by

$$u = \frac{\text{Number of nonconformities } (c)}{\text{Number in subsample } (n)}$$

The calculation for the first subsample is

$$u = \frac{c_1}{n_1} = \frac{38}{27} = 1.4$$

The average number of nonconformities, the $u$-bar ($\bar{u}$), is calculated as

$$\bar{u} = \frac{\text{Total nonconformities}}{\text{Total units inspected}} = \frac{\Sigma c}{\Sigma n} = \frac{970}{675} = 1.4$$

The capability of the process is 1.4 nonconformities per unit.

The standard deviation of the Poisson distribution (another approximation of the binomial) is used to calculate the control limits for the $u$ and $c$ charts. The standard deviation is the square root of the $\bar{u}$ divided by $n$. Since the subsample number $n$ in this example changes, it is considered in the control limit calculations for the $u$ chart. The standard deviation of the Poisson distribution (as applied to the $u$ chart) is

$$\sqrt{\frac{\bar{u}}{n}}$$

Three standard deviations each side of the average ($\bar{u}$) are used to establish the control limits, so the control limits are calculated by

$$\text{UCL}_u = \bar{u} + 3\sqrt{\frac{\bar{u}}{n}}$$

$$\text{LCL}_u = \bar{u} - 3\sqrt{\frac{\bar{u}}{n}}$$

EXAMPLE 7.5

Using the short-cut demonstrated with the $p$ chart, calculate the control limits for all subsample sizes $n$ within 25 percent of the average subsample size $\bar{n}$.

*Solution:* First, the average $n$ is calculated.

$$\bar{n} = \frac{\Sigma n}{\text{Number of subgroups } (k)} = \frac{675}{25} = 27$$

The range of the subsample sizes within 25 percent is

$$27 \pm (27 \times 0.25) = 27 \pm 6.75 = 20.25 \text{ to } 33.75$$

Only one subgroup, subgroup number 24 ($n$ = 19), requires different control limits.

Calculating the control limits for all of the subgroups except subgroup 24,

$$UCL_u = 1.4 + 3\sqrt{\frac{1.4}{27}} = 2.1$$

$$LCL_u = 1.4 - 3\sqrt{\frac{1.4}{27}} = 0.72$$

It is left for you to calculate the control limits for subgroup 24. Your result may be verified by checking the position of the control limits on the chart.

Confirm the scaling of the chart and the correct placement of the average line ($\bar{u}$ line) on the chart. The chart scale factor for the major divisions is 0.25, in order to use even numbers on the major divisions and use a fair proportion of the chart space. The scale factor for the minor divisions (one-fifth of that for the major divisions) is noted on the left-hand side of the chart. Finally, confirm the correct placement of the upper and lower control limits and the plotting of the individual subsample values for $u$.

Chart interpretation is facilitated by the notes on the bottom section of the form. After problems with engine seals (points 1 and 2), the chief executive officer (CEO) for the plant visits the seal supplier. The problem apparently was solved with the seals, resulting in the out-of-control point to the good (point 3). A shortage of belts causes the final out-of-control point on the chart.

The quality team on this project should look for reasons why so many of the points are near or beyond the control limits (rule of thirds). It looks as if the distribution is bimodal, meaning that there are two different shifts of assemblers or two different assembly lines on the same chart. What other reasons could account for the apparent bimodality?

# The c Chart

The *c* **chart** is used to chart nonconformities when the subsample size is constant. Even if the subsample size *n* is not one, the group may be considered as one. Typical applications of *c* charts are

- carpet weave imperfections in a sample area of 100 square yards;
- number of defective machine screws in a box of 1,000;
- count of paint flaws on an auto body.

The final application example in this chapter will be used to discuss the construction of the *c* chart. Figure 7.12 is a *c* chart of the flaws per paint job in a medium-sized auto body shop in the Midwest.

The data show a constant subsample size of one automobile ($n = 1$). Since the *c* chart is to be used, the line for the fraction defective ($u$) is not completed. The average number of nonconformities ($\bar{c}$) is calculated with the scientific calculator.

$$\bar{c} = \frac{\text{Total nonconformities}}{\text{Total units inspected}} = \frac{c}{n}$$

Confirm the average number as 6.8 (accurate to two significant figures). The capability of the process is the $\bar{c}$ or 6.8 nonconformities per unit. (If *n* is greater than one, the total of nonconformities does not have to be divided by the constant *n* for each subsample. The average from the chart can be interpreted later as *X* number of nonconformities per *Y* number of units.)

The standard deviation of the Poisson for the *c* chart is simply $\sqrt{\bar{c}}$. Please note that the constant subsample size *n* is already a part of the *c*. Therefore, *n* is not used in the standard deviation term. Since three standard deviations each side of the average are required, the control limits are calculated by

$$\text{UCL}_c = \bar{c} + 3\sqrt{\bar{c}}$$

$$\text{LCL}_c = \bar{c} - 3\sqrt{\bar{c}}$$

EXAMPLE 7.6 Calculate the control limits for this example, the number of flaws per paint job.

**ATTRIBUTE CONTROL CHART**

| Part Name | Chart type | Process Number | Chart No. |
|---|---|---|---|
| Auto | p ☐ np ☐ c ☒ u ☐ | N/A | 1 |

| Department | Operation (Process) | Type of Nonconformance |
|---|---|---|
| Paint shop | Paint auto | Flaw |

Avg. = 6.8 UCL = 14.6 LCL = 0 Average sample size (n) = 1

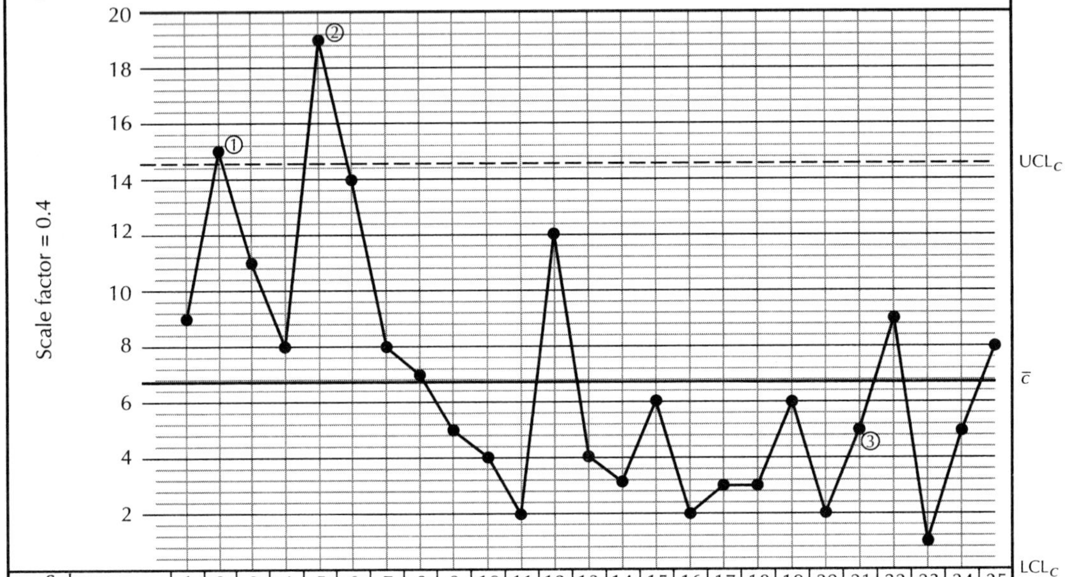


| Subgroup | 1 | 2 | 3 | 4 | 5 | 6 | 7 | 8 | 9 | 10 | 11 | 12 | 13 | 14 | 15 | 16 | 17 | 18 | 19 | 20 | 21 | 22 | 23 | 24 | 25 |
|---|---|---|---|---|---|---|---|---|---|---|---|---|---|---|---|---|---|---|---|---|---|---|---|---|---|
| Subsample (n) | 1 → | | | | | | | | | | | | | | | | | | | | | | | | → 1 |
| Nonconforming: Number (np, Ⓒ) | 9 | 15 | 11 | 8 | 19 | 14 | 8 | 7 | 5 | 4 | 2 | 12 | 4 | 3 | 6 | 2 | 3 | 3 | 6 | 2 | 5 | 9 | 1 | 5 | 8 |
| Nonconforming: Fraction (p, u) | | | | | | | | | | | | | | | | | | | | | | | | | |
| Date | 3/4 | 5 | 6 | 7 | 10 | 11 | 12 | 13 | 14 | 17 | 18 | 19 | 20 | 21 | 23 | 24 | 25 | 26 | 27 | 30 | 31 | 4/1 | 2 | 3 | 5 |

Notes

1. Changed paint supplier.
2. New spray guns.
3. Boss on vacation.

**FIGURE 7.12**
A *c* chart is used to control flaws per paint job.

*Solution:* The calculations are as follows:

$$UCL_c = 6.8 + 3\sqrt{6.8} = 14.6$$

$$LCL_c = 6.8 - 3\sqrt{6.8} = -1.02 \text{ (or zero)}$$

After confirming all calculations, confirm the scaling of the chart and the correct placement of the average line ($\bar{c}$ line) on the chart. The chart scale factor for the major divisions is 2, in order to use even numbers on the major divisions and use a fair proportion of the chart space. The scale factor for the minor divisions (one-fifth of that for the major divisions) is noted on the left-hand side of the chart. Finally, confirm the correct placement of the upper and lower control limits and the plotting of the individual values of $c$. (Do you know why the lower control limit for $c$, or $LCL_c$, is zero in this case?)

Chart interpretation is facilitated by the notes on the bottom section of the form. After some problems with a brand of paint (point 1) and the paint sprayers (point 2), the chart seems in control. So why is point 3 noted? This question and the interpretation of the cause of the out-of-control condition are left for your interpretation.

## Section 7.3 Review Problems

1. The $u$ chart is used to track ____________________
   (nonconforming units, nonconformities), with a ______________ (varying, constant) subsample size.
2. The $c$ chart is used to track ____________________
   (nonconforming units, nonconformities), with a ______________ varying, constant) subsample size.
3. See Figure 7.12, and solve for the $\bar{c}$ and the upper control limit for $c$ if the subsample size was actually 2 ($n = 2$) instead of 1. The $UCL_c$ is calculated as
   a. 6.8
   b. 7.3
   c. 8.9
   d. 14.6

# Summary

- The four steps preparatory to attribute control charting are selecting the product attribute to be studied, determining the type of chart to use (Figure 7.1), deciding on how the nonconforming unit or nonconformity is to be judged, and data collection techniques.
- Constructing the $p$ chart, for nonconforming units with changing sample size, involves the following equations:

$$p = \frac{\text{Number nonconforming units in subsample } (np)}{\text{Number inspected in subgroup } (n)}$$

$$\overline{p} = \frac{\text{Total nonconforming}}{\text{Total inspected}} = \frac{\Sigma\, np}{\Sigma n}$$

$$\text{UCL}_p, \text{LCL}_p = \overline{p} \pm 3\sqrt{\frac{\overline{p}(1-\overline{p})}{n}}$$

- Constructing the $np$ chart, for nonconforming units with constant sample size, involves the same calculation for $\overline{p}$ as the preceding and the following equations for the control limits:

$$\text{UCL}_{np}, \text{LCL}_{np} = n\overline{p} \pm 3\sqrt{n\overline{p}(1-\overline{p})}$$

- Use the following equations to construct the $u$ chart (changing subsample size):

$$\overline{u} = \frac{\text{Total nonconformities}}{\text{Total inspected}} = \frac{\Sigma\, c}{\Sigma\, n}$$

$$\text{UCL}_u, \text{LCL}_u = \overline{u} \pm 3\sqrt{\frac{\overline{u}}{n}}$$

- The $c$ chart is constructed by using the average number of nonconformities per subsample ($\overline{c}$), and since the subsample size is constant, the control limits are calculated by

$$\text{UCL}_c, \text{LCL}_c = \overline{c} \pm 3\sqrt{\overline{c}}$$

- The capability of any attribute chart is the average of the particular chart. The $\overline{p}$ and $\overline{u}$ are expressed as fractions or percentages, and the $n\overline{p}$ and $\overline{c}$ are expressed as number of units.

- Chart interpretation involves the same rules for out-of-control as apply to the $\overline{X}$ and $R$ chart (variable chart)—point out, rule of thirds, and rule of sevens.

## CASE STUDY · CHAPTER 7

The Reeaally Kleen Car Wash chain has received a number of complaints on damage to automobiles. Damage claims settled in and out of court have amounted to a sizable loss in profits for the company. At this point, no one in the company knows the proper way to attack the problem.

You are the maintenance manager for the company. After receiving quality instruction as part of your degree work in manufacturing technology, you have decided that there is a sound approach to solving the problem. You suggest to the owner that an attribute study be performed, keeping a count of the units (number of cars) cleaned and the nonconformities per unit—for example, number of nicks, mars, scratches, unremoved tar, and waterspots. You tell the owner that simply doing an inspection of each unit—before and after the cleaning process—will involve the staff, and this feedback will result in a decrease in nonconformities.

You recommend the following specifics for a quality system for the Reeaally Kleen Car Wash.

- The use of a $u$ chart to track nonconformities per automobile. Subsample size should be based on a daily basis, resulting in a changing subsample size, hence the need for the $u$ chart rather than the $c$ chart.
- Thorough notes must be kept on the chart form to explain assignable causes that are operating, such as stiff brushes scratching cars when the ambient temperature drops below a certain level.
- A Pareto chart will be maintained that categorizes the causes for the nonconformities.
- Quality costs (discussed in Section 10.1) should be tracked to either prove or disprove that the effort is worthwhile. Some considerations are cost of training employees; time spent inspecting; and cost of materials, such as check sheets used to collect data on the location and type of nonconformity.
- Quality costs during the period must be compared to quality savings (e.g., the historical costs of lawsuits and repair costs). If there is

increased business, it should be reflected as a cost saving. Perhaps employees will find inexpensive ways to repair minor nonconformities. Other cost benefits from increased employee involvement should be noted and cost benefits estimated and compared with overall costs.

- Staff should brainstorm ways to determine nonconformities. The manager should clearly explain to all employees that they must not fear reporting nonconformities, that over the long run the increased business will more than make up for the increased costs of repair of nonconformities.
- Management must decide how damages are to be rectified—for example, on site, under contract with a local body shop, or by a private contractor.

## Chapter Review Problems

1. For the following quality problems, determine the most appropriate type of control chart to use ($\overline{X}$ and *R*, *p*, *np*, *c*, or *u*). Note that the $\overline{X}$ and *R* chart (from Chapter 4) is a choice.
   a. The percent of housekeeping nonconformities per day in a hospital—the sample size varies day by day.
   b. The number of engine nonconformities—sample size is one engine.
   c. Time of service in a restaurant—data are in minutes.
   d. Time of service in an airport—0 to 7 minutes is considered good service and anything over 7 minutes is considered a nonconforming unit; sample size is variable.
   e. The percent of nonconforming two-cycle engines built per day.
   f. The number of data errors, each one considered a nonconforming unit—subsample size is 200.
2. The following data (page 259) represent number of units produced per day by a manufacturer of toasters. The nonconforming units per day are also shown.

   Calculate the percent of nonconforming units for each subsample ($p$). Calculate the $p$-bar ($\overline{p}$). Calculate the average subsample size ($\overline{n}$). List the subsample sizes within ±25 percent of the average subsample size ($\overline{n}$).

| Subgroup | Sample Size | Nonconforming Units | Subgroup | Sample Size | Nonconforming Units |
|---|---|---|---|---|---|
| 1 | 425 | 17 | 13 | 727 | 16 |
| 2 | 364 | 26 | 14 | 1,000 | 19 |
| 3 | 379 | 22 | 15 | 1,200 | 19 |
| 4 | 706 | 24 | 16 | 800 | 8 |
| 5 | 600 | 30 | 17 | 1,000 | 8 |
| 6 | 500 | 35 | 18 | 1,010 | 23 |
| 7 | 417 | 15 | 19 | 521 | 20 |
| 8 | 562 | 19 | 20 | 410 | 18 |
| 9 | 575 | 23 | 21 | 325 | 13 |
| 10 | 243 | 18 | 22 | 526 | 20 |
| 11 | 417 | 15 | 23 | 520 | 14 |
| 12 | 808 | 21 | 24 | 387 | 17 |
| | | | 25 | 512 | 16 |

3. Calculate the control limits for all other subsamples in problem 2. Prepare a chart for the process by constructing the average line and the control limit lines. Chart the process.
4. How many out-of-control points do you find on the chart for problem 3? What is the capability of the process? Could the rule of sevens be used between subsample points 10 and 17? If so, why? How could it help the company to recognize the pattern of the subsample points between 10 and 17?
5. See Figure 7.7. Calculate the control limits for the eleven subsamples not calculated in the accompanying discussion and Examples 7.2 and 7.3. Confirm that the control limit lines on the chart are correctly located.
6. Sheet-metal screws in large lots are ordered from a single supplier and 100 percent inspected. The process flow is pictured in Figure 7.13. The data shown below indicate the number inspected ($n$) in each lot and the number nonconforming ($np$).

| Lot Number | $n$ | $np$ |
|---|---|---|
| 1 | 6,000 | 183 |
| 2 | 6,000 | 131 |
| 3 | 10,000 | 184 |
| 4 | 1,783 | 23 |

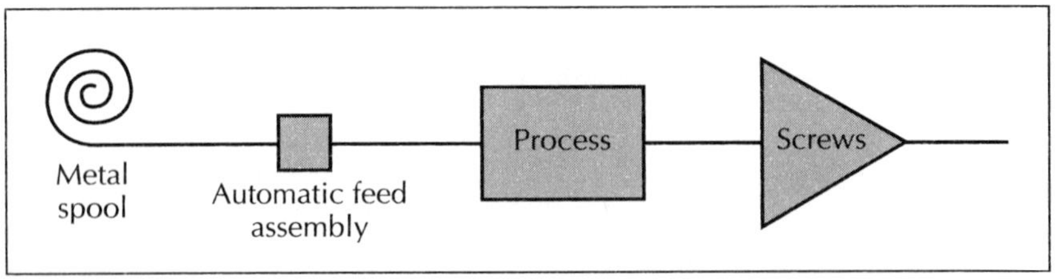


**FIGURE 7.13**
The process flow diagram for producing sheet-metal screws (problem 6).

A depleted metal spool is replaced with a full one.

| | | |
|---|---|---|
| 5 | 3,089 | 171 |
| 6 | 5,774 | 65 |

As you can see, these are only a few subsamples, offering a snapshot of the process around the material change (the new metal spool). Construct the appropriate control chart for the data. Indicate with an X any points out of control.

7. From Figure 7.13 and the attribute chart constructed in problem 6, recommend a possible study that could result in a more stable process. Explain your rationale.
8. A large motel is maintaining a control chart on the housekeeping function. Every day 315 rooms are inspected. If more than two nonconformities (e.g., no towels, dirty tub, hair on carpet) are found in any one room, the room is treated as a nonconforming unit. The following data were collected for 20 days:

| *Day* | *np* | *Day* | *np* |
|---|---|---|---|
| 1 | 14 | 11 | 17 |
| 2 | 10 | 12 | 5 |
| 3 | 16 | 13 | 6 |
| 4 | 11 | 14 | 10 |
| 5 | 12 | 15 | 9 |
| 6 | 15 | 16 | 9 |
| 7 | 28 | 17 | 10 |
| 8 | 13 | 18 | 8 |
| 9 | 17 | 19 | 27 |
| 10 | 11 | 20 | 18 |

Calculate the process average and control limits, and construct the appropriate chart. Mark any points above or below the control limits with an X.

9. In the chart for problem 8 identify out-of-control signals due to the rule of sevens and the rule of thirds. How could the company make use of the assignable cause operating between subsamples 12 and 18?

10. An airline is conducting a study on baggage defects (rips or tears, lost baggage, sent to wrong destination, etc.). Each day, for 18 days, the airline tries to check 10 to 15 flights at randomly selected locations across the country. The following data have been collected:

| *Day* | *n (bags)* | *c* | *Day* | *n (bags)* | *c* |
|---|---|---|---|---|---|
| 1 | 1700 | 30 | 10 | 800 | 12 |
| 2 | 900 | 16 | 11 | 1300 | 21 |
| 3 | 1300 | 26 | 12 | 900 | 16 |
| 4 | 1000 | 23 | 13 | 1400 | 28 |
| 5 | 1000 | 14 | 14 | 700 | 17 |
| 6 | 1200 | 23 | 15 | 1200 | 20 |
| 7 | 900 | 22 | 16 | 1000 | 12 |
| 8 | 1000 | 12 | 17 | 900 | 18 |
| 9 | 1300 | 17 | 18 | 1300 | 17 |

Calculate the percent of nonconformities for each subsample ($u$). Calculate the $u$-bar ($\overline{u}$). Calculate the average subsample size ($\overline{n}$). Be sure to use the method of keeping the control limits constant for all subsample sizes within $\pm 25$ percent of the average subsample size ($\overline{n}$). Calculate the control limits for all other subsamples. Chart the process. Mark with an X any out-of-control points.

11. In problem 10, identify out-of-control subsamples that are due to the rule of sevens and rule of thirds. What would you recommend the airline look for between days 8 and 15?

12. What is the capability of the process in problem 10?

13. See Example 7.5. Calculate the control limits for the remaining point, number 24. Confirm your calculation with the position of the control limits on the $u$ chart in Figure 7.11.

14. In Figure 7.11, the later stages of the process (points 16 to 25) show out of control by the rule of thirds. The author's explanation

involves a bimodal distribution. What could this mean? Explain the rule of thirds in terms of bimodality.

15. A textile manufacturer is charting nonconformities per 100 square yards of carpet. The following data have been collected:

| *Piece No.* | *Nonconformities (c)* | *Piece No.* | *Nonconformities (c)* |
|---|---|---|---|
| 1 | 11 | 11 | 11 |
| 2 | 8 | 12 | 5 |
| 3 | 9 | 13 | 7 |
| 4 | 12 | 14 | 12 |
| 5 | 4 | 15 | 13 |
| 6 | 14 | 16 | 8 |
| 7 | 5 | 17 | 23 |
| 8 | 8 | 18 | 11 |
| 9 | 21 | 19 | 9 |
| 10 | 10 | 20 | 10 |

Calculate the average and the control limits and construct the appropriate chart. Mark any points out of control with an X.

16. An automobile assembler records the nonconformities in subsamples of 100 automobiles, as reported by owners in the first three months of use. The data are as follows:

| *Reporting Period* | *Nonconformities (c)* | *Reporting Period* | *Nonconformities (c)* |
|---|---|---|---|
| 1 | 111 | 9 | 33 |
| 2 | 153 | 10 | 130 |
| 3 | 181 | 11 | 71 |
| 4 | 157 | 12 | 52 |
| 5 | 192 | 13 | 122 |
| 6 | 113 | 14 | 37 |
| 7 | 139 | 15 | 76 |
| 8 | 81 | | |

Prepare the appropriate control chart.

17. Prior to reporting period number 8, the assembler implemented companywide SPC training. Would the control chart in problem 16 encourage management to continue SPC training? Explain why or why not.

18. See Figure 7.12. Why is point 3 noted with an assignable cause? What would you recommend the manager do?

# LABORATORY · CHAPTER 7
## Attribute Charting Using the Computer

*Materials needed:* Computer with SQC software, or a manufacturer using SQC software. The instructor will provide attribute data to be charted.

After becoming oriented to the computer system, enter attribute data into the system by using the keyboard. Any of the data sets presented in this chapter may be used. You may then practice using the SQC software to construct *p, np, u,* or *c* charts. After the various charts have been examined on the screen, the charts should be printed out. Affirm by calculation at least one of the parameters calculated by the computer.

# 8

# Acceptance Sampling and Inspection

**Acceptance Sampling** 8.1
**MIL–STD–105** 8.2
**Nondestructive Testing (NDT)** 8.3

## 8.1 *Acceptance Sampling*

Statistical process control (SPC) has been the focus of previous chapters. SPC has been shown to be an effective quality tool—a major tool that is essential to the improvement of quality in both manufacturing and service industries.

This chapter will discuss acceptance sampling. Recently, acceptance sampling has been much maligned. It has been described as a system that encourages the production of nonconforming goods.

Hopefully, you will discover that acceptance sampling is another major quality tool—a tool that may be used when SPC is absent or impossible to implement.

In Chapter 1, control charting or SPC was shown to be but one of the two major components of **statistical quality control (SQC).** The other component is acceptance sampling. Figure 8.1 shows the two major components of SQC. Both components, acceptance sampling and SPC or control charting, are divided into the same major elements derived from the type of data collected—attribute data (go/no-go) or variable data (measurable).

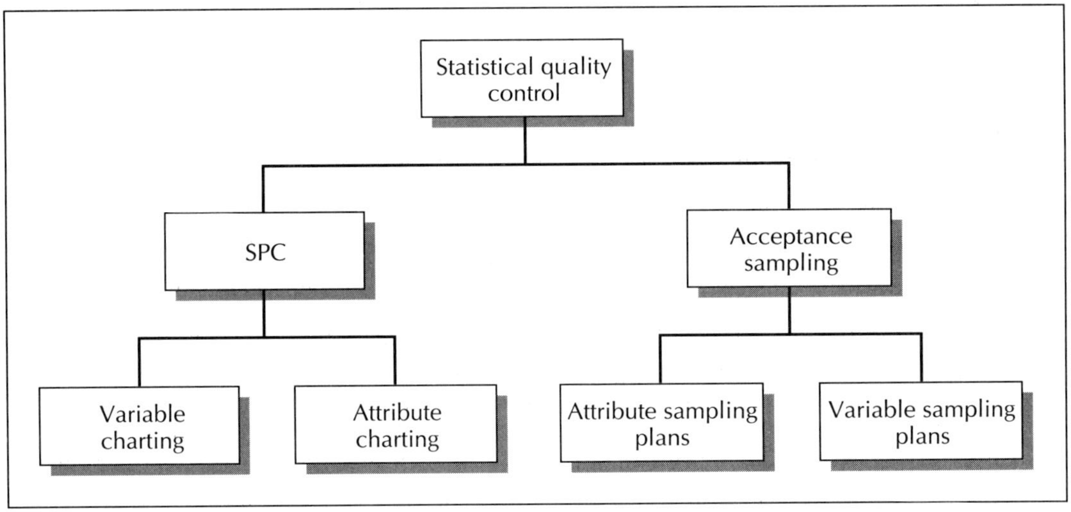


**FIGURE 8.1**
The two components of SQC—SPC and acceptance sampling.

**Acceptance sampling** is the inspection of **lots** of product for specified quality characteristics after the product has been produced. After inspection of a sample drawn from the lot, the lot is either accepted or rejected.

Implementing acceptance sampling may be visualized by considering Figure 8.2, in which an inspector is seen examining a random sample of parts drawn from the lot. Each part in the sample is inspected, and the lot is either accepted or rejected based on the number of nonconforming units in the sample. (Several quality characteristics may be inspected.) The quality characteristics to be used to accept or reject

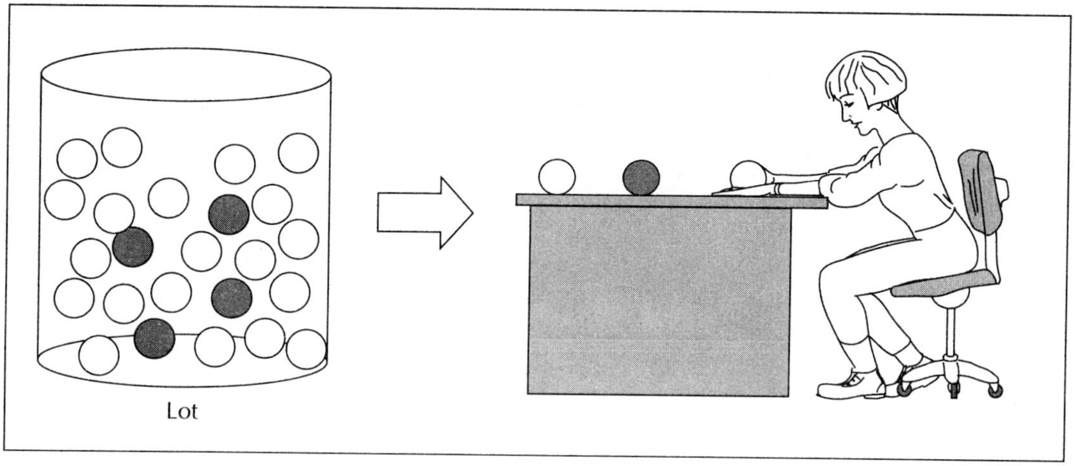


**FIGURE 8.2**
Acceptance sampling involves a random selection of parts from a lot.

the lot must be identified and described. Each quality characteristic must be given a **seriousness classification,** which includes such descriptive terms as these:

- **Critical**—a critical nonconformance will cause failure of the product, injury, or illness (i.e., food industry).
- **Major**—A major nonconformance will cause loss of customers or not conform to government regulations or standards.
- **Minor**—A minor nonconformance includes none of the above characteristics and should be considered for elimination as a nonconformance.

EXAMPLE 8.1

A large assembly plant for trucks receives thousands of parts from hundreds of suppliers every day. Similar parts arrive from more than one supplier, even though **single sourcing** is strongly encouraged by some quality leaders. Some parts arrive from new suppliers, and the quality of the parts may be uncertain. One way to cost-effectively monitor the quality of the incoming parts from the suppliers is to sample their product (as contrasted with 100 percent inspection) to make sure it contains few nonconforming parts. This method is known as acceptance sampling—to accept what is being delivered.

Imagine further that the plant is only receiving enough parts to build a day's worth of trucks. This practice of keeping low inventories saves money and is known as **just-in-time** manufacturing or **JIT.** If the quality of any of the incoming parts is not predictable, considerable downtime will result and profits will suffer. How can the company assure that suppliers are providing acceptable product?

*Solution:* The management feels that it cannot afford to set up and monitor quality programs for its suppliers. Not being able to afford what many consider the best type of quality assurance, and needing to depend heavily on the incoming quality of the product from their suppliers, they elect to use acceptance sampling plans for the necessary inspection of incoming parts and supplies.

The use of the term "acceptance sampling" is not meant to imply that this method is used only at the front end of a manufacturing operation or in the parts receiving area. Acceptance sampling may be used anywhere within the plant. It may be used between two departments when one department supplies parts to the other. Acceptance sampling, which may then be more appropriately referred to as **final inspection,** is also used for the final product.

The major difference between the use of "run charts" to control processes and acceptance sampling is that acceptance sampling is used after the process has been run. Process control (SPC) is used as the process produces goods. Another way to look at the difference between SPC and acceptance sampling is to consider the two flowcharts in Figure 8.3.

The similarities of the two systems are the following:

- Each system has the same inputs to the process—MMMMEP (Chapter 1).
- Both involve some type of inspection.
- There are two possible outcomes of inspection in both cases.

The acceptance sampling model shown in the figure differs from the SPC model in the following ways:

- Acceptance sampling is primarily designed and used for manufacturing (SPC applies to service industries as well).
- Inspection is integral to the process in SPC and may be carried out by operators, thereby building pride in their work. (Accep-

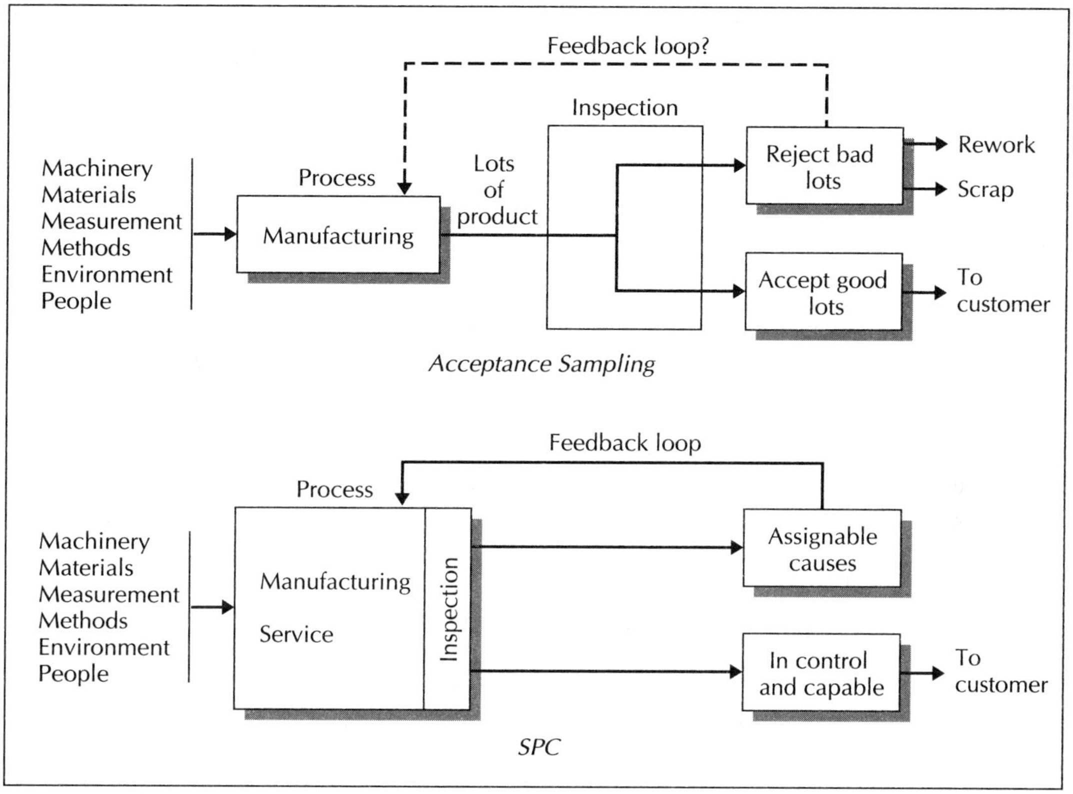


**FIGURE 8.3**
SQC involves SPC and acceptance sampling.

tance sampling inspection is by lots of product drawn from the process and performed by people designated as inspectors.)

- The feedback loop from inspection back to the process is longer with acceptance sampling and may not occur at all.

One of the major differences involves the outputs of the two systems. With SPC—using variable charting techniques—once a process is in control and capable, there should be no scrap or rework. Inspecting lots of product results in some nonconforming product always getting through to the customer. (Why this is so will become more apparent later when an important component of the acceptance sampling system, the acceptable quality level, or AQL, is discussed.) In fact, the U.S. Department of Defense, a longtime believer in the effectiveness of acceptance sampling, has recently moved to requiring the increased use of SPC by suppliers to the armed forces.

Regardless of the obvious advantages of SPC, acceptance sampling still has its advantages, and quality control technologists or quality managers must have all of the tools available in their toolbox in order to enjoy flexibility in problem solving and to be maximally effective in their careers.

The advantages of acceptance sampling are as follows:

- It may be an inexpensive way to inspect product or to provide quality assurance, especially if the part being accepted or produced is not critical to the final product. *Acceptance sampling is not recommended for critical quality characteristics.*
- It is at times the only feasible way to inspect product. Take, for example, the manufacture of shells for weapons, or any other product where conformance testing must be destructive **(destructive testing).**
- It usually is superior to 100 percent inspection, mainly because inspectors become bored with simply inspecting the product, which is often perceived as a negative part of manufacturing.
- SPC requires training of the entire workforce. Acceptance sampling plans may be managed by a specific group called inspectors.

## *100 Percent Sampling*

Acceptance sampling plans have been shown to produce a lower level of nonconformances than 100 percent sampling practices. We should convince ourselves that this statement is true before continuing on to explore the statistics of published acceptance sampling plans.

**EXAMPLE 8.2** Count the number of occurrences of the letter *f* on this page. Include this example. Do it quickly, because the inspection function in a plant is expected to be as productive as possible. To simulate realistic in-plant conditions, you might watch a ball game on television—turned way up so you can hear the crowd noise—while you complete this assignment.

*Solution:* Check your results with others. If in a classroom setting, calculate the average results for the class. Next, sample only 10 randomly sampled lines from the page and average your results. You will probably find that the sample results agree more closely with the actual number than the results of 100 percent inspection. If you performed this experiment over the entire chapter or over the entire text, you would certainly find that sampling is both more efficient and more accurate than 100 percent inspection.

Often, 100 percent inspection is perceived as the only way to produce high-quality product. But, a manufacturing plant that uses it is not focusing on making the product *right the first time.* Instead, the focus is on the use of inspectors to make sure it is right later. This approach leads to very expensive and inefficient manufacture (see the Case Study—Chapter 8).

One hundred percent inspection is more costly than other methods. There is more handling damage, and more inspectors are required. Products that require destructive testing cannot be 100 percent inspected.

## No Inspection

The other extreme from 100 percent inspection is none at all. Figure 8.4 shows what happens with no inspection. The manufacturer ships to a customer or sends to another department exactly what the incoming quality is. If 10 percent of product is nonconforming, then 10 percent of that product is passed on to the consumer—the customer—or used in making another product.

## Constant Percentage Sampling

Another sampling method commonly used in industry is the constant percentage sample. The **constant percentage sample** is a sample that is a fixed percentage of the lot size. This type of plan will not offer adequate protection for small lots and offers too much protection (requires too much inspection) for large lots. *Statistically determined acceptance sampling plans do not depend on lot size* and are much more efficient and less costly than constant percentage plans.

## Collecting Random Samples

In this chapter we will be selecting samples of product. The sample will be *assumed* to be representative of the population. The best way to assure that the sample is representative of the population is to draw

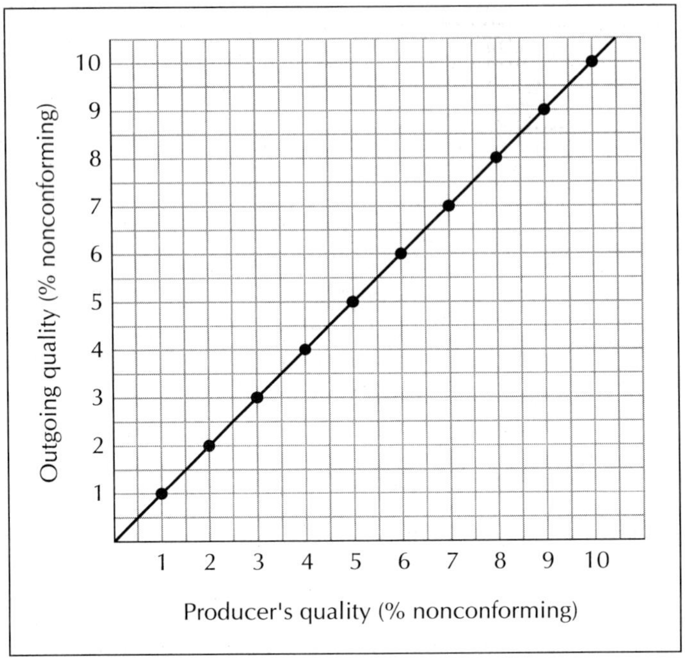


**FIGURE 8.4**
The result of no inspection on outgoing quality.

a random set of parts or products. **Random sampling** means that each part from a finite population has an equal probability of being selected. For instance, if a lot contains 1,000 parts (the finite population) and 50 parts are to be selected (the sample) to represent the population, how would you select the parts to ensure a random selection? The method most acceptable to statisticians is to number each part and use a table of random numbers to determine which parts to select. (Your calculator will probably have a random number generator.)

EXAMPLE 8.3

Use your calculator's random number generator to determine which of 20 parts to select from a lot of product having the serial numbers 1024300 to 1028300.

*Solution:* Check your calculator's manual for the use of the random number generator. The range of the generated random numbers for most calculators is 0.000 to 0.999 or three significant figures. For this example, which requires four significant figures, you could generate the first number of each of the 20 parts to be sampled by using any of the three numbers generated. Always use a consistent method—say, always use the first number. If the number is out of the range of the part numbers, 4 to 8 in this case, then another number must be generated. The remaining three numbers may then be generated in the same way. Random numbers may also be selected using random number tables.

If you wanted to select a sample of 300 screws from a lot of 10,000, the method in Example 8.3 would be too laborious (it is difficult to put a serial number on a screw). A method next most preferred to assure randomness is to do a systematic sample. A **systematic sample** uses samples occurring in equally spaced locations or at equally spaced periods of time. One type of systematic sample is illustrated in Figure 8.5.

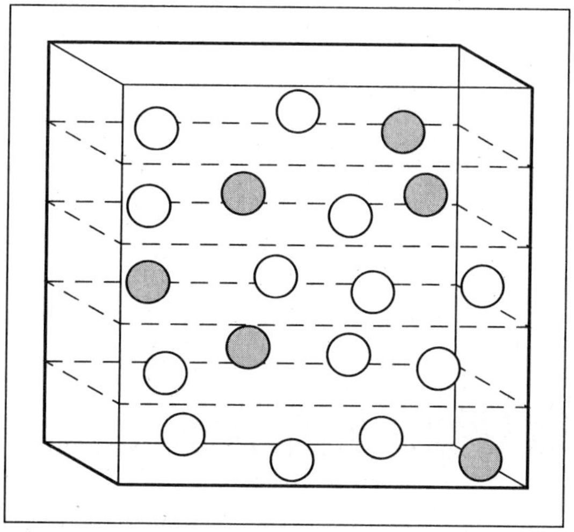

**FIGURE 8.5**
One type of systematic sample is a selection of parts from different levels (strata) of the lot.

Another systematic sample can be created by selecting every 50th name from a telephone directory to prepare a sample of survey respondents. Yet another type of systematic sampling is to collect samples at prescribed intervals from a continuous process, such as an assembly line.

Too often in business and industry, faulty sampling techniques invalidate any conclusions. Take, for example, the case where an inspector is afraid to report low quality. When taking a smoothness-of-bore measurement with the profilometer, the first reading may be high (see Example 2.1). Consequently, the measurement is not recorded, but another is taken. After five tries a low enough reading is obtained for the inspection report. This practice of multiple readings is known as "editing the data." Of course, this example describes a selected sample and not a random sample of the smoothness of the bore. It is a search for the least rough area of the bore, and the results are invalid and useless. Deming, an expert on how good management leads to high quality, would say that the inspector's action is due to "management by fear." The inspector is afraid to report poor quality. Elimination of fear prevents such a waste of human energy.

## Operating Characteristic Curves

The best way for you to understand an acceptance sampling plan is to examine the risks associated with a particular plan by using a graph called the operating characteristic curve, or simply the OC curve. The **OC curve** or graph is a plot of the probabilities of accepting a lot of product given the largest number of nonconformances allowed $c$ in a sample size $n$.

Figure 8.6 shows the type of graph we will consider throughout our discussion of OC curves. You will note the $X$ axis is labeled $\overline{p}$. The symbol $\overline{p}$ represents the fraction nonconforming or average fraction nonconforming in a lot. In other words, it represents the lot quality. The $Y$ axis is the probability of accepting the lot, or the probability of acceptance $P(A)$, which will be expressed as a percentage.

Figure 8.6a shows a perfect sampling plan for only passing lots that are 3 percent nonconforming ($\overline{p} = 0.03$) or less. This sampling plan is considered perfect or ideal because it could yield the exact quality break point for the lots studied with 100 percent accuracy.

Figure 8.6b is just the opposite in its effectiveness. This situation could be due to the inspectors not being well trained and to quality characteristics not being well defined in terms of quality standards. The graph shows that no matter what the quality level of the lots, 50 percent of them pass. In other words, all of the lots tested have a 50/50 chance of being accepted regardless of the quality level. Needless

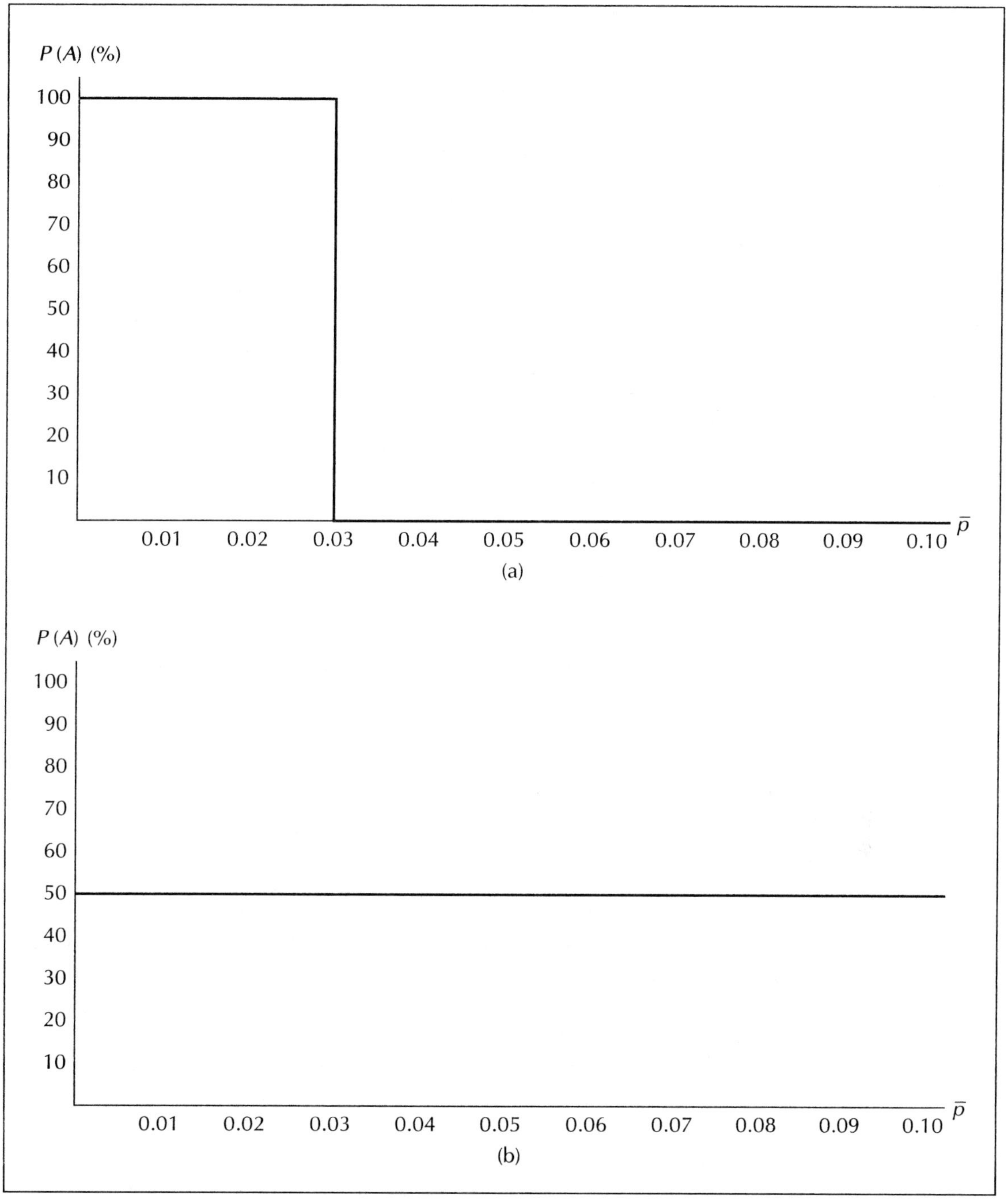


**FIGURE 8.6**
A perfect sampling plan (a) and an undocumented or "hit or miss" plan (b).

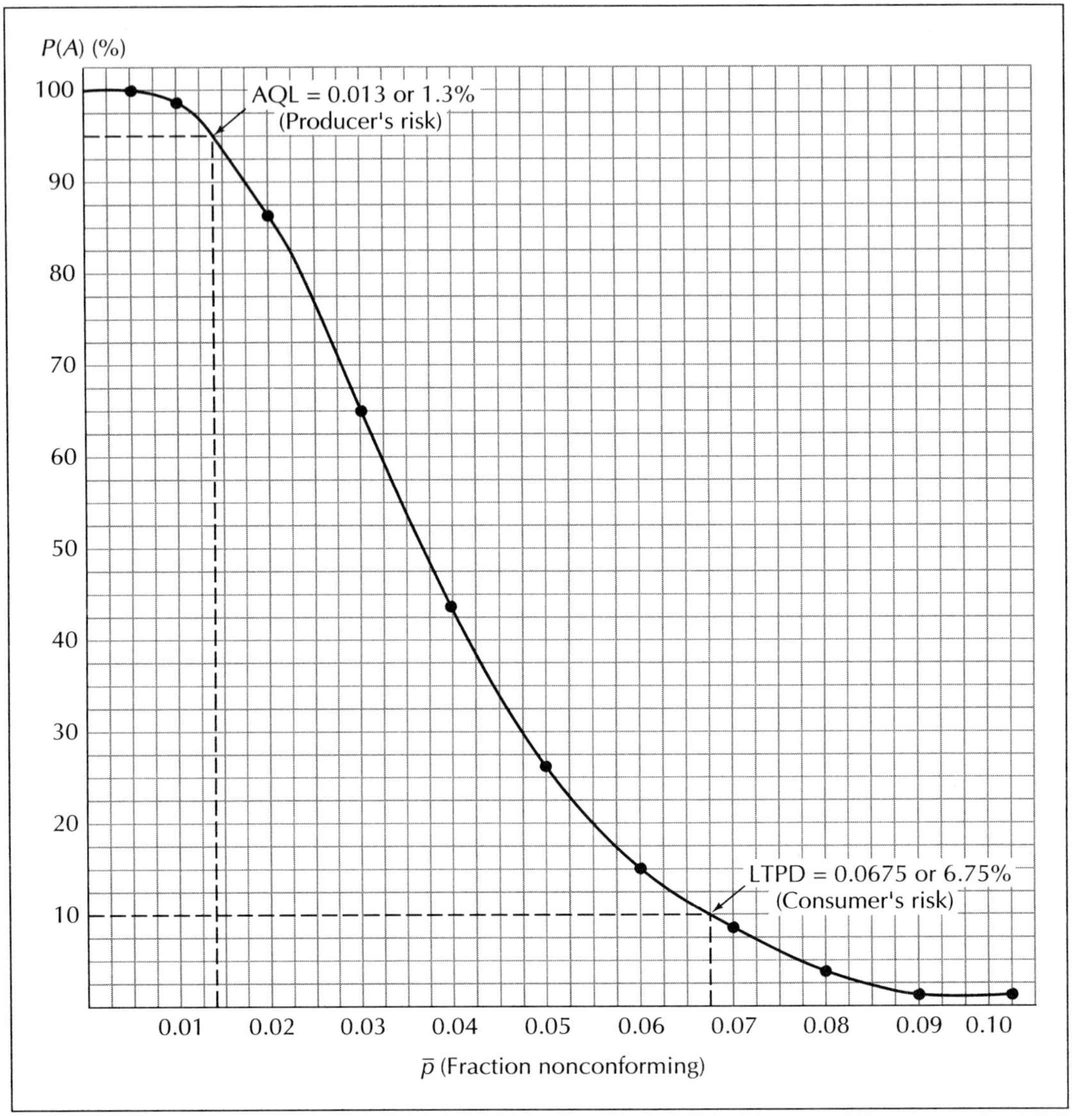


**FIGURE 8.7**
The OC curve for $n = 100$, $c = 3$.

to say a *real* sampling plan must be somewhere in between the vertical line of Figure 8.6a and the horizontal line of Figure 8.6b.

An OC curve will be constructed by using the Poisson distribution (Appendices B and D). The Poisson is an approximation of the binomial that may be used if we assume a steady stream of product or long runs and if the $\overline{p}$ is small ($\overline{p} < 0.1$; see the probability flowchart, Figure 6.7). The main reason to use the Poisson to estimate the binomial is that it is easier to use a table than calculate the binomial probabilities. You may wish to review the material in Chapter 6 at this time.

To discuss any particular sampling plan we only have to know two things:

- the sample size $n$;
- the number of nonconformances allowed to accept the lot $c$.

Note that we do not need to know the size of the lot $N$, but a lot size significantly greater than the sample size is assumed. *For our purposes in this chapter we will assume the lot size* N *to be at least ten times the sample size* n ($N > 10n$).

Assume a sampling plan with a sample size of 100 ($n = 100$) and an acceptance number of three ($c = 3$). Using the cumulative values in parentheses from the Poisson table (Appendix D) and the $\overline{p}$ values used in Figure 8.7, we construct a table of values (Table 8.1). The probability of acceptance $P(A)$ has been rounded to a unit's accuracy. Please verify the probabilities in Table 8.1.

Figure 8.7 shows a graph of the OC curve generated from Table 8.1. Confirm the correct plotting of the points at this time. A smooth curve has been drawn through the points.

Assuming lot sizes $N$ significantly larger than sample sizes $n$, the OC curve will tell us our chances of accepting lots of differing $\overline{p}$.

TABLE 8.1

**Values Used in Constructing the OC Curve**

| $\overline{p}$ | $n\overline{p}$ ($n = 100$) | $P(A)$ % from Poisson (using the row for $c = 3$) |
|---|---|---|
| 0.005 | 0.5 | 100 |
| 0.01 | 1 | 98 |
| 0.02 | 2 | 86 |
| 0.03 | 3 | 65 |
| 0.04 | 4 | 43 |
| 0.05 | 5 | 27 |
| 0.06 | 6 | 15 |
| 0.07 | 7 | 8 |
| 0.08 | 8 | 4 |
| 0.09 | 9 | 2 |
| 0.10 | 10 | 1 |

EXAMPLE 8.4

Find the probability of accepting lots having the following quality levels *p*: 0.025, 0.048, 0.062, 0.085, 0.093.

*Solution:* Use the graph in Figure 8.7. Draw vertical lines from the *X* axis up to the curve. Connect a horizontal line over to the *Y* axis. The result is the *P*(*A*) for that particular $\overline{p}$. The respective probabilities of acceptance are as follows: 75 percent, 30 percent, 10.4 percent, 2.5 percent, 1.2 percent. A way to view the percentages just calculated is to take the case of $\overline{p} = 0.025$. The probability of accepting a lot with a quality level of 2.5 percent nonconforming product is 75 percent. This means there are three chances in four that the lot will be accepted.

Most sampling plans derived directly from the OC curve are based on the probability of acceptance *P*(*A*) of 95 percent and 10 percent. The *P*(*A*) of 95 percent is called the **acceptable quality level (AQL).** The *P*(*A*) of 10 percent (or the probability of rejection of 90 percent) is called the **limiting quality level (LQL).** Sampling plans of the LQL type are known variously as **rejectable quality level (RQL), lot tolerance percent defective (LTPD),** or **unacceptable quality level (UQL)** plans.

The AQL is that point where there is a 95 percent chance of accepting a lot. For the sampling plan described by the OC curve in Figure 8.7, the AQL is at 0.013 on the *X* axis. The value 0.013 means the producer has only a 5 percent chance of rejecting a lot with a quality level less than 1.3 percent nonconforming ($\overline{p} = 0.013 \times 100$). Because the protection is defined for the producer (95% chance of accepting a lot), AQL plans are referred to as producer's risk plans.

The LTPD is the point on the OC curve where there is only a 10 percent chance of accepting a lot. For the sampling plan $n = 100$, $c = 3$, that *X* axis point is $\overline{p} = 0.0675$. Confirm that there is only a 10 percent chance of accepting a lot of 6.75 percent nonconforming material under this sampling plan. Because the protection is defined for the customer or consumer (90% chance of rejecting the lot), LQL plans are referred to as consumer's risk plans.

## *Type I and Type II Errors*

The AQL is said to protect the producer against rejecting good lots. This is known as a **Type 1 error.** The LQL protects the customer (or consumer) against receiving bad lots. This is known as a **Type 2 error.**

EXAMPLE 8.5

You are serving on a jury at a murder trial. The death penalty may be invoked if the judge wishes. The evidence points to a guilty verdict, but to

| | | The truth is | |
|---|---|---|---|
| | | Guilty | Not guilty |
| The jury decides | Not guilty | Incorrect Type II | Correct |
| | Guilty | Correct | Incorrect Type I |

**FIGURE 8.8**
Type I and Type II errors (also known as $\alpha$ and $\beta$ errors). The jury is still out.

your way of thinking it is somewhat shaky, and the defense did a poor job of defending the accused.

Others on the jury agree with you on the shaky evidence and the poor defense.

*Solution:* To set him free when he is guilty is like accepting a low-quality lot—in this case not a lot but a very dangerous person who could harm others. This decision is analogous to a Type II error.

You ask that a flip chart with paper be furnished to the jury and you draw upon your quality control experience to draw and explain Figure 8.8. You point out that there are two kinds of errors that can be made regarding the jury's decision. You add, "We can decide he is guilty when he is not, or we can decide he is not guilty when he really is."

Is there any wonder that "hung juries" result from having to make such a decision? Consider the decision "He is found guilty when he really is not" to be a Type I error. It is like rejecting a good lot!

As far as I know, the jury got hung on this one!

Some sampling plans have evolved from a third type of statistic that is not directly derived from the OC curve. The statistic is called the **average outgoing quality limit** or **AOQL.***

* For any acceptance sampling plan, the AOQL is the maximum average outgoing quality over all possible levels of incoming quality.

The AOQL sampling plan is diagrammed in Figure 8.9. Incoming product is submitted in lots to acceptance sampling inspection. Rejected lots are 100 percent inspected. All of the good parts and those that can be reworked are combined with all of the accepted lots. The mix of product used to calculate the AOQL includes

- no nonconforming items from the sorting of the rejected lots;
- the mix of conforming and nonconforming items from the accepted lots.

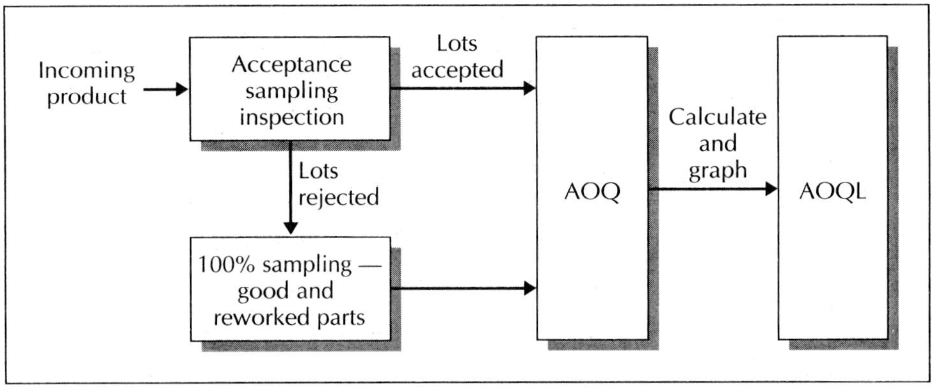


**FIGURE 8.9**
An AOQL system.

We will use the OC curve in Figure 8.7 to calculate the points to graph for the AOQL curve in Figure 8.10. The process is to calculate the average outgoing quality for lots with incoming quality $\overline{p}$ across the range of percent nonconforming ($\overline{p}$ = 0.01, 0.02, 0.03, etc.).

EXAMPLE 8.6

Calculate the average outgoing quality of lots containing 0.01 or 1 percent nonconforming product, using the AOQL system.

*Solution:*

*Step 1:* The percentage of lots accepted under the plan $n$ = 100, $c$ = 3 can be determined from the OC curve in Figure 8.7 or from the accompanying table. The value is $P(A)$ = 98 percent.

*Step 2:* Only 2 percent will fail. Since these are 100 percent inspected and sorted lots, the average outgoing quality level will be 0 (meaning there are no nonconforming units from these lots).

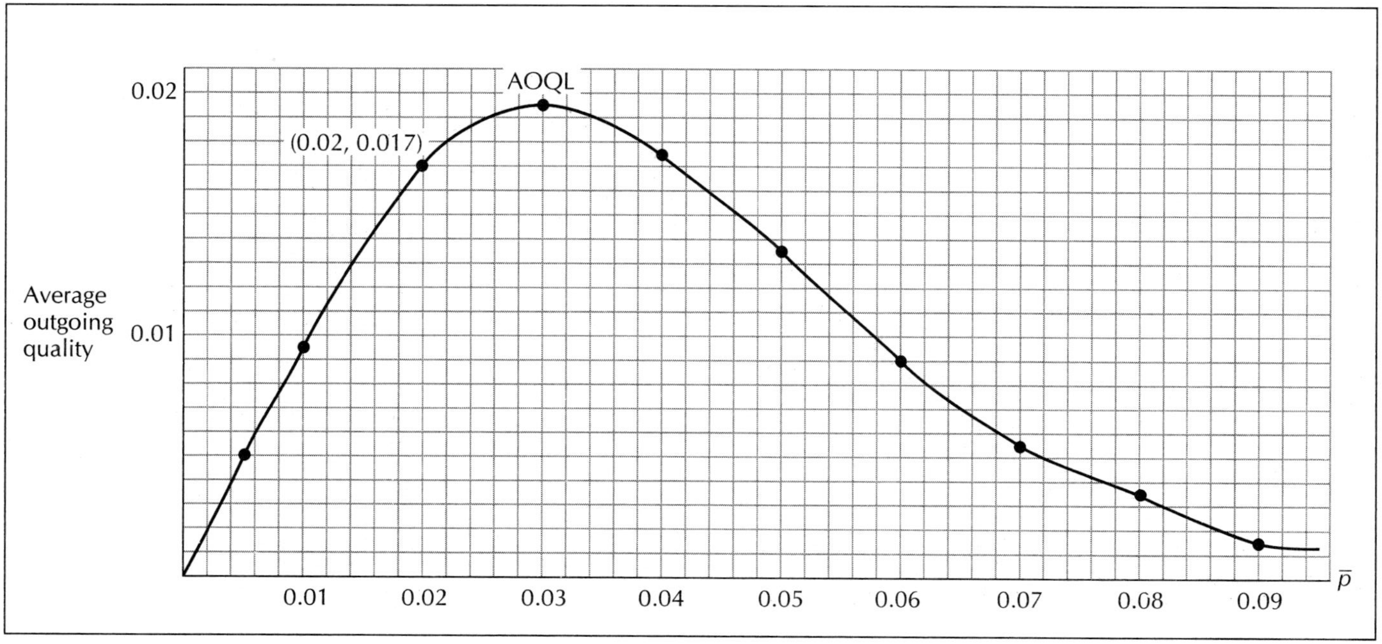


**Figure 8.10**
Finding the AOQL.

*Step 3:* The average outgoing quality of the accepted lots will be

$$\text{AOQL} = P(A) \times \overline{p} = 0.98 \times 0.01 = 0.0098$$

*Step 4:* The average outgoing quality of product from both accepted and rejected lots at the incoming quality level of 0.01 then is 0 + 0.0098, or simply the calculation in Step 3.

From the example, we can see that for every $\overline{p}$ value the AOQL will be equal to

$$\text{AOQL} = P(A) \times \overline{p}$$

You may confirm the calculations for the graph of the AOQL in Figure 8.10. For instance, at 0.02 the average outgoing quality level is

$$\text{AOQL} = P(A) \times \overline{p} = 0.86 \times 0.02 = 0.0172$$

The AOQL is discovered from the graph. The peak of the graph is the AOQL, and its value in this case is a little less than 2 percent.

To summarize, the average outgoing quality limit or AOQL (2% in the example) is the average outgoing quality limit of the product shipped under an acceptance sampling plan where rejected lots are 100 percent inspected and only conforming product is shipped.

## *Section 8.1 Review Problems*

1. SQC includes the use of which of the following?
   a. variable control charts
   b. variable sampling plans
   c. attribute sampling plans
   d. all of the above
2. In sampling inspection, the maximum percent nonconforming in a lot that will be shipped to the customer is known as the
   a. LQL
   b. AQL
   c. RQL
   d. UQL
3. The AOQL includes product from rejected and accepted lots. True or false?

# 8.2 *MIL–STD–105*

The following text is taken or adapted from Military Standard 105 (MIL–STD–105). Also refer to the sampling tables (Figures 8.11 to 8.15). Examples 8.7 and 8.8 only illustrate the use of attribute sampling plans. (Variable sampling plans are beyond the scope of this textbook.)

To use the sampling tables, the following procedures must be followed:

- The AQLs listed across the tops of the tables represent the average level of the percent nonconforming, or the nonconformities per 100 units of product.
- The sample size for the plan (listed in the first column of each table) will be accessed by using Figure 8.11a, the sample size code letters. Special inspection levels are used only in the case of small sample sizes, where large sampling risks may be tolerated. The general inspection levels are the most frequently used, and level II is used if the inspection level is unspecified. (Figure 8.11b depicts the OC curves and the probabilities of acceptance for the general inspection levels, with single sampling.)
- **Normal inspection** will be used at the beginning of inspection, unless management specifies otherwise.
- **Tightened inspection** will be implemented when two out of five consecutive lots have been rejected under normal inspection. Normal inspection will be resumed when five consecutive lots have been accepted under tightened inspection.
- **Reduced inspection** may be implemented when 10 lots have been accepted under normal inspection. If production becomes irregular or delayed, normal inspection should be reinstituted.
- The types of sampling plans are single, double, and multiple. Only the single and double plans are presented here.

EXAMPLE 8.7

For a lot size of 1,400, an AQL of 0.40 percent, and an inspection level of II, determine the single sampling plans for normal tightened and reduced inspection.

*Solution:* For all tables, use sample-size code letter *K* (Figure 8.11a). From Figure 8.12, *normal inspection:* $n = 125$, $Ac = 1$, $Re = 2$, which means to randomly draw a sample of 125 from the lot, accept the lot if only 1 or fewer nonconforming parts are found, and reject the lot if 2 or more nonconformities are found.

| | Special Inspection Levels | | | | General Inspection Levels | | |
|---|---|---|---|---|---|---|---|
| Lot or Batch Size | S-1 | S-2 | S-3 | S-4 | I | II | III |
| 2–8 | A | A | A | A | A | A | B |
| 9–15 | A | A | A | A | A | B | C |
| 16–25 | A | A | B | B | B | C | D |
| 26–50 | A | B | B | C | C | D | E |
| 51–90 | B | B | C | C | C | E | F |
| 91–150 | B | B | C | D | D | F | G |
| 151–280 | B | C | D | E | E | G | H |
| 281–500 | B | C | D | E | F | H | J |
| 501–1,200 | C | C | E | F | G | J | K |
| 1,201–3,200 | C | D | E | G | H | K | L |
| 3,201–10,000 | C | D | F | G | J | L | M |
| 10,001–35,000 | C | D | F | H | K | M | N |
| 35,001–150,000 | D | E | G | J | L | N | P |
| 150,001–500,000 | D | E | G | J | M | P | Q |
| 500,001and over | D | E | H | K | N | Q | R |

(a)

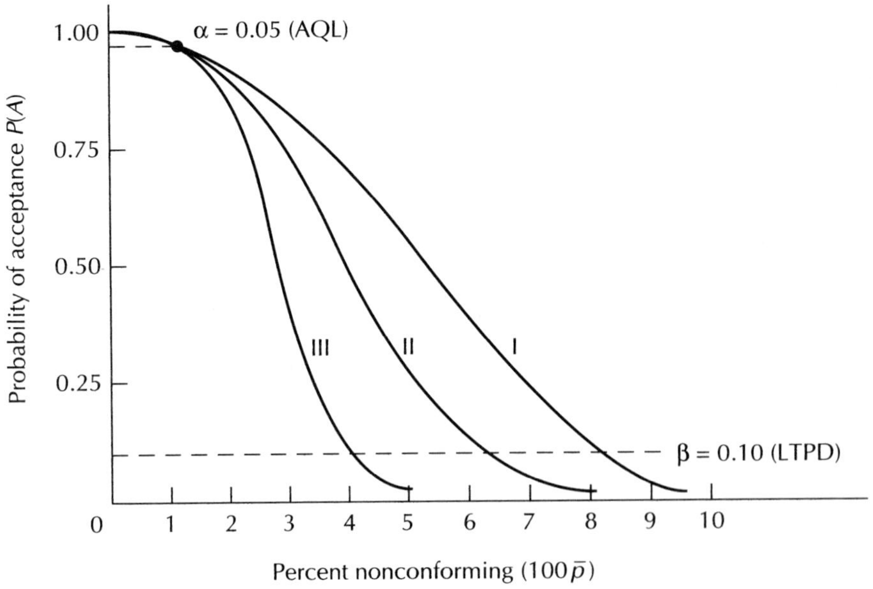


(b)

**FIGURE 8.11**
(a) Sample size code letters. (b) The OC curves for MIL–STD–105.

| Sample Size Code Letter | Sample Size | Acceptable Quality Levels (normal inspection) | | | | | | | | | | | | | | | | | | | | | | | | | |
|---|---|---|---|---|---|---|---|---|---|---|---|---|---|---|---|---|---|---|---|---|---|---|---|---|---|---|---|
| | | 0.010 | 0.015 | 0.025 | 0.040 | 0.065 | 0.10 | 0.15 | 0.25 | 0.40 | 0.65 | 1.0 | 1.5 | 2.5 | 4.0 | 6.5 | 10 | 15 | 25 | 40 | 65 | 100 | 150 | 250 | 400 | 650 | 1,000 |
| | | Ac Re | Ac Re | Ac Re | Ac Re | Ac Re | Ac Re | Ac Re | Ac Re | Ac Re | Ac Re | Ac Re | Ac Re | Ac Re | Ac Re | Ac Re | Ac Re | Ac Re | Ac Re | Ac Re | Ac Re | Ac Re | Ac Re | Ac Re | Ac Re | Ac Re | Ac Re |
| A | 2 | ↓ | ↓ | ↓ | ↓ | ↓ | ↓ | ↓ | ↓ | ↓ | ↓ | ↓ | ↓ | ↓ | ↓ | 0 1 | ↓ | ↓ | 1 2 | 2 3 | 3 4 | 5 6 | 7 8 | 10 11 | 14 15 | 21 22 | 30 31 |
| B | 3 | ↓ | ↓ | ↓ | ↓ | ↓ | ↓ | ↓ | ↓ | ↓ | ↓ | ↓ | ↓ | ↓ | 0 1 | ↑ | ↓ | 1 2 | 2 3 | 3 4 | 5 6 | 7 8 | 10 11 | 14 15 | 21 22 | 30 31 | 44 45 |
| C | 5 | ↓ | ↓ | ↓ | ↓ | ↓ | ↓ | ↓ | ↓ | ↓ | ↓ | ↓ | ↓ | 0 1 | ↑ | ↓ | 1 2 | 2 3 | 3 4 | 5 6 | 7 8 | 10 11 | 14 15 | 21 22 | 30 31 | 44 45 | ↑ |
| D | 8 | ↓ | ↓ | ↓ | ↓ | ↓ | ↓ | ↓ | ↓ | ↓ | ↓ | ↓ | 0 1 | ↑ | ↓ | 1 2 | 2 3 | 3 4 | 5 6 | 7 8 | 10 11 | 14 15 | 21 22 | 30 31 | 44 45 | ↑ | ↑ |
| E | 13 | ↓ | ↓ | ↓ | ↓ | ↓ | ↓ | ↓ | ↓ | ↓ | ↓ | 0 1 | ↑ | ↓ | 1 2 | 2 3 | 3 4 | 5 6 | 7 8 | 10 11 | 14 15 | 21 22 | 30 31 | 44 45 | ↑ | ↑ | ↑ |
| F | 20 | ↓ | ↓ | ↓ | ↓ | ↓ | ↓ | ↓ | ↓ | ↓ | 0 1 | ↑ | ↓ | 1 2 | 2 3 | 3 4 | 5 6 | 7 8 | 10 11 | 14 15 | 21 22 | ↑ | ↑ | ↑ | ↑ | ↑ | ↑ |
| G | 32 | ↓ | ↓ | ↓ | ↓ | ↓ | ↓ | ↓ | ↓ | 0 1 | ↑ | ↓ | 1 2 | 2 3 | 3 4 | 5 6 | 7 8 | 10 11 | 14 15 | 21 22 | ↑ | ↑ | ↑ | ↑ | ↑ | ↑ | ↑ |
| H | 50 | ↓ | ↓ | ↓ | ↓ | ↓ | ↓ | ↓ | 0 1 | ↑ | ↓ | 1 2 | 2 3 | 3 4 | 5 6 | 7 8 | 10 11 | 14 15 | 21 22 | ↑ | ↑ | ↑ | ↑ | ↑ | ↑ | ↑ | ↑ |
| J | 80 | ↓ | ↓ | ↓ | ↓ | ↓ | ↓ | 0 1 | ↑ | ↓ | 1 2 | 2 3 | 3 4 | 5 6 | 7 8 | 10 11 | 14 15 | 21 22 | ↑ | ↑ | ↑ | ↑ | ↑ | ↑ | ↑ | ↑ | ↑ |
| K | 125 | ↓ | ↓ | ↓ | ↓ | ↓ | 0 1 | ↑ | ↓ | 1 2 | 2 3 | 3 4 | 5 6 | 7 8 | 10 11 | 14 15 | 21 22 | ↑ | ↑ | ↑ | ↑ | ↑ | ↑ | ↑ | ↑ | ↑ | ↑ |
| L | 200 | ↓ | ↓ | ↓ | ↓ | 0 1 | ↑ | ↓ | 1 2 | 2 3 | 3 4 | 5 6 | 7 8 | 10 11 | 14 15 | 21 22 | ↑ | ↑ | ↑ | ↑ | ↑ | ↑ | ↑ | ↑ | ↑ | ↑ | ↑ |
| M | 315 | ↓ | ↓ | ↓ | 0 1 | ↑ | ↓ | 1 2 | 2 3 | 3 4 | 5 6 | 7 8 | 10 11 | 14 15 | 21 22 | ↑ | ↑ | ↑ | ↑ | ↑ | ↑ | ↑ | ↑ | ↑ | ↑ | ↑ | ↑ |
| N | 500 | ↓ | ↓ | 0 1 | ↑ | ↓ | 1 2 | 2 3 | 3 4 | 5 6 | 7 8 | 10 11 | 14 15 | 21 22 | ↑ | ↑ | ↑ | ↑ | ↑ | ↑ | ↑ | ↑ | ↑ | ↑ | ↑ | ↑ | ↑ |
| P | 800 | ↓ | 0 1 | ↑ | ↓ | 1 2 | 2 3 | 3 4 | 5 6 | 7 8 | 10 11 | 14 15 | 21 22 | ↑ | ↑ | ↑ | ↑ | ↑ | ↑ | ↑ | ↑ | ↑ | ↑ | ↑ | ↑ | ↑ | ↑ |
| Q | 1,250 | 0 1 | ↑ | ↓ | 1 2 | 2 3 | 3 4 | 5 6 | 7 8 | 10 11 | 14 15 | 21 22 | ↑ | ↑ | ↑ | ↑ | ↑ | ↑ | ↑ | ↑ | ↑ | ↑ | ↑ | ↑ | ↑ | ↑ | ↑ |
| R | 2,000 | ↑ | ↑ | 1 2 | 2 3 | 3 4 | 5 6 | 7 8 | 10 11 | 14 15 | 21 22 | ↑ | ↑ | ↑ | ↑ | ↑ | ↑ | ↑ | ↑ | ↑ | ↑ | ↑ | ↑ | ↑ | ↑ | ↑ | ↑ |

↓ = Use first sampling plan below arrow. If sample size equals, or exceeds, lot or batch size, do 100 percent inspection.

↑ = Use first sampling plan above arrow.

*Ac* = Acceptance number.

*Re* = Rejection number.

**FIGURE 8.12**
Single sampling plans for normal inspection (MIL–STD–105).

| Sample Size Code Letter | Sample Size | Acceptable Quality Levels (tightened inspection) | | | | | | | | | | | | | | | | | | | | | | | | | |
|---|---|---|---|---|---|---|---|---|---|---|---|---|---|---|---|---|---|---|---|---|---|---|---|---|---|---|---|
| | | 0.010 | 0.015 | 0.025 | 0.040 | 0.065 | 0.10 | 0.15 | 0.25 | 0.40 | 0.65 | 1.0 | 1.5 | 2.5 | 4.0 | 6.5 | 10 | 15 | 25 | 40 | 65 | 100 | 150 | 250 | 400 | 650 | 1,000 |
| | | Ac Re | Ac Re | Ac Re | Ac Re | Ac Re | Ac Re | Ac Re | Ac Re | Ac Re | Ac Re | Ac Re | Ac Re | Ac Re | Ac Re | Ac Re | Ac Re | Ac Re | Ac Re | Ac Re | Ac Re | Ac Re | Ac Re | Ac Re | Ac Re | Ac Re | Ac Re |
| A | 2 | | | | | | | | | | | | | | | ↓ | | | ↓ | 1 2 | 2 3 | 3 4 | 5 6 | 8 9 | 12 13 | 18 19 | 27 28 |
| B | 3 | | | | | | | | | | | | | | ↓ | 0 1 | | ↓ | 1 2 | 2 3 | 3 4 | 5 6 | 8 9 | 12 13 | 18 19 | 27 28 | 41 42 |
| C | 5 | | | | | | | | | | | | | ↓ | 0 1 | | ↓ | 1 2 | 2 3 | 3 4 | 5 6 | 8 9 | 12 13 | 18 19 | 27 28 | 41 42 | ↑ |
| D | 8 | | | | | | | | | | | | ↓ | 0 1 | | ↓ | 1 2 | 2 3 | 3 4 | 5 6 | 8 9 | 12 13 | 18 19 | 27 28 | 41 42 | ↑ | |
| E | 13 | | | | | | | | | | | ↓ | 0 1 | | ↓ | 1 2 | 2 3 | 3 4 | 5 6 | 8 9 | 12 13 | 18 19 | 27 28 | 41 42 | ↑ | | |
| F | 20 | | | | | | | | | | ↓ | 0 1 | | ↓ | 1 2 | 2 3 | 3 4 | 5 6 | 8 9 | 12 13 | 18 19 | ↑ | ↑ | ↑ | | | |
| G | 32 | | | | | | | | | ↓ | 0 1 | | ↓ | 1 2 | 2 3 | 3 4 | 5 6 | 8 9 | 12 13 | 18 19 | ↑ | | | | | | |
| H | 50 | | | | | | | | ↓ | 0 1 | | ↓ | 1 2 | 2 3 | 3 4 | 5 6 | 8 9 | 12 13 | 18 19 | ↑ | | | | | | | |
| J | 80 | | | | | | | ↓ | 0 1 | | ↓ | 1 2 | 2 3 | 3 4 | 5 6 | 8 9 | 12 13 | 18 19 | ↑ | | | | | | | | |
| K | 125 | | | | | | ↓ | 0 1 | | ↓ | 1 2 | 2 3 | 3 4 | 5 6 | 8 9 | 12 13 | 18 19 | ↑ | | | | | | | | | |
| L | 200 | | | | | ↓ | 0 1 | | ↓ | 1 2 | 2 3 | 3 4 | 5 6 | 8 9 | 12 13 | 18 19 | ↑ | | | | | | | | | | |
| M | 315 | | | | ↓ | 0 1 | | ↓ | 1 2 | 2 3 | 3 4 | 5 6 | 8 9 | 12 13 | 18 19 | ↑ | | | | | | | | | | | |
| N | 500 | | | ↓ | 0 1 | | ↓ | 1 2 | 2 3 | 3 4 | 5 6 | 8 9 | 12 13 | 18 19 | ↑ | | | | | | | | | | | | |
| P | 800 | | ↓ | 0 1 | | ↓ | 1 2 | 2 3 | 3 4 | 5 6 | 8 9 | 12 13 | 18 19 | ↑ | | | | | | | | | | | | | |
| Q | 1,250 | ↓ | 0 1 | | ↓ | 1 2 | 2 3 | 3 4 | 5 6 | 8 9 | 12 13 | 18 19 | ↑ | | | | | | | | | | | | | | |
| R | 2,000 | 0 1 | ↑ | ↓ | 1 2 | 2 3 | 3 4 | 5 6 | 8 9 | 12 13 | 18 19 | ↑ | | | | | | | | | | | | | | | |
| S | 3,150 | | | 1 2 | | | | | | | | | | | | | | | | | | | | | | | |

↓ = Use first sampling plan below arrow. If sample size equals or exceeds lot or batch size, do 100 percent inspection.

↑ = Use first sampling plan above arrow.

*Ac* = Acceptance number.

*Re* = Rejection number.

**FIGURE 8.13**
Single sampling plans for tightened inspection (MIL–STD–105).

| Sample Size Code Letter | Sample Size | Acceptable Quality Levels (reduced inspection)† | | | | | | | | | | | | | | | | | | | | | | | | | |
|---|---|---|---|---|---|---|---|---|---|---|---|---|---|---|---|---|---|---|---|---|---|---|---|---|---|---|---|
| | | 0.010 | 0.015 | 0.025 | 0.040 | 0.065 | 0.10 | 0.15 | 0.25 | 0.40 | 0.65 | 1.0 | 1.5 | 2.5 | 4.0 | 6.5 | 10 | 15 | 25 | 40 | 65 | 100 | 150 | 250 | 400 | 650 | 1,000 |
| | | *Ac Re* | *Ac Re* | *Ac Re* | *Ac Re* | *Ac Re* | *Ac Re* | *Ac Re* | *Ac Re* | *Ac Re* | *Ac Re* | *Ac Re* | *Ac Re* | *Ac Re* | *Ac Re* | *Ac Re* | *Ac Re* | *Ac Re* | *Ac Re* | *Ac Re* | *Ac Re* | *Ac Re* | *Ac Re* | *Ac Re* | *Ac Re* | *Ac Re* | *Ac Re* |
| A | 2 | | | | | | | | | | | | | | ↓ | 0 1 | | ↓ | 1 2 | 2 3 | 3 4 | 5 6 | 7 8 | 10 11 | 14 15 | 21 22 | 30 31 |
| B | 2 | | | | | | | | | | | | | ↓ | 0 1 | ↑ | ↓ | 0 2 | 1 3 | 2 4 | 3 5 | 5 6 | 7 8 | 10 11 | 14 15 | 21 22 | 30 31 |
| C | 2 | | | | | | | | | | | | ↓ | 0 1 | ↑ | ↓ | 0 2 | 1 3 | 1 4 | 2 5 | 3 6 | 5 8 | 7 10 | 10 13 | 14 17 | 21 24 | ↑ |
| D | 3 | | | | | | | | | | | ↓ | 0 1 | ↑ | ↓ | 0 2 | 1 3 | 1 4 | 2 5 | 3 6 | 5 8 | 7 10 | 10 13 | 14 17 | 21 24 | | |
| E | 5 | | | | | | | | | | ↓ | 0 1 | ↑ | ↓ | 0 2 | 1 3 | 1 4 | 2 5 | 3 6 | 5 8 | 7 10 | 10 13 | 14 17 | 21 24 | | ↑ | |
| F | 8 | | | | | | | | | ↓ | 0 1 | ↑ | ↓ | 0 2 | 1 3 | 1 4 | 2 5 | 3 6 | 5 8 | 7 10 | 10 13 | ↑ | ↑ | ↑ | ↑ | | |
| G | 13 | | | | | | | | ↓ | 0 1 | ↑ | ↓ | 0 2 | 1 3 | 1 4 | 2 5 | 3 6 | 5 8 | 7 10 | 10 13 | ↑ | | | | | | |
| H | 20 | | | | | | | ↓ | 0 1 | ↑ | ↓ | 0 2 | 1 3 | 1 4 | 2 5 | 3 6 | 5 8 | 7 10 | 10 13 | ↑ | | | | | | | |
| J | 32 | | | | | | ↓ | 0 1 | ↑ | ↓ | 0 2 | 1 3 | 1 4 | 2 5 | 3 6 | 5 8 | 7 10 | 10 13 | ↑ | | | | | | | | |
| K | 50 | | | | | ↓ | 0 1 | ↑ | ↓ | 0 2 | 1 3 | 1 4 | 2 5 | 3 6 | 5 8 | 7 10 | 10 13 | ↑ | | | | | | | | | |
| L | 80 | | | | ↓ | 0 1 | ↑ | ↓ | 0 2 | 1 3 | 1 4 | 2 5 | 3 6 | 5 8 | 7 10 | 10 13 | ↑ | | | | | | | | | | |
| M | 125 | | | ↓ | 0 1 | ↑ | ↓ | 0 2 | 1 3 | 1 4 | 2 5 | 3 6 | 5 8 | 7 10 | 10 13 | ↑ | | | | | | | | | | | |
| N | 200 | | ↓ | 0 1 | ↑ | ↓ | 0 2 | 1 3 | 1 4 | 2 5 | 3 6 | 5 8 | 7 10 | 10 13 | ↑ | | | | | | | | | | | | |
| P | 315 | ↓ | 0 1 | ↑ | ↓ | 0 2 | 1 3 | 1 4 | 2 5 | 3 6 | 5 8 | 7 10 | 10 13 | ↑ | | | | | | | | | | | | | |
| Q | 500 | 0 1 | ↑ | ↓ | 0 2 | 1 3 | 1 4 | 2 5 | 3 6 | 5 8 | 7 10 | 10 13 | ↑ | | | | | | | | | | | | | | |
| R | 800 | ↑ | | 0 2 | 1 3 | 1 4 | 2 5 | 3 6 | 5 8 | 7 10 | 10 13 | ↑ | | | | | | | | | | | | | | | |

↓ = Use first sampling plan below arrow. If sample size equals or exceeds lot or batch size, do 100 percent inspection.

↑ = Use first sampling plan above arrow.

*Ac* = Acceptance number.

*Re* = Rejection number.

† If the acceptance number has been exceeded, but the rejection number has not been reached, accept the lot, but reinstate normal inspection.

**FIGURE 8.14**
Single sampling plans for reduced inspection (MIL–STD–105).

| Sample Size Code Letter | Sample | Sample Size | Cumulative Sample Size | Acceptable Quality Levels (double sampling, normal inspection) | | | | | | | | | | | | | | | | | | | | | | | | | |
|---|---|---|---|---|---|---|---|---|---|---|---|---|---|---|---|---|---|---|---|---|---|---|---|---|---|---|---|---|---|
| | | | | 0.010 | 0.015 | 0.025 | 0.040 | 0.065 | 0.10 | 0.15 | 0.25 | 0.40 | 0.65 | 1.0 | 1.5 | 2.5 | 4.0 | 6.5 | 10 | 15 | 25 | 40 | 65 | 100 | 150 | 250 | 400 | 650 | 1,000 |
| | | | | Ac Re | Ac Re | Ac Re | Ac Re | Ac Re | Ac Re | Ac Re | Ac Re | Ac Re | Ac Re | Ac Re | Ac Re | Ac Re | Ac Re | Ac Re | Ac Re | Ac Re | Ac Re | Ac Re | Ac Re | Ac Re | Ac Re | Ac Re | Ac Re | Ac Re | Ac Re |
| A | | | | ↓ | ↓ | ↓ | ↓ | ↓ | ↓ | ↓ | ↓ | ↓ | ↓ | ↓ | ↓ | ↓ | ↓ | * | ↓ | ↓ | * | * | * | * | * | * | * | * | * |
| B | First | 2 | 2 | ↓ | ↓ | ↓ | ↓ | ↓ | ↓ | ↓ | ↓ | ↓ | ↓ | ↓ | ↓ | ↓ | * | ↑ | ↓ | 0 2 | 0 3 | 1 4 | 2 5 | 3 7 | 5 9 | 7 11 | 11 16 | 17 22 | 25 31 |
| | Second | 2 | 4 | | | | | | | | | | | | | | | | | 1 2 | 3 4 | 4 5 | 6 7 | 8 9 | 12 13 | 18 19 | 26 27 | 37 38 | 56 57 |
| C | First | 3 | 3 | ↓ | ↓ | ↓ | ↓ | ↓ | ↓ | ↓ | ↓ | ↓ | ↓ | ↓ | ↓ | * | ↑ | ↓ | 0 2 | 0 3 | 1 4 | 2 5 | 3 7 | 5 9 | 7 11 | 11 16 | 17 22 | 25 31 | ↑ |
| | Second | 3 | 6 | | | | | | | | | | | | | | | | 1 2 | 3 4 | 4 5 | 6 7 | 8 9 | 12 13 | 18 19 | 26 27 | 37 38 | 56 57 | |
| D | First | 5 | 5 | ↓ | ↓ | ↓ | ↓ | ↓ | ↓ | ↓ | ↓ | ↓ | ↓ | ↓ | * | ↑ | ↓ | 0 2 | 0 3 | 1 4 | 2 5 | 3 7 | 5 9 | 7 11 | 11 16 | 17 22 | 25 31 | ↑ | ↑ |
| | Second | 5 | 10 | | | | | | | | | | | | | | | 1 2 | 3 4 | 4 5 | 6 7 | 8 9 | 12 13 | 18 19 | 26 27 | 37 38 | 56 57 | | |
| E | First | 8 | 8 | ↓ | ↓ | ↓ | ↓ | ↓ | ↓ | ↓ | ↓ | ↓ | ↓ | * | ↑ | ↓ | 0 2 | 0 3 | 1 4 | 2 5 | 3 7 | 5 9 | 7 11 | 11 16 | 17 22 | 25 31 | ↑ | ↑ | ↑ |
| | Second | 8 | 16 | | | | | | | | | | | | | | 1 2 | 3 4 | 4 5 | 6 7 | 8 9 | 12 13 | 18 19 | 26 27 | 37 38 | 56 57 | | | |
| F | First | 13 | 13 | ↓ | ↓ | ↓ | ↓ | ↓ | ↓ | ↓ | ↓ | ↓ | * | ↑ | ↓ | 0 2 | 0 3 | 1 4 | 2 5 | 3 7 | 5 9 | 7 11 | 11 16 | ↑ | ↑ | ↑ | ↑ | ↑ | ↑ |
| | Second | 13 | 26 | | | | | | | | | | | | | 1 2 | 3 4 | 5 4 | 6 7 | 8 9 | 12 13 | 18 19 | 26 27 | | | | | | |
| G | First | 20 | 20 | ↓ | ↓ | ↓ | ↓ | ↓ | ↓ | ↓ | ↓ | * | ↑ | ↓ | 0 2 | 0 3 | 1 4 | 2 5 | 3 7 | 5 9 | 7 11 | 11 16 | ↑ | ↑ | ↑ | ↑ | ↑ | ↑ | ↑ |
| | Second | 20 | 40 | | | | | | | | | | | | 1 2 | 3 4 | 4 5 | 6 7 | 8 9 | 12 13 | 18 19 | 26 27 | | | | | | | |
| H | First | 32 | 32 | ↓ | ↓ | ↓ | ↓ | ↓ | ↓ | ↓ | * | ↑ | ↓ | 0 2 | 0 3 | 1 4 | 2 5 | 3 7 | 5 9 | 7 11 | 11 16 | ↑ | ↑ | ↑ | ↑ | ↑ | ↑ | ↑ | ↑ |
| | Second | 32 | 64 | | | | | | | | | | | 1 2 | 3 4 | 4 5 | 6 7 | 8 9 | 12 13 | 18 19 | 26 27 | | | | | | | | |
| J | First | 50 | 50 | ↓ | ↓ | ↓ | ↓ | ↓ | ↓ | * | ↑ | ↓ | 0 2 | 0 3 | 1 4 | 2 5 | 3 7 | 5 9 | 7 11 | 11 16 | ↑ | ↑ | ↑ | ↑ | ↑ | ↑ | ↑ | ↑ | ↑ |
| | Second | 50 | 100 | | | | | | | | | | 1 2 | 3 4 | 4 5 | 6 7 | 8 9 | 12 13 | 18 19 | 26 27 | | | | | | | | | |
| K | First | 80 | 80 | ↓ | ↓ | ↓ | ↓ | ↓ | * | ↑ | ↓ | 0 2 | 0 3 | 1 4 | 2 5 | 3 7 | 5 9 | 7 11 | 11 16 | ↑ | ↑ | ↑ | ↑ | ↑ | ↑ | ↑ | ↑ | ↑ | ↑ |
| | Second | 80 | 160 | | | | | | | | | 1 2 | 3 4 | 4 5 | 6 7 | 8 9 | 12 13 | 18 19 | 26 27 | | | | | | | | | | |
| L | First | 125 | 125 | ↓ | ↓ | ↓ | ↓ | * | ↑ | ↓ | 0 2 | 0 3 | 1 4 | 2 5 | 3 7 | 5 9 | 7 11 | 11 16 | ↑ | ↑ | ↑ | ↑ | ↑ | ↑ | ↑ | ↑ | ↑ | ↑ | ↑ |
| | Second | 125 | 250 | | | | | | | | 1 2 | 3 4 | 4 5 | 6 7 | 8 9 | 12 13 | 18 19 | 26 27 | | | | | | | | | | | |
| M | First | 200 | 200 | ↓ | ↓ | ↓ | * | ↑ | ↓ | 0 2 | 0 3 | 1 4 | 2 5 | 3 7 | 5 9 | 7 11 | 11 16 | ↑ | ↑ | ↑ | ↑ | ↑ | ↑ | ↑ | ↑ | ↑ | ↑ | ↑ | ↑ |
| | Second | 200 | 400 | | | | | | | 1 2 | 3 4 | 4 5 | 6 7 | 8 9 | 12 13 | 18 19 | 26 27 | | | | | | | | | | | | |
| N | First | 315 | 315 | ↓ | ↓ | * | ↑ | ↓ | 0 2 | 0 3 | 1 4 | 2 5 | 3 7 | 5 9 | 7 11 | 11 16 | ↑ | ↑ | ↑ | ↑ | ↑ | ↑ | ↑ | ↑ | ↑ | ↑ | ↑ | ↑ | ↑ |
| | Second | 315 | 630 | | | | | | 1 2 | 3 4 | 4 5 | 6 7 | 8 9 | 12 13 | 18 19 | 26 27 | | | | | | | | | | | | | |
| P | First | 500 | 500 | ↓ | * | ↑ | ↓ | 0 2 | 0 3 | 1 4 | 2 5 | 3 7 | 5 9 | 7 11 | 11 16 | ↑ | ↑ | ↑ | ↑ | ↑ | ↑ | ↑ | ↑ | ↑ | ↑ | ↑ | ↑ | ↑ | ↑ |
| | Second | 500 | 1000 | | | | | 1 2 | 3 4 | 4 5 | 6 7 | 8 9 | 12 13 | 18 19 | 26 27 | | | | | | | | | | | | | | |
| Q | First | 800 | 800 | * | ↑ | ↓ | 0 2 | 0 3 | 1 4 | 2 5 | 3 7 | 5 9 | 7 11 | 11 16 | ↑ | ↑ | ↑ | ↑ | ↑ | ↑ | ↑ | ↑ | ↑ | ↑ | ↑ | ↑ | ↑ | ↑ | ↑ |
| | Second | 800 | 1,600 | | | | 1 2 | 3 4 | 4 5 | 6 7 | 8 9 | 12 13 | 18 19 | 26 27 | | | | | | | | | | | | | | | |
| R | First | 1,250 | 1,250 | ↑ | ↑ | 0 2 | 0 3 | 1 4 | 2 5 | 3 7 | 5 9 | 7 11 | 11 16 | ↑ | ↑ | ↑ | ↑ | ↑ | ↑ | ↑ | ↑ | ↑ | ↑ | ↑ | ↑ | ↑ | ↑ | ↑ | ↑ |
| | Second | 1,250 | 2,500 | | | 1 2 | 3 4 | 4 5 | 6 7 | 8 9 | 12 13 | 18 19 | 26 27 | | | | | | | | | | | | | | | | |

↓ = Use first sampling plan below arrow. If sample size equals or exceeds lot or batch size, do 100 percent inspection.

↑ = Use first sampling plan above arrow.

*Ac* = Acceptance number.

*Re* = Rejection number.

* = Use corresponding single sample plan (or alternatively, use double sampling plan below, where available).

**FIGURE 8.15**
Double sampling plans for normal inspection (MIL–STD–105).

For *tightened inspection,* use Figure 8.13: $n = 200$ (vertical arrow down changes lot size as well as *Ac* and *Re* values), $Ac = 1$, $Re = 2$. In this case, you are forced to examine more items with the same *Ac* and *Re* values.

For *reduced inspection,* use Figure 8.14: $n = 50$, $Ac = 0$, $Re = 2$. If 1 nonconforming unit is found, the lot is accepted, but the type of inspection changes from *reduced* to *normal.*

EXAMPLE 8.8

For a lot size *N* of 300,000, an AQL of 0.65 percent, and an inspection level of II, determine the double sampling plan for normal inspection.

*Solution:* Use sample-size code letter *P* (Figure 8.11a). From Figure 8.15: $n = 500$ for both the first and second samples, $Ac = 5$, $Re = 9$ for the first sample. If the number of nonconforming items is 5 or less the lot is accepted. If the number nonconforming is 9 or more the lot is rejected. If the number of nonconforming units is 6, 7, or 8, the second sample of 500 is drawn from the lot, and the *Ac* and *Re* values are 12 and 13, respectively.

Before leaving the subject of acceptance sampling, we need to consider what to do with rejected lots. This is such an important consideration that the new European standard, **ISO 9000** (Chapter 10), discusses it at length. The parts from rejected lots not suitably identified often find their way into the work flow. Therefore, rejected lots must be recorded and placed in areas designated as scrap or rework. Rework should be accomplished by production.

## *Section 8.2 Review Problems*

1. Using MIL–STD–105, determine the sample size for a lot size of 100, AQL = 0.1 percent, level II, normal inspection, single sampling.
   a. 30
   b. 50
   c. 125
   d. 200
2. A sampling plan for reduced inspection accepts 3 and rejects 7. If 2 nonconformities are found and production falls off, you should
   a. accept the lot and return to normal inspection
   b. draw a second sample
   c. reinspect the sample
   d. move to the next sample lot

3. Using MIL–STD–105, determine the sample size for a lot size of 400, AQL = 0.4 percent, level II, normal inspection, double sampling.
   a. 32
   b. 50
   c. 125
   d. 200

## 8.3 *Nondestructive Testing (NDT)*

If acceptance sampling proves costly or otherwise ineffective, a company might consider a form of **nondestructive testing (NDT).** With NDT, the product is not damaged or destroyed, and the cost of inspection of more samples may not be prohibitive. SPC may become more practical when a product can be conformance tested and not destroyed. Acceptance sampling costs can be lowered by using nondestructive testing in incoming or final inspection. When we think of NDT we usually consider the use of specialized equipment and technology; obviously any type of visual examination and many types of measurement are nondestructive tests, but these are usually not considered under NDT.

Some of the most popular types of NDT technology include the following, in the order of their most frequent use: liquid penetrant dyes, magnetic particles, radiography (e.g., X rays), other electromagnetic waves, ultrasonic waves, and eddy currents. The technologist and manager benefit from knowing the principles of operation and applications, as well as the advantages and limitations of each.

The user of **liquid penetrant dyes** uses special lighting, usually ultraviolet (UV) light with the dyes to detect nonconformities on the surface of parts—for example, cracks, holes, porosity. The advantages of such testing are that all nonabsorbent materials may be tested, the technology is simple and easily understood, and the equipment is low in cost and portable (does not have to be used only in a laboratory). The main limitation is that only surface nonconformities are detected by liquid penetrant inspection. Also, parts must be cleaned before and after inspection.

Blowholes, cavities, forging cracks, laps, grinding cracks, and stress cracks can be detected by **magnetic particle testers.** The technology involves the use of small magnetic particles placed on ferromagnetic materials (iron, steel, nickel, cobalt). The tested materials must be placed in a magnetic field (magnetism must be established in at least two directions to see all discontinuities). The particles align so as to detect flaws up to ¼ inch below the surface of the part. Advantages of the use of magnetic particles are the use of portable equipment and the simple ability to detect subsurface nonconformities. Magnetic particle testing is limited to ferromagnetic materials.

The use of small (short-wavelength) and vigorous electromagnetic waves to test for nonconformities or correct thickness is classified as **radiography.** The most frequently used waves, as well as the most familiar to all of us, are **X rays.** Gamma and beta waves are also used. Limitations involve the use of specialized equipment and the potential for hazardous operation. Specially trained technicians are necessary, and the results are not immediate. The principal advantages of radiographic testing include the testing of a great number of materials and the minute analysis of subsurface nonconformities.

Electromagnetic waves in the light spectrum are provided by lasers. High-powered lasers, combined with holographic techniques, are known as **optical NDT** systems (Figure 8.16). Optical NDT can measure thicknesses and find weaknesses in installed materials, such as oil pipelines, that are only partially accessible. By focusing the laser beam on the surface of the material and causing it to vibrate, another laser can measure the small vibrations. The object under test does not have to be placed in water or be in contact with transducers or detectors.

We are all familiar with the use of ultrasonic waves in medicine (sonograms). Sound waves are less vigorous (less penetrating) than electromagnetic waves, but they are safer. **Ultrasonic testing** begins at frequencies above the threshold of hearing, above 20 kilohertz. The waves are reflected or scattered by nonconformities such as cracks, cast inclusions, poor brazing or other bonds, and improper laminations. The reflected or transmitted waves can also be used to determine thickness. Parts often must be submerged in water. It is difficult to use this technology with irregularly shaped parts, and highly skilled operators are needed.

**Eddy current testing** involves placing a coil in contact with a conductive surface. The coil impedance (impedance is similar to DC resistance properties) will change as the material properties change. Changes in impedance expose discontinuities and nonhomogeneous materials. Impedance may also be used to check surface finish and part dimensions. Eddy current testers are used on automated production lines because of their ability to measure material properties rapidly. The inability to measure nonconductive materials limits the use of

**FIGURE 8.16**
Optical NDT uses light to study materials that cannot be touched.
(*Source:* The Johns Hopkins University School of Engineering.)

eddy current testers mainly to metals, and even these materials can be measured only in a qualitative sense. Absolute measurements cannot be accomplished with eddy current testers.

There are many other forms of NDT available, such as capacitive and fiber optical testing. Entire books are available on the subject, and new equipment is being developed at a rapid rate. The technologist and manager may save their company a great deal of money by keeping up with the latest developments.

## *Section 8.3 Review Problems*

1. Of those listed below, which is considered nondestructive testing (NDT)?
   a. radiographic
   b. ultrasonic

c. eddy current
d. all are forms of NDT

2. Liquid penetrant dyes will detect
   a. subsurface weld flaws
   b. subsurface cracks
   c. both a and b are correct
   d. only surface nonconformities

3. X rays, gamma rays, and beta waves are used in

   ______________________________ (radiographic, magnetic) testing.

# Summary

The **system for sampling** consists of the following steps:

- The product to be inspected is identified.
- The quality characteristics to be inspected are identified.
- The lot size is determined.
- The inspection level is determined.
- The acceptable quality level (AQL), limiting quality level (LQL, RQL, or LTPD), and/or the average outgoing quality level (AOQL) are determined.
- A sampling plan is selected.
- How the lot is to be sampled is decided.
- Samples are drawn and inspected for the quality characteristics of interest.
- The inspector needs three things:
  a. a sample size *n*;
  b. an acceptance number *c*;
  c. a list and thorough description of the quality characteristics to be checked.
- Each lot is accepted or rejected according to an established quality level (AQL, LQL, or AOQL).
- Rejected lots must be recorded and placed in areas designated as scrap or rework. Rework should be accomplished by production.
- If the use of the AOQL is desired, the rejected lots are 100 percent inspected and sorted, and good products must be returned to shipping.

- Nondestructive testing (NDT) can cut costs of inspection—in acceptance sampling or SPC.
- Technical managers should be aware of the advantages and limitations of the most frequently used types of NDT: liquid penetrant dyes, magnetic particles, radiography (e.g., X rays), other electromagnetic waves, ultrasonic waves, and eddy currents.

## CASE STUDY · CHAPTER 8

You have just been promoted to quality technologist at a small plant that supplies a rather complicated small plastic part to the automotive industry. The product is a combination of three parts, all molded plastic, which must be free of nonconformances such as scratches, mold marks, and other visual imperfections. The parts are produced by the production department with approximately 20 employees.

You know that SPC was tried once in the production department, but soon dispensed with as, according to the plant manager, "too hard to implement with the production workers." Of course, no training classes were offered during the trial period, and inspectors were expected to learn SPC on their own time with books provided by management. The bottom line is that SPC is "out of the question."

The only inspection occurs in a separately managed department called final inspection. Fifteen employees perform 100 percent inspection. The nonconformances result in attribute-type data—scratches, mold marks, and so on. The parts found to be "out of spec" are reworked by the inspectors or scrapped if they are "considered" not repairable.

The plant manager asks that you look into quality problems at the plant. Specifically, you are to see why some of the inspectors are scrapping good product and others are passing poor product, when all they have to do is to "look at the damn things" (again a quote from the manager). The manager is also concerned with the throughput of the department. It can be a real bottleneck.

With this limited information, what steps might you take to seek a solution to the problem? What recommendations might you make to the plant manager?

From the information given, you may surmise that the plant manager's mind-set is that "employees usually do not think for themselves, and that training is a waste of resources and more importantly represents an unwanted increase in indirect costs (costs not associated with producing goods)." It follows that the plant manager will not hear of another attempt at statistical process control (SPC) in the production department, and SPC for attributes may not be effective anyhow. You also feel that the production department should have some responsibility for the quality of the product.

You decide to first work with the final inspection department to assure yourself that the quality characteristics have been well defined and each inspector knows which parts should be scrapped and which should be reworked. This action may be accomplished with a few good check sheets and a little "undercover training." An appropriate report to the manager should follow this action.

After the plant manager has seen the report and reacted to it, it might be time to suggest a sampling-plan approach to reduce bottlenecks. During your discussion you should be prepared to address the benefits of acceptance sampling—for example, fewer inspectors, fewer bottlenecks, and more efficient and, therefore, less costly inspection.

Further, *suggest that the rework be returned to the production department where operators can be made aware of the type of nonconformances.* The feedback will result in solutions to some nonconformances, while increasing pride in their work. (Production workers will say, "I can finally see what kind of work I am producing so that I may produce better work.") This suggestion may create a potentially explosive situation with the production manager, so the plant manager should be familiar with what you are about to suggest. If the rework is performed in production, a few of the 15 inspectors can be assigned to production to help with the material handling.

Next, you may suggest that a **material review board** be established. The board, composed of you, the design engineer, and the marketing manager or a customer, could better define what is a nonconformity, decide on disposition of lots not conforming, and better fix responsibilities for losses incurred.

These are the most positive steps you can take under the circumstances. Perhaps the plant manager will be exposed to and adopt a more modern management philosophy. Perhaps he or she will attend a workshop on **total quality management (TQM)** that will be of further aid in improving quality while building pride in the people doing the work.

# *Chapter Review Problems*

1. List the two major types of acceptance sampling plans.
2. Compare and contrast the two flowcharts in Figure 8.3 (two or three paragraphs or a listing of characteristics—instructor's determination).
3. Discuss the advantages of acceptance sampling when compared

with SPC. What are the advantages of SPC over acceptance sampling? Answer in two paragraphs.

4. What are the advantages of acceptance sampling over other sampling schemes, such as
   - no inspection;
   - inspecting a constant percentage of the lot size;
   - 100 percent inspection.

   Brief answers will suffice.
5. If you believe there is ever a case for using 100 percent sampling, state at least one example.
6. As a production worker you are responsible for placing parts produced from your machine in a tray. The tray goes on to an inspector who immediately reports all negative findings to your manager. You notice that he systematically samples the parts from the corner areas of the tray. How will you fill the tray if you can quickly select out conforming units during production and if you want to get this inspector "off your back"? Briefly report on how you would feel as a worker facing the described circumstance of an inspector reporting directly to your supervisor. What management improvements could you suggest?
7. From a lot of 200 radios you are to select 10 for inspection and testing. Each radio carries a three-digit number from 001 to 200. Use your calculator's random number generator to select the 10 parts. What is the total number of random numbers required to obtain the 10 numbers between 001 and 200?
8. Coal is being carried from a centralized mine to three local power plants by separate conveyor belts. Describe the most efficient way to collect samples and still give each piece of coal an equal chance of being sampled and inspected for sulfur content.
9. From a deck of cards take out the two red aces. Draw samples of five from the lot size of 50. Shuffle the deck thoroughly before drawing the random sample. The only nonconforming unit in the deck is the ace of spades ($\overline{p} = 2\%$). Using an acceptance number of $c = 0$, experimentally determine, from at least 150 trials, the probability of acceptance $P(A)$. Compare your answer with the probability from the Poisson table in Appendix D ($n = 5, \overline{p} = 0.02$, and $c = 0$).
10. What three things must be provided to an acceptance sampling inspector?

11. Using MIL–STD–105, determine the sample size and the *Ac* and *Re* values for a lot size of 1,300, AQL = 0.15 percent, level II, normal inspection, single sampling.
12. Using MIL–STD–105, determine the sample size and the *Ac* and *Re* values for a lot size of 350,000, AQL = 0.1 percent, level I, normal inspection, double sampling.
13. Using MIL–STD–105, determine the sample size and the *Ac* and *Re* values for a lot size of 100, AQL = 1.0 percent, level II, tightened inspection, single sampling.
14. Graph the OC curve for $n = 50$, $c = 0$. From the curve estimate the value of $\overline{p}$ at the AQL and the LQL.
15. One of the most cost-effective sampling plans for chaotic processes is the plan $n = 200$, $c = 0$. Lot sizes should be at least 2,000 ($n$ = 2,000) under the plan. Draw the OC curve for the plan. Use a range of 0.0002 to 0.02 for the fraction nonconforming ($\overline{p}$). From the curve estimate the value of the $\overline{p}$ at the AQL and the LQL.
16. From the OC curve in problem 14, draw the curve for and estimate the AOQL for the plan $n = 50$, $c = 0$.
17. Write a one-page paper describing an NDT test for a specific product that you are familiar with. Describe the materials making up the product.

# LABORATORY · CHAPTER 8
## Acceptance Sampling

*Materials needed:* A deck of cards or examples of inspections from industry. Use the deck of cards in class to perform the experiment outlined in problem 9.

*or:*

The instructor will provide inspection reports from industry that will be used with the MIL–STD–105 tables (Figures 8.11 to 8.15).

# 9

# Computer-Aided SQC

**Quality Control Software** 9.1
**Computer Analysis** 9.2
**Automated Process Control** 9.3

## 9.1 Quality Control Software

Whether you will be employed in a manufacturing or service industry, it is virtually certain that a computer is in your future. The computer is especially useful in quality control, and many companies currently produce software specifically geared to QC problem solving. Today's quality software is easily learned and easy to use, and most companies are using it.

You have already realized how difficult it is to lay out a Pareto chart or a cause-and-effect diagram neatly. This section covers the ways quality software may be used in QC problem solving. Section 9.1 also describes how gauge calibration by repeatability and reproducibility (R&R) studies is aided by using computer software.

The computer may be helpful in solving quality problems by using two of the seven tools of quality: Pareto analysis and the cause-and-effect diagram. A Pareto chart taken from the scrap reports of a manufacturer of axle assemblies is shown in Figure 9.1.

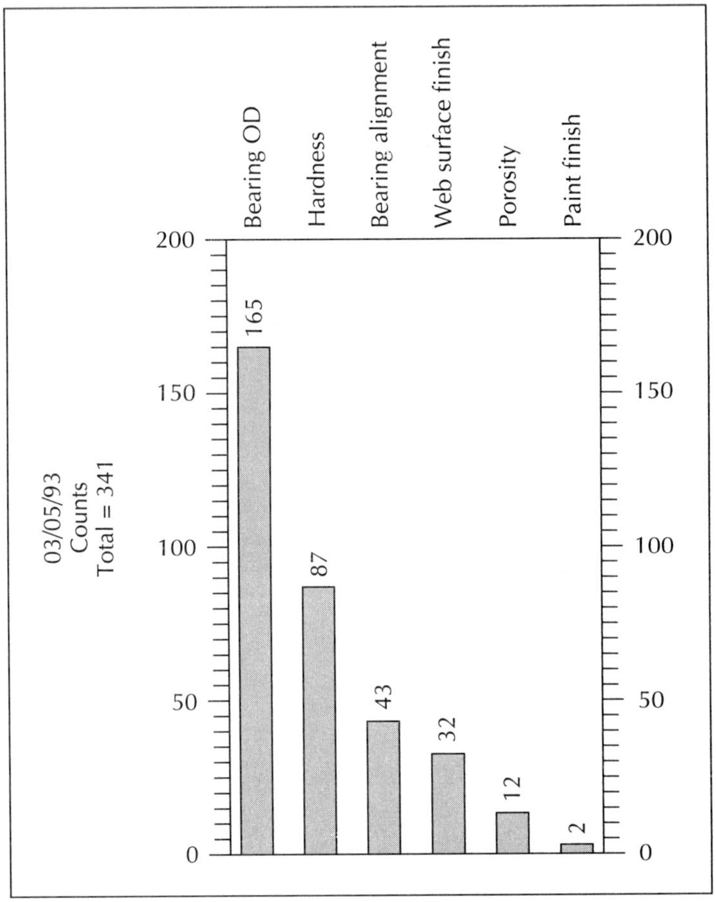


**FIGURE 9.1**
A computer-generated Pareto analysis.

The resulting charts present a clear picture to the technologist or manager of which problems are causing most of the losses. The axle assemblies are scrapped largely because a bearing's outer diameter is out of specification. To interpret the chart, you may wish to review Section 2.5, "Pareto's Principle."

In order for the computer software to construct the chart, several decisions must be made by the user. For example, the softwear might query the user as to the following:

- number of digits in largest count;
- number of whole digits in cost data;
- number of letters in category information;
- maximum number of categories.

Some of the categories may be optional, but most must be answered for the computer to know how to display the findings.

This example clearly shows the value of using QC software with the personal computer. If you first tried to do the graphing by hand, you would become aware of the ease with which the computer can prepare the chart while improving neatness and readability. The computer can also accept new information on the process as improvements are made and quickly yield a different Pareto chart that may show new areas to work on.

The QC software produced by most companies may also include cause-and-effect diagramming. The cause-and-effect diagram is another of the seven tools of quality credited to Kaoru Ishikawa and discussed in Chapter 1, "1960—Quality Problem Solving." The example shown in Figure 9.2 concerns the axle bearing's outer diameter, a

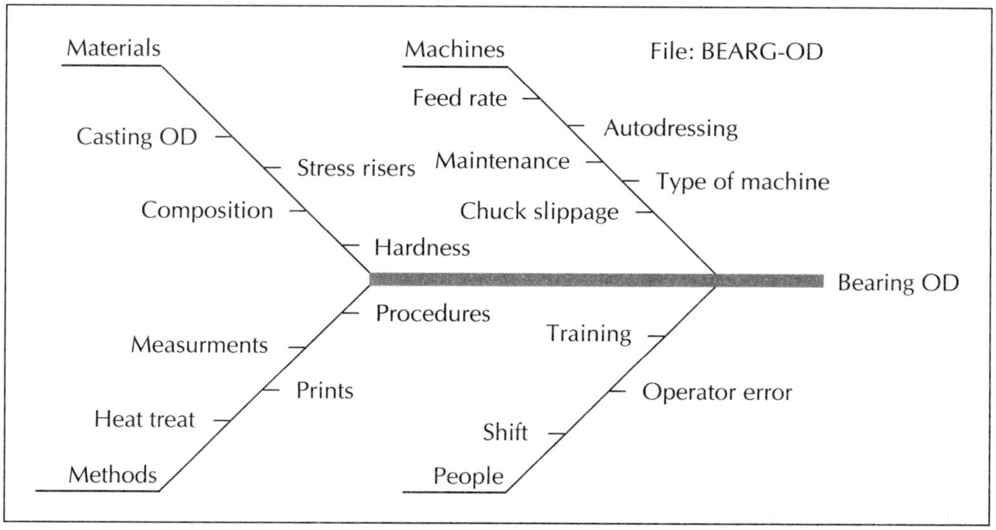


**FIGURE 9.2**
A computer-generated cause-and-effect diagram.

continuation of the major problem identified by the Pareto chart in Figure 9.1. You will appreciate how neat and clear the cause-and-effect diagram constructed on the computer is.

## R&R Studies

Calibration is critical before data are collected. Most commercial QC software can help with this need, too. Repeatability and reproducibility (R&R) studies have already been addressed (see the laboratories at the

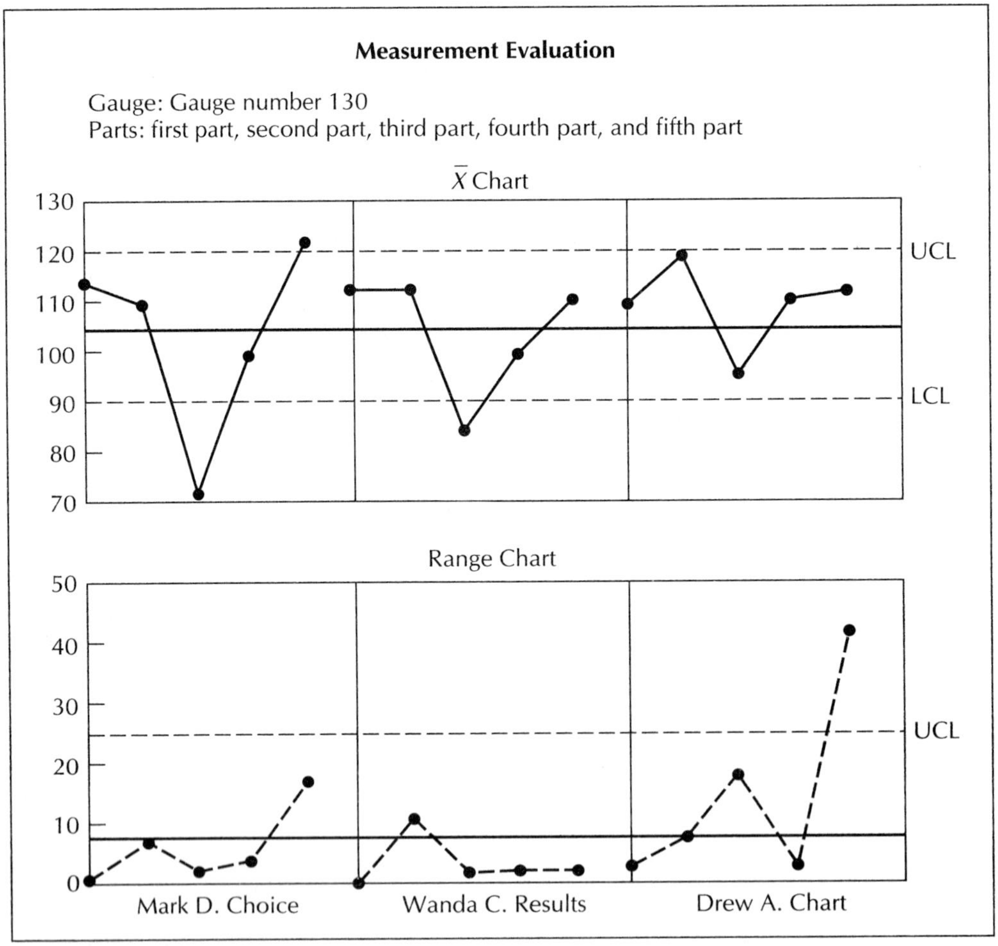


**FIGURE 9.3**
Gauge variability (R&R) for three operators. (*Source:* By permission of PQ Systems.)

ends of Chapters 3 and 4). To review, R&R studies determine the percent of process spread consumed by equipment or inspector error. R&R studies can be enhanced and facilitated by computer programs based on the same principles.

Figure 9.3 shows an $\overline{X}$ and $R$ chart of the variability of a gauge (Gauge Number 130) used by three operators to measure the same five parts. Each part is measured up to five times, using a randomized order for the parts each time, in order to obtain each plotted point. The measurement variation for each part is reflected by the value of the range parameters ($\overline{R}$ and $UCL_R$). When these data are combined with the $\overline{X}$ chart, which shows part-to-part variation, the effect of gauge ($R$ chart) and inspector ($\overline{X}$ chart) measurement variation may be determined.

The same data reveal the accuracy and variability of each operator with a given gauge. Figure 9.4 takes the results of the $\overline{X}$ and $R$ study and presents the results in median chart form (Section 5.2), along with the true average value of the parts. The "true value" is used to deter-

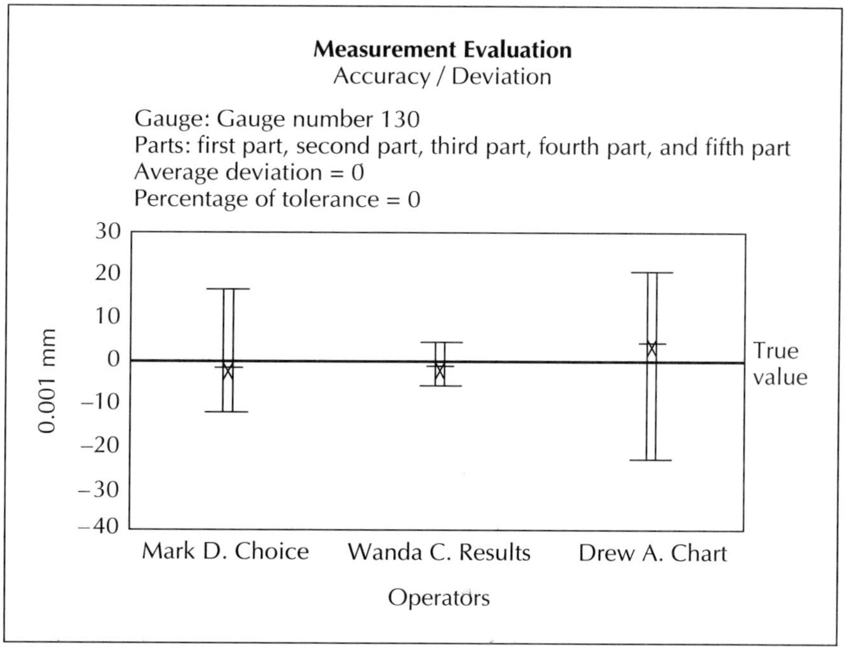


**FIGURE 9.4**
Accuracy and range report for three operators. (*Source:* By permission of PQ Systems.)

mine whether gauges are accurate and precise enough. The width of the deviations of each operator's measurements shows when operators require additional measurement training.

**EXAMPLE 9.1** If you were the QC manager, how would you act on the results of the R&R study?

*Solution:* Wanda's measurements reflect the lowest variation and the most accurate results. The three inspectors will be asked to work as a team, with Wanda as the team coordinator, in order to come up with the best method of measuring the parts from the production line. All three employees should practice until the variation is as close as possible to, and perhaps even better than, Wanda's initial results.

## Section 9.1 Review Problems

1. How many digits are in the "largest count" in the Pareto example in Figure 9.1?
   a. 1
   b. 2
   c. 3
   d. 4
2. What *minimum amount* would satisfy the maximum number of letters in the category information in Figure 9.1?
   a. 10
   b. 20
   c. 30
   d. 40
3. What type of chart is used in Figure 9.4 to study the accuracy and variability of each operator with a given gauge?
   a. $\overline{X}$ and $R$ chart
   b. individuals chart
   c. precontrol chart
   d. median chart
4. The gauge variation in Figure 9.3 is revealed by the ________ chart(s).
   a. $\overline{X}$
   b. $R$
   c. both
   d. neither

5. Which operator (see Figure 9.4) may require more training in measurement techniques?
   a. Drew
   b. Wanda
   c. Mark
   d. the chart does not yield such information

# 9.2 *Computer Analysis*

The computer handles large amounts of data with ease. It can collect data in millisecond intervals if necessary. The computer can analyze the data in seconds or minutes at most, yielding such statistics as the mean, range, standard deviation, and control limit calculations and charting the results. This section addresses the capability of QC software in collecting and analyzing data. *If a computer with QC software is available, this section should be used with the supporting equipment.*

In order for QC computer software to analyze data, the data must first be placed in the computer's memory. The importing of data to the computer is known as **data acquisition.** One method of data acquisition is to simply use the keyboard to enter the data that have already been collected by hand. However, hand recording of data is not easy. In fact, anytime a computer is at hand the technologist should feel inefficient when recording anything by hand. Fortunately, there is equipment available that allows for electronic data collection.

## *Electronic Data Collectors*

Suppliers of measurement equipment have many types of gauges that convert the analog world being measured into the digital world of the computer. These devices are generally referred to as **electronic data collectors.** (Converting electronic signals from analog to digital is technically known as **A/D conversion**—read as A to D conversion. You perform A/D conversion when reading data from a continuous scale and rounding it to the value recorded.) These digital-readout instruments have various names, such as "digimatic" or "electronic," as the supplier chooses. Figure 9.5 shows a digital-readout device for weight measurement.

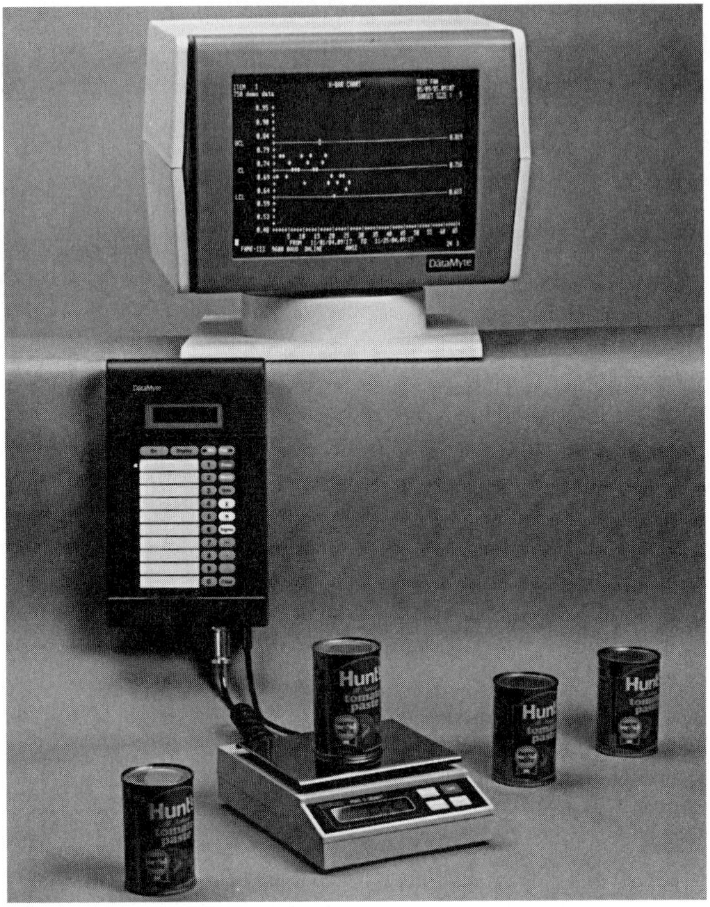


**FIGURE 9.5**
A data collection system for weight measurement. (*Source:* Reprinted by permission from the *DataMyte Handbook—A Practical Guide to Computerized Data Collection for Statistical Process Control,* 4th ed., DataMyte division of Allen-Bradley Company, Inc., 1989.)

The data from a digital instrument can be transferred electronically by way of data switches to small computers known as **data collectors** or **gauge talkers** (Figures 9.5 and 9.6). Compared to data recorded by hand, data transmitted electronically will be faster and more free of mistakes.

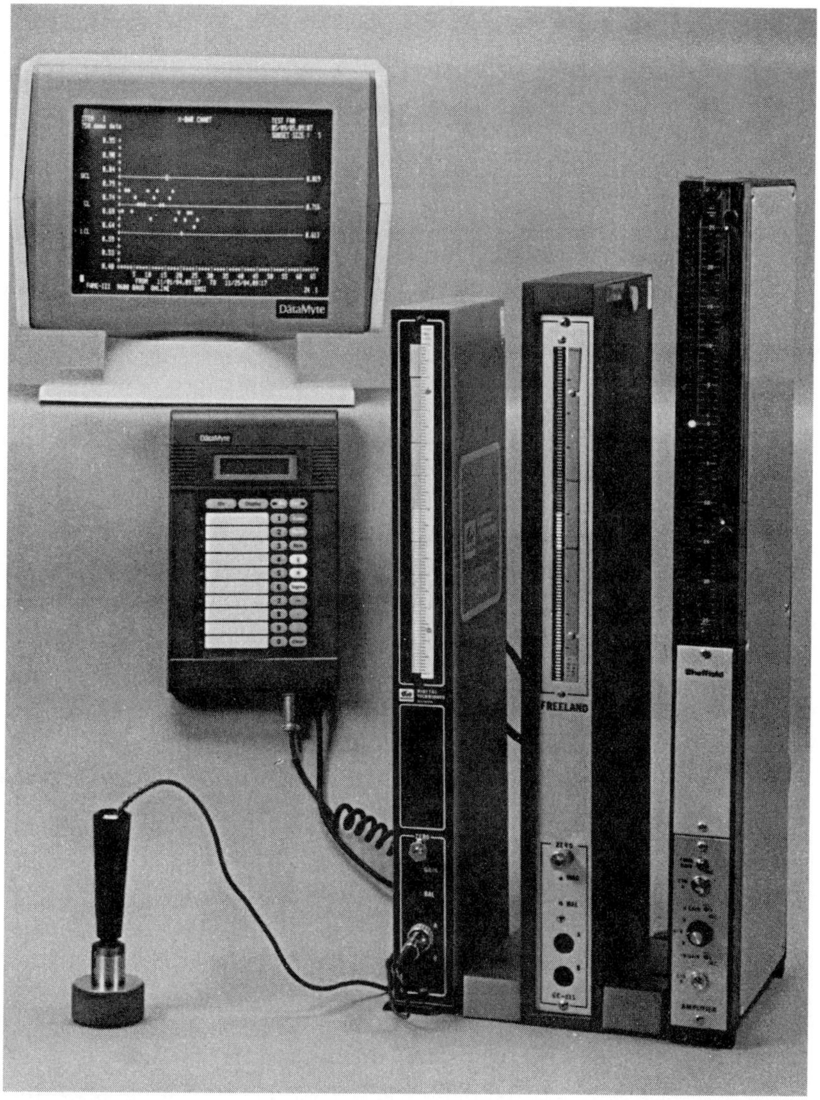


**FIGURE 9.6**
The portable data collector or gauge talker connected to an air gauge for dimensional measurement. (*Source:* Reprinted by permission from the *DataMyte Handbook—A Practical Guide to Computerized Data Collection for Statistical Process Control,* 4th ed., DataMyte division of Allen-Bradley Company, Inc., 1989.)

Many industries are using data collectors at the workstations. These data collectors have features such as automatic date stamping, an alarm that alerts the inspector to a data input higher or lower than expected, and some analysis of data—even $\overline{X}$ and $R$ charts and capability calculations. The use of multiple gauges is another advantage. One data collector can have a micrometer, vernier caliper, and bar code reader connected with inputs to separate programs in the unit. This capability makes it possible to measure multiple quality characteristics with one data collector. Differing connection schemes with a large range of hardware choices are required.

Data collectors often communicate directly to a computer, usually a personal computer. The personal computer's larger memory and faster processing can make the analysis of the data, such as constructing $\overline{X}$ and $R$ charts, go much smoother.

The data collector must be connected to the computer by a special cable. Special **communication software** is used whenever data are to be transferred from one computer to another. Most quality software suppliers provide transparent communications software that the user does not directly interact with, except for having to make the following choices when setting up the system:

- a baud rate;
- the number of data bits to control each character;
- the number of stop bits;
- the type of parity used for error checking as data are downloaded.

These communications software specific terms will not be defined here, for the computer software you work with will have specific definitions.

Data from several digital instruments may be entered into the computer simultaneously. Special hardware is necessary. This interfacing hardware is known as a multiplexer. A **multiplexer** allows the transmission of data from several data sending devices to be sent over a single, series communication port on the computer. *When multiple gauges are connected to a computer, a computer file must be established for each gauge.* In this way, each measurement will produce separate charts, histograms, and so on.

No matter how the data are entered or what communications software you use, the QC software must be prepared to accept it. **QC software** is software specifically written for the quality practitioner.

Software packages are regularly advertised in such quality control journals as *Quality Progress,* published by the ASQC.

The QC software, once you select the "enter" or "edit data" part of the program, will query you with such questions as "How many digits go before and after the decimal point?" and "Are there any minus numbers?" These are questions you have already answered as you completed prior studies without the computer. See Figure 9.7 for a screen from a particular QC software package that is used to prepare the software to collect data on a particular quality characteristic.

```
Your Company Name
SQCpack/PLUS Release 2.0                                File: POND
Variables Data—File Header                              15 Samples

Characteristic        =
Plant                 =
Department            =
Machine               =
Part name             =
Part number           =
Operation             =
Sample frequency      =
Units                 =

Number of digits left of decimal     =    Any negative values? N
Decimal places in the raw values     =    ... in the computed values =
Specs: Upper =              Lower =       Nominal dimension =
Sample size =                             Identifier length =
Maximum number of samples (opt) =         Number of historical samples =

F1 = Quit   F2 = Edit data   F3 = Values   F4 = Edit headers   F5 = Edit causes
```

**FIGURE 9.7**
A typical screen used to prepare the QC software to enter or edit data. (*Source:* By permission of PQ Systems.)

## *Computer Analysis*

When the data have been loaded into the computer, the analysis can begin. Once you have used the computer to do $\overline{X}$ and $R$ charts or any of the attribute charts, you will never wish to do them by hand again.

**Nevertheless, the practice you have already acquired through doing charts by hand is considered to be necessary in order to effectively interpret the chart designed and plotted by the computer.**

EXAMPLE 9.2

Enter the data by way of the keyboard into the QC software, prepare an $\overline{X}$ and $s$ chart on the screen, and print out the result. The data from the bearing OD (Figures 9.1 and 9.2) will be used for this study.

*Solution:* First, the queries on the screen must be answered before data entry begins. For this study (see Figure 9.7), the "Number of digits left of the decimal" is 1. After preparing the software for the data you are to enter, you must select the enter or edit data part of the program. Use function key F2, as instructed in Figure 9.7. Data entry is very easy, because the decimal point is automatically entered and the cursor is automatically sequenced to the next data set. All you must do is type the significant figures!

After data entry, you must select another part of the program to set up the chart output. This screen with the correct choices made for this study is shown in Figure 9.8.

The user may select an output to the screen or the printer. Selecting "Screen" for the output gives the user a chance to make sure the chart instructions yield the desired result before the printout. The output selected in Figure 9.8 is "Printer," so that a printout may be obtained. With the chart instructions completed, the command is entered to do the charts (F2 in this example). The final product—the $\overline{X}$ and $s$ chart—is shown in Figure 9.9.

Confirm the control limits on the $\overline{X}$ and $s$ chart by (see Appendix C)

$$\text{UCL}_{\overline{X}} = \overline{\overline{X}} + (A_3 \times \overline{s})$$
$$\text{LCL}_{\overline{X}} = \overline{\overline{X}} - (A_3 \times \overline{s})$$
$$\text{UCL}_s = B_4 \times \overline{s}$$
$$\text{LCL}_s = B_3 \times \overline{s}$$

**Using another part of the QC software's program allows you to compute capability. The printout includes a great deal of information, including the following: a histogram of the data, whether or not the distribution of the data is normal, and the capability ($C_p$ and $C_{pk}$).**

EXAMPLE 9.3

With the computer and the quality software, construct a histogram of the distribution of the data in Figure 9.9, and determine the following:

- whether or not the distribution is normal;
- the percent "out of spec";

PQ Systems
SQCpack/PLUS Release 2.0
Variables Data-Charting

File: POND
20 Samples
of 7 observations

Control limits — Set 1 — Set 2 — Set 3

Tabular (3*e*) 1-10

| | *E* Upper | Lower | Mean | *E* Upper | Lower | Mean | *E* Upper | Lower |
|---|---|---|---|---|---|---|---|---|
| $\overline{X}$ (R) | 1.623927 | 1.623693 | 1.6238 | | | | | |
| Range | 0.000539 | 0.000021 | 0.0002 | | | | | |
| $\overline{X}$ (S) | 1.623929 | 1.623691 | 1.6238 | | | | | |
| Sigma | 0.000189 | 0.000012 | 0.000 | | | | | |
| Median | 1.623947 | 1.623663 | 1.62380 | | | | | |

Output = Printer, graphics chart.
Chart direction = Across the page, paper size = 8 1/2 x 11.
Charts: $\overline{X}$, Sigma.
Chart all samples.
Chart scaling is based on the data.
Control limits for charting: chosen by SQCpack/PLUS.
Recompute control limits = No.
Change chart title(s): No.

F1 = Quit charting F2 = Do chart(s)

**Figure 9.8**
A typical screen used to prepare for charting. (*Source:* By permission of PQ Systems.)

- product above and below the specification limits;
- the capability ($C_p$ and $C_{pk}$) of the process.

*Solution:* Upon selecting the capability analysis of the program, you must alert the software to the parameters necessary for the particular study as outlined in Example 9.2. A screen is used for the selection. The user may define the parameters for the cells of the histogram or allow the software to best determine the cell interval and limits.

With the appropriate chart instructions entered, the computer command to do the histogram with capability calculations is entered. The result is shown in Figure 9.10.

Confirm the estimated capability by (see Appendix C)

$$\hat{\sigma} = \frac{\overline{s}}{c_4}$$

Part name: Knuckle
Part number: G–2165
Operation: Grinding
Characteristic: Bearing OD
Sample frequency: 7
Units: Inches

| | 1 | 2 | 3 | 4 | 5 | 6 | 7 | 8 | 9 | 10 | 11 | 12 | 13 | 14 | 15 | 16 | 17 | 18 | 19 | 20 |
|---|---|---|---|---|---|---|---|---|---|---|---|---|---|---|---|---|---|---|---|---|
| 1 | 1.62395 | 1.62395 | 1.62370 | 1.62380 | 1.62385 | 1.62380 | 1.62380 | 1.62355 | 1.62355 | 1.62380 | 1.62380 | 1.62380 | 1.62375 | 1.62395 | 1.62395 | 1.62375 | 1.62355 | 1.62360 | 1.62385 | 1.62385 |
| 2 | 1.62395 | 1.62385 | 1.62395 | 1.62395 | 1.62370 | 1.62390 | 1.62380 | 1.62365 | 1.62370 | 1.62380 | 1.62375 | 1.62380 | 1.62375 | 1.62385 | 1.62395 | 1.62385 | 1.62355 | 1.62365 | 1.62385 | 1.62390 |
| 3 | 1.62415 | 1.62400 | 1.62370 | 1.62380 | 1.62375 | 1.62375 | 1.62380 | 1.62370 | 1.62375 | 1.62385 | 1.62390 | 1.62375 | 1.62375 | 1.62400 | 1.62380 | 1.62390 | 1.62375 | 1.62355 | 1.62390 | 1.62390 |
| 4 | 1.62405 | 1.62395 | 1.62380 | 1.62385 | 1.62400 | 1.62385 | 1.62360 | 1.62360 | 1.62400 | 1.62380 | 1.62385 | 1.62385 | 1.62375 | 1.62405 | 1.62400 | 1.62375 | 1.62370 | 1.62360 | 1.62390 | 1.62375 |
| 5 | 1.62395 | 1.62395 | 1.62375 | 1.62375 | 1.62375 | 1.62385 | 1.62385 | 1.62350 | 1.62375 | 1.62385 | 1.62385 | 1.62390 | 1.62380 | 1.62395 | 1.62395 | 1.62385 | 1.62360 | 1.62365 | 1.62385 | 1.62395 |
| 6 | 1.62395 | 1.62415 | 1.62400 | 1.62380 | 1.62375 | 1.62390 | 1.62370 | 1.62355 | 1.62375 | 1.62355 | 1.62375 | 1.62390 | 1.62380 | 1.62395 | 1.62385 | 1.62385 | 1.62365 | 1.62365 | 1.62395 | 1.62390 |
| 7 | 1.62400 | 1.62375 | 1.62380 | 1.62395 | 1.62385 | 1.62375 | 1.62365 | 1.62375 | 1.62355 | 1.62385 | 1.62390 | 1.62375 | 1.62375 | 1.62395 | 1.62395 | 1.62390 | 1.62355 | 1.62365 | 1.62400 | 1.62380 |

| | 1 | 2 | 3 | 4 | 5 | 6 | 7 | 8 | 9 | 10 | 11 | 12 | 13 | 14 | 15 | 16 | 17 | 18 | 19 | 20 |
|---|---|---|---|---|---|---|---|---|---|---|---|---|---|---|---|---|---|---|---|---|
| $\overline{X}$: | 1.624000 | 1.623943 | 1.623814 | 1.623843 | 1.623807 | 1.623829 | 1.623743 | 1.623614 | 1.623721 | 1.623786 | 1.623829 | 1.623822 | 1.623764 | 1.623957 | 1.623921 | 1.623836 | 1.623621 | 1.623621 | 1.623900 | 1.623864 |
| UCLs: | 1.623929 | 1.623929 | 1.623929 | 1.623929 | 1.623929 | 1.623929 | 1.623929 | 1.623929 | 1.623929 | 1.623929 | 1.623929 | 1.623929 | 1.623929 | 1.623929 | 1.623929 | 1.623929 | 1.623929 | 1.623929 | 1.623929 | 1.623929 |
| Means: | 1.623810 | 1.623810 | 1.623810 | 1.623810 | 1.623810 | 1.623810 | 1.623810 | 1.623810 | 1.623810 | 1.623810 | 1.623810 | 1.623810 | 1.623810 | 1.623810 | 1.623810 | 1.623810 | 1.623810 | 1.623810 | 1.623810 | 1.623810 |
| LCLs: | 1.623691 | 1.623691 | 1.623691 | 1.623691 | 1.623691 | 1.623691 | 1.623691 | 1.623691 | 1.623691 | 1.623691 | 1.623691 | 1.623691 | 1.623691 | 1.623691 | 1.623691 | 1.623691 | 1.623691 | 1.623691 | 1.623691 | 1.623691 |

| | | | | | | | | | | | | | | | | | | | | |
|---|---|---|---|---|---|---|---|---|---|---|---|---|---|---|---|---|---|---|---|---|
| Sigmas: | 0.000076 | 0.000124 | 0.000118 | 0.000079 | 0.000102 | 0.000064 | 0.000093 | 0.000090 | 0.000152 | 0.000107 | 0.000064 | 0.000064 | 0.000024 | 0.000061 | 0.000070 | 0.000063 | 0.000081 | 0.000039 | 0.000058 | 0.000069 |
| UCLs: | 0.000189 | 0.000189 | 0.000189 | 0.000189 | 0.000189 | 0.000189 | 0.000189 | 0.000189 | 0.000189 | 0.000189 | 0.000189 | 0.000189 | 0.000189 | 0.000189 | 0.000189 | 0.000189 | 0.000189 | 0.000189 | 0.000189 | 0.000189 |
| Means: | 0.000100 | 0.000100 | 0.000100 | 0.000100 | 0.000100 | 0.000100 | 0.000100 | 0.000100 | 0.000100 | 0.000100 | 0.000100 | 0.000100 | 0.000100 | 0.000100 | 0.000100 | 0.000100 | 0.000100 | 0.000100 | 0.000100 | 0.000100 |
| LCLs: | 0.000012 | 0.000012 | 0.000012 | 0.000012 | 0.000012 | 0.000012 | 0.000012 | 0.000012 | 0.000012 | 0.000012 | 0.000012 | 0.000012 | 0.000012 | 0.000012 | 0.000012 | 0.000012 | 0.000012 | 0.000012 | 0.000012 | 0.000012 |

$\overline{X}$ Chart
LCL = 1.623691 Mean = 1.62381 UCL = 1.623929
Using tabular constants, based on samples 1–10

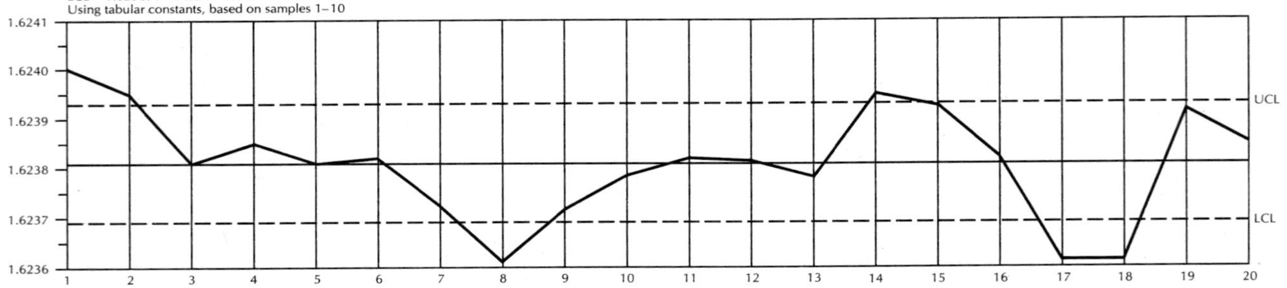


Sigma Chart
LCL = 0.000012 Mean = 0.0001 UCL = 0.000189

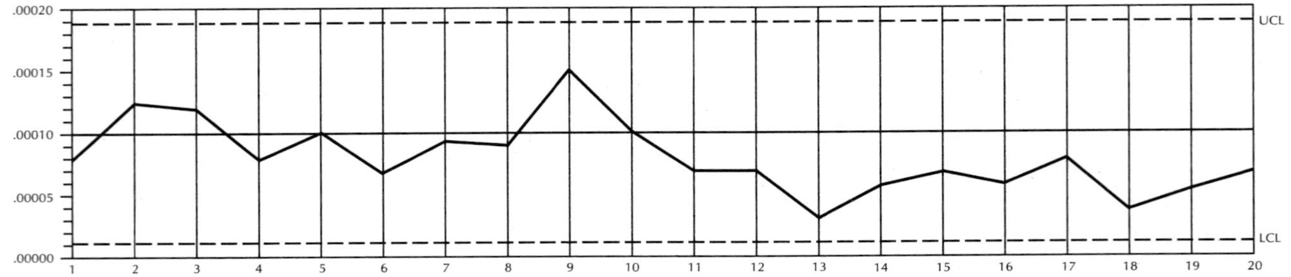


**FIGURE 9.9**
The computer-designed $\overline{X}$ and $s$ chart.

**Bearing OD**

| | | | |
|---|---|---|---|
| Operation | : Grinding | Part name | : Knuckle |
| Characteristic | : Bearing OD | Part number | : G–2165 |
| | | Sample frequency | : 7 |
| | | Units | : Inches |

**Descriptive Statistics**

| | | |
|---|---|---|
| All samples ($n$ = 7) | Min. value = 1.62350 | Interval = 0.000050 |
| 140 data points | Max. value = 1.62415 | lower boundary |
| Mean = 1.623812 | Kurtosis = –0.264 | Chi Squared = 21.309 |
| Sigma indiv. = 0.000133 | Skewness = –0.196 | deg. free = 8 |
| Est. Sigma = 0.000104 | Actual % | Conf. level = 95% |
| Coeff. var. = 0.000082 | Above spec. = 0.00 | Not Normal |
| Capability | Below spec. = 0.00 | Mean + 3s = 1.624210 |
| Using Sigma indiv. | Out of spec. = 0.00 | Mean – 3s = 1.623414 |
| $C_{pk}$ = 0.78 | Theoretical | |
| $C_p$ = 1.25 | Above spec. = 0.00 | |
| $C_r$ = 0.80 | Below spec. = 0.94 | |
| Z upper = 5.19 | Out of spec. = 0.94 | |
| Z lower = 2.35 | | |

**Histogram Capability**

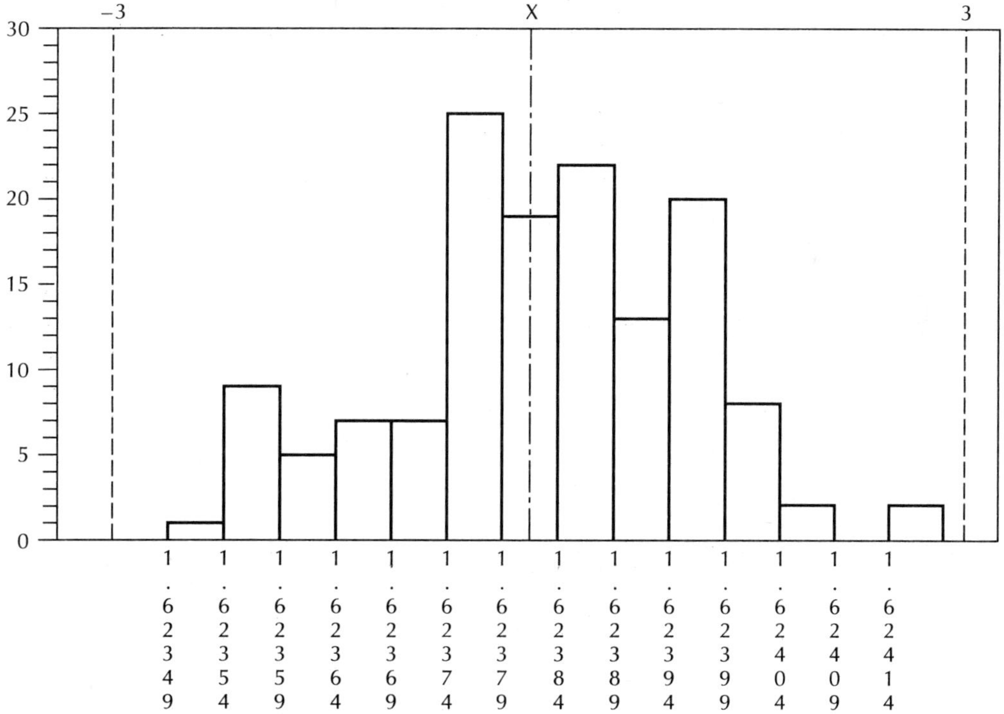


**FIGURE 9.10**
The computer-generated histogram with capability.

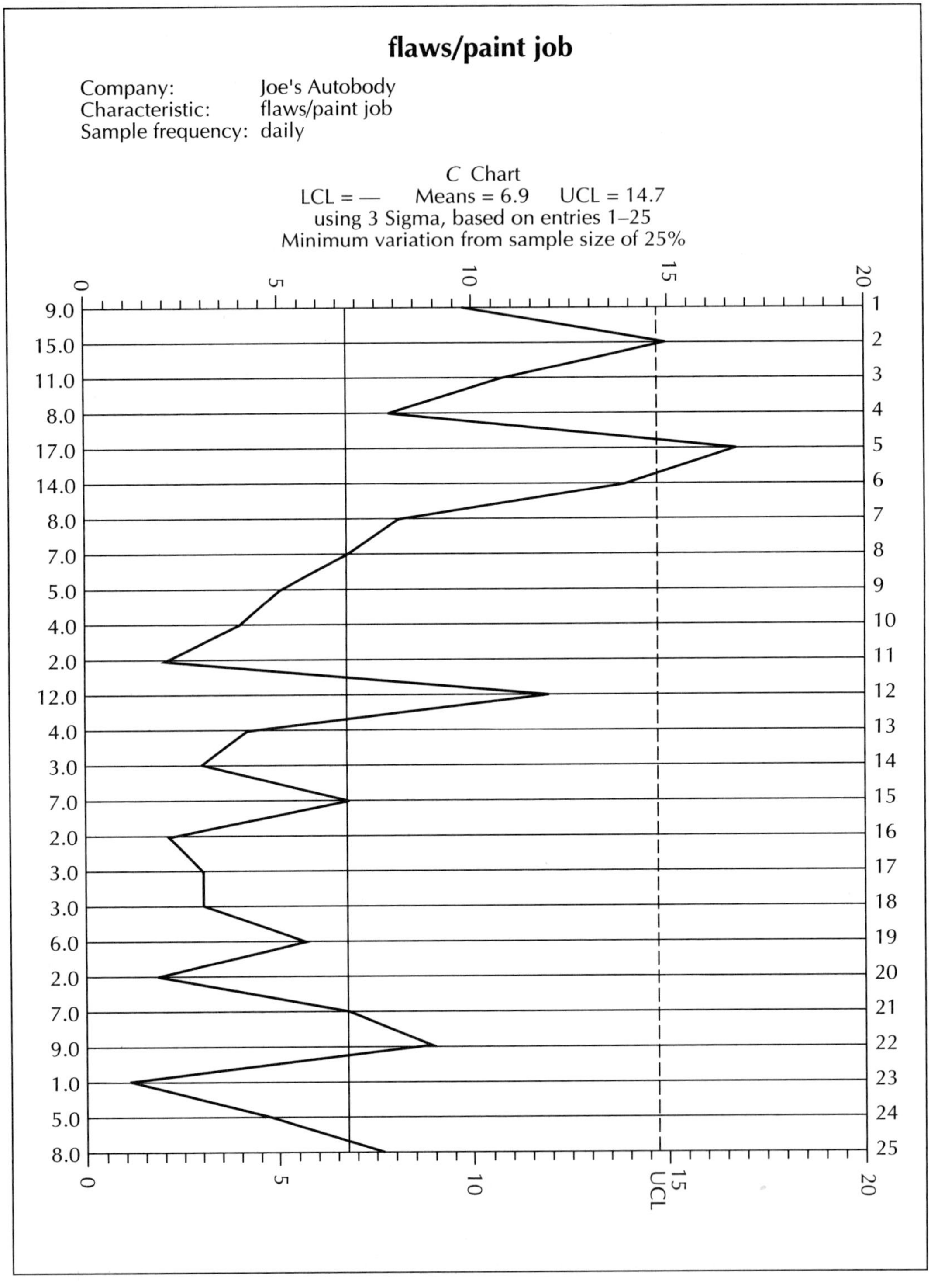


**FIGURE 9.11**
One example of the many different types of computer-generated charts, a *c* chart.

Confirm the $C_p$ by using the Sigma indiv. and the specification limits as below:

$$C_p = \frac{1.6245 - 1.6235}{6(0.000133)} = 1.25$$

(The multimodality evidenced by the distribution is typical of automated grinding operations lacking SPC optimization [Section 9.3]. In this case, the grinding wheel is dressed periodically, regardless of the need for dressing. A good example of overcontrol.)

Examples 9.2 and 9.3 represent a few of the most important analysis functions accomplished by currently available QC software. Most QC software will also construct a broad range of other charts, such as individuals charts, trend charts, all types of attribute charts (Figure 9.11), and CuSum charts (Section 9.3).

In addition, the software will evaluate more than one variable. Cause-and-effect relationships between variables may be plotted on a scatter diagram (see Section 2.2).

## *Purchasing Computer Equipment*

Whenever you purchase computer equipment to aid in any specialized task, it is important that you consider your needs carefully. Begin with the software: Will it work on an inexpensive personal computer? Can it perform necessary tasks? Does the documentation explain the way the software works? Is the software friendly enough for the people that will be using it? Is the communication software easily set up and used? And now for some important hardware decisions: Will the computer support the software? Are there enough ports and cards to allow the necessary connections for data collection? Is the system capable of being upgraded? For both hardware and software decisions: Does the supplier offer service and support after the sale?

These are a few of the questions you must ask yourself when considering computerized data collection and analysis for your company. Many of the questions cannot be answered by technologists or managers not having charting and process capability experience. In fact, **most of a company's processes should be brought into control by more traditional methods before computer analysis of the processes is seriously considered.** Finally, it will be helpful for you to keep current in this volatile field by reading the technical journals and attending professional meetings.

## *Section 9.2 Review Problems*

1. See Figure 9.7 and the data in Figure 9.9. How many "Decimal places in the raw values"?
   a. 1
   b. 3
   c. 5
   d. 7
2. With your calculator confirm some of the sample standard deviation calculations in Figure 9.9. Do you use the sigma ($\sigma$) or $s$ ($\sigma_{n-1}$) key? (Note that the software rounds off to the sixth decimal place in the computed values.)
3. When using quality software to collect data from a data collector with two gauges attached
   a. only one control chart may be generated
   b. two control charts are generated
   c. one or two control charts may be generated
   d. two gauges cannot be connected to one data collector
4. What out-of-control signals are revealed in the $\overline{X}$ and $s$ chart in Figure 9.9?
5. Other than the points out (points 2 and 5), is there an out-of-control region in the $c$ chart in Figure 9.11? How should Joe use the information in this region of the chart?

# 9.3 *Automated Process Control*

The computer may be part of a feedback loop that can effectively control a process. The same statistical "run" charts you have already constructed are used with the computer to prevent overcontrol in automatically controlled processes. This section discusses computer-controlled processes utilizing statistical process control.

Figure 9.12a shows the flowchart of a process controlled by an operator in what is described as open-loop control. **Open-loop control** is operator control. As material flows through the process and is formed

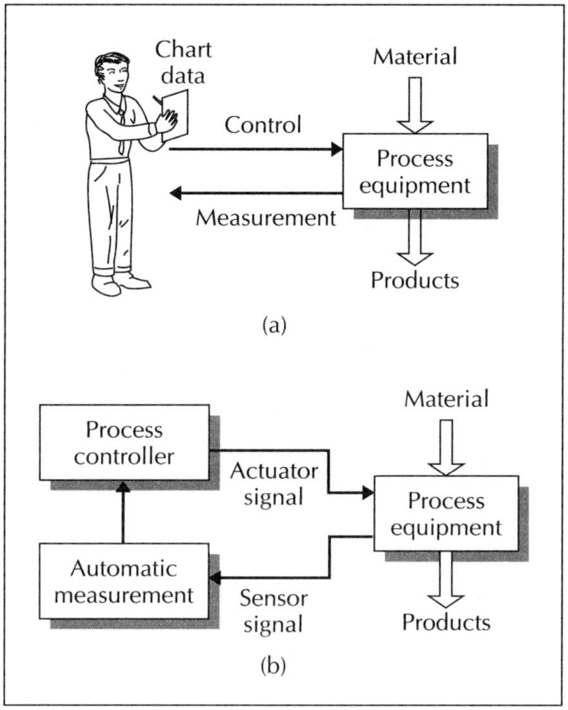


**FIGURE 9.12**
Open-loop control (a) is compared with closed-loop control (b).

into products, measurements are taken on specific quality characteristics by the operator, and decisions are made to adjust the process or leave it alone. Process adjustments are not made automatically, but by the operator.

In contrast, **closed-loop control,** shown in Figure 9.12b, is automatic control. Factory automation uses closed-loop control. In the flowchart we see the same process, but now the quality characteristics are automatically being measured by a sensor, such as a tachometer or thermocouple, and a process controller is used to signal motors, heaters, or other actuators to make changes in the process and compensate for errors.

Each process controller is set or "tuned" to control one process quality characteristic at a time. Technicians set the controller's zero point (the target) and the gain (the response of the feedback signal through the closed loop to overcome a process deviation to target) to control the process in the most effective way.

But, as you will recall from discussions of **overcontrol** (Section 4.2), continuous adjustment of a process can cause the system to overshoot. Some processes are especially sensitive to continuous control. These are the processes in which there are controllable long-term variations (such as trends), which we have learned to call **assignable causes,** as well as the always present uncontrollable short-term variations or **chance causes.** Temperature control is a good example. The long-term variation may be a heater that when turned on will cause the temperature to rise (trend upward) in the oven. Drafts or air currents within the oven would be the chance causes that affect the sensor's measurement.

The flowchart in Figure 9.13 shows two additional blocks. The SPC data analysis block represents a computer that can collect data in real time (as they are generated) and plot the data in a run-chart format, or in some other problem-solving format. The computer can also have alarms set at the statistically computed control limits, with an output to the process controller (Trends [rule of sevens] and other out-of-control signals [rule of thirds] can also trigger alarms.) The operator interface block can also be used to allow the operator to monitor the automatic controlled process and intercede when necessary. Operator interaction may be necessary, even with automatic control. Note that the use of a control chart does not take the place of a closed-loop controller. The SPC chart serves to enhance or optimize the action of the controller.

One way to view how SPC charts aid in the control of a process is to view the process as having a significant chance cause region that

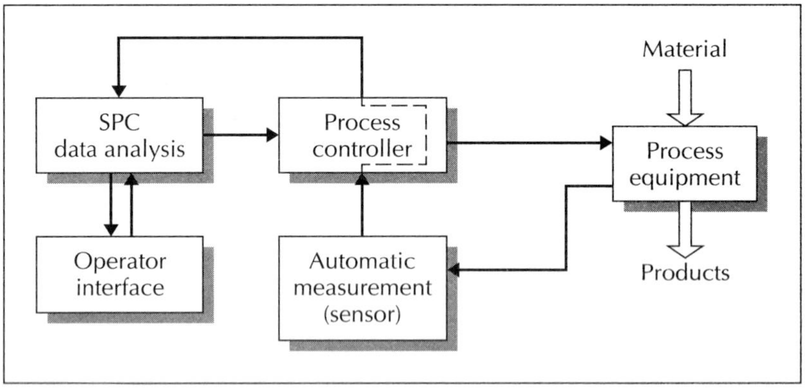


**FIGURE 9.13**
Closed-loop control assisted by QC software and interactive with the operator.

expands the uncontrollable or chance cause distribution of the quality characteristic of interest. (After all, a control chart's limits are based on three standard deviations of the normal distribution, which defines the chance cause region.) Figure 9.14a shows a process where the output on the controller is adjusting for the last part measured and is not centering the entire distribution of chance causes between the LSL

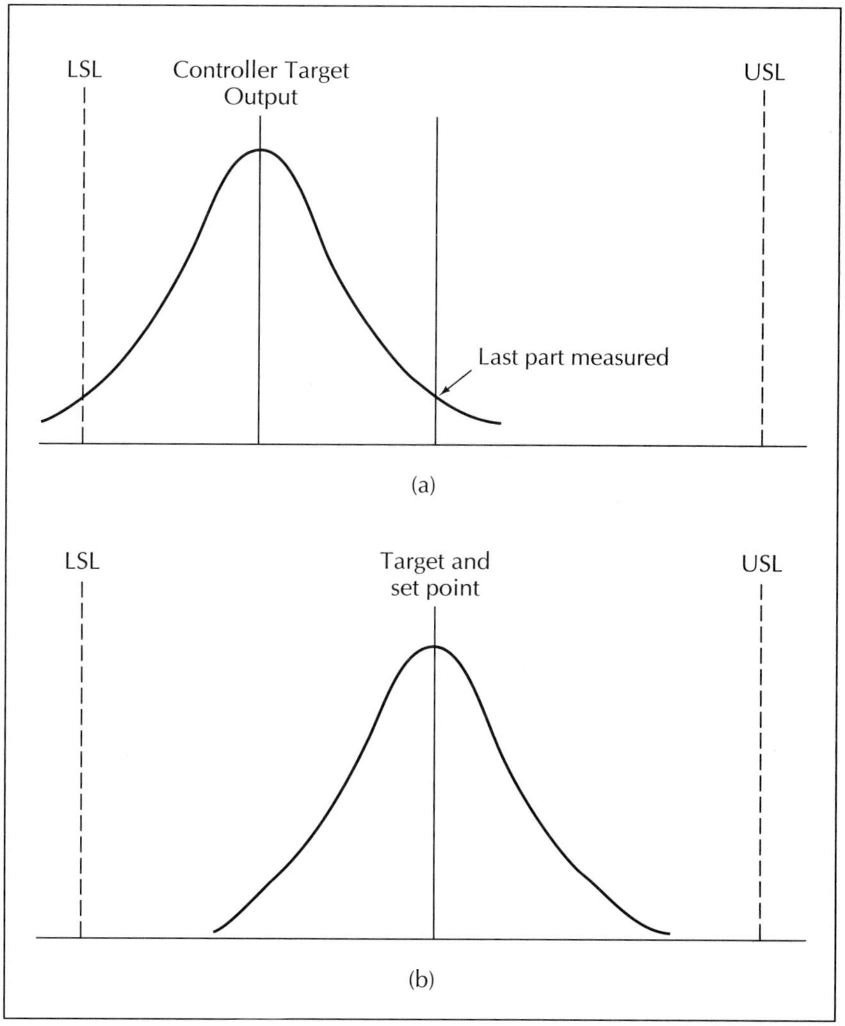


**FIGURE 9.14**
Controlling a process with chance causes operating—without SPC (a) and with SPC (b).

and USL. The controller is continually adjusting for the most recent measurement. Overcontrol problems result.

In Figure 9.14b we see the effects of the control loop using the variable SPC chart. Now the process is tracked as it runs, and the average of the product's quality characteristic (in this case length) is adjusted to the mean of the distribution. With the chance causes of the process understood, the set point may be more accurately determined, overcontrol is minimized, and savings will result.

Statistical process control, then, is used to monitor processes and to use the variable data from the process to build a control chart with control limits. The control limits are used to establish the natural spread or range of the process so the controller will not signal changes in the process until the natural control limits (or other out-of-control signals) are exceeded. In this way, the system makes "intelligent adjustments." Hence, the system is referred to as **artificial intelligence (AI)** control, or control by **expert system.** Other descriptors include **predictive control, SPC optimization,** and **intelligent alarms.**

One advantage of the computer systems described is the operator interface (Figure 9.15). The operator can see what is happening with the process in real time or while the event is still taking place. A real time Pareto analysis that identifies the most frequent or most costly causes of variation can also be shown, in order to assist the operator in making process adjustment decisions. The expert system, then, acts as an adviser to the operator; it does not make all the decisions. The operator must be well trained to interpret the control chart and Pareto information, determine the exact cause, and take prompt action. The computer helps the operator by reducing the load of data gathering and remembering which are the most frequent assignable causes of variation.

The main advantage of an automated system with SPC is the same as the advantage of SPC as described for processes that are not automated. **SPC allows the operator to see what is happening as it happens and not later when it is too late to do anything about it.** The expert system acts as an aid, not as a replacement, to the operator.

EXAMPLE 9.4 Research and describe a typical application of the use of an expert system utilizing SPC control limit alarms.

*Solution:* The Instrument Society of America's journal, *Intech,* has just such an example in the June 1991 issue, "SPC + Expert Systems = Intelligent Alarms" by Donald K. Frerichs. The article begins on page 22.

**FIGURE 9.15**
A machine operator in a paper mill uses a distributed control system with specialized software to capture an overview of the quality of paper being made. If preset quality parameters vary, the operator takes corrective action on the computer without leaving the control room. (*Source:* By permission of ABB Process Automation, Inc.)

The expert system described on page 24 utilizes SPC control limit alarms to control an industrial, steam-heated air dryer.

Operators found that if alarms occur from the control limits for the range (*R*) chart the dryer control is considered to be erratic. On the other hand, $\overline{X}$ chart alarms indicate that the process—the environment to be dried—is changing too fast to be controlled.

Fault conditions are described by the computer, but the operator must take corrective actions based on his or her best judgment. Cost factors dictate that process shutdowns for maintenance can only be implemented as a last resort.

The chemical industries have used the cumulative sum chart (CuSum chart) to control batch processes effectively. The **batch process** involves the mixing of batches of ingredients. The brewing industry is one example of an industry requiring batch processing. Other examples of batch-processing industries are the food, drug, and paper industries. A typical batch process may include such process quality characteristics as process time and temperature, agitator RPM, and amount and feed rate of additives.

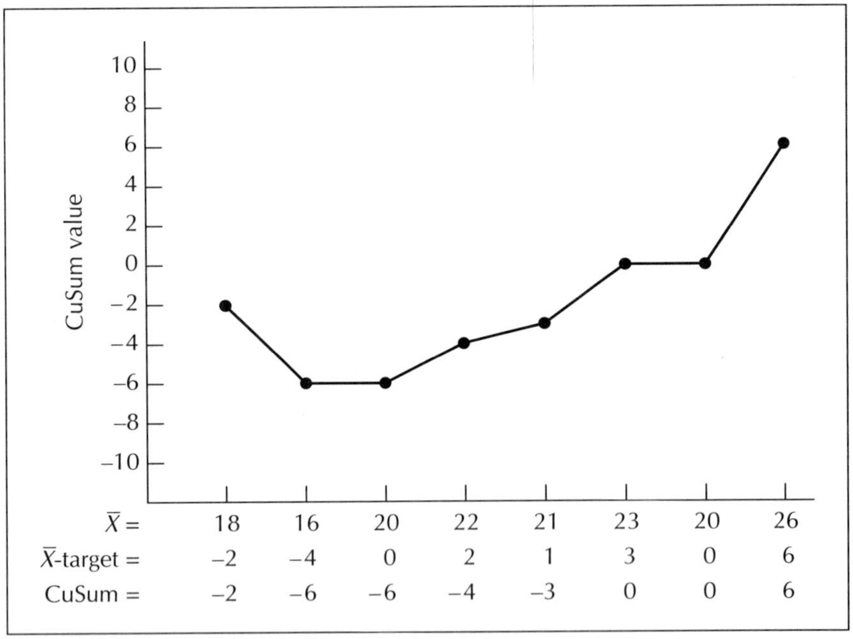


**FIGURE 9.16**
A CuSum chart (target = 20).

A **CuSum chart** plots the cumulative deviations of each measurement (an $\overline{X}$ for variable data) from the nominal or target value. The slope of the function plotted represents where the process really is in respect to target and the amount of deviation from target. The CuSum chart triggers an alarm based on established control limits, and the amount of change desired is downloaded to the controller. See Figure 9.16, a CuSum chart constructed using simplified data.

In Figure 9.16, the first mean or $\overline{X}$ of the data is 18. The 18 is subtracted from the target value of 20 to find the cumulative sum or CuSum value of $-2$. The next $\overline{X}$ value of 16 results in a $\overline{X}$ − target = $-4$, and the CuSum value is $(-2) + (-4) = -6$. Note that the slope of the chart is 0 or horizontal between the earlier $\overline{X}$ values of 16 and 20, and the later $\overline{X}$ values of 23 and 20. The slope of 0 means the process is stable in its relationship to the target value, and not that it is at target. Control limits (not shown) would be placed according to experience with the process. The CuSum chart is often used with an $\overline{X}$ and *R* chart—the CuSum exaggerates shifts from the target value and is best at detecting small but significant shifts in the process, while the $\overline{X}$ and *R* chart is best at detecting large shifts in the process.

## *Computer-Integrated Manufacturing (CIM)*

Perhaps the greatest advantage of the use of computers and expert systems to control processes is that a database may be established for each process. Processes then may be linked through a **distributed control system.** Linking of process control systems to business systems, known as **computer-integrated manufacturing (CIM),** can be quite advantageous. The information available from a batch system on a paint product is as follows:

- Pareto analysis of customer complaints as to color, viscosity, and chips in the product;
- the history of each batch, including such information as the reactor's (containment vessel's) temperature and pressure;
- $\overline{X}$ and *R* or $\overline{X}$ and *s* charts on individual quality characteristics;
- CuSum charts on individual quality characteristics;
- scatter diagrams and other correlation tools;
- cost data, including losses due to scrap or rework.

The last item, linking cost data to technical data, needs explanation. Bringing process control into business systems is not easy. The first hurdle is to bring management to the realization that business and engineering should be working together (Chapter 10). Next, the difference between business systems and process control systems must be acknowledged. One of the most important differences is that a business system is always open looped, never closed loop. People need to be involved with the business data—that is, costs, sales volumes, production figures. All of these data need people to interpret them effectively. In spite of the differences between the way business and engineering people think and use data, there are enormous advantages to linking the databases. Up-to-date quality reporting and meaningful forecasting are a couple of the great advantages of using CIM.

The Pareto chart in Figure 2.11b is a good example of how important it is to let production personnel know cost data. Without cost data, they would assume that gauge (the thickness of the film) is the most important problem to work on. But, in looking at costs, bubble breaks represent a far more serious problem. Giving production people business information builds pride and allows them to accomplish much more than if they only know process data—kilograms of scrap, degrees of heat, and milliseconds of process time.

## When to Automate

In closing this section on process control, keep in mind that not all processes are easily automated. Automation is also costly, but the cost of computer equipment is decreasing. The technology will continue to improve, further driving down the costs of implementation. Industrially hardened systems will become more available. Quality control departments should be involved in decisions on automating particular processes. They should also be involved with the amount of operator control that should be available on automated systems and how much training is necessary. The rule of thumb for automating processes is that *the process should already be running well and be well understood before attempting automation.*

The same is true of implementation of CIM. A factory should be running efficiently before anything like a CIM system is introduced. In all cases, plantwide automation should include training of operators, management, and maintenance so that quality can be maintained. Total quality management (Chapter 10) includes the planning and implementation of automated systems.

## *Section 9.3 Review Problems*

1. A process controller is used in
   a. open-loop systems
   b. closed-loop systems
   c. both types of systems
2. The chart that uses the nominal or target value in each point plotted is the
   a. median chart
   b. individuals chart
   c. $\overline{X}$ and $s$ chart
   d. CuSum chart
3. Automating a process should never be considered unless
   a. it can be shown to be cost effective
   b. the process is already running well
   c. staff are trained to maintain the automated system
   d. all of the above are true

# *Summary*

- The quality control technologist or manager will find the computer to be indispensable in problem solving (Pareto analysis and cause-and-effect diagramming), calibrating gauges and collecting data, preparing quality charts and reports, and assisting automatic systems in controlling processes (expert systems).
- Quality software is available and must be purchased to use the computer for specialized quality control data collection and reporting.
- Digital instruments can be purchased that will enter measurements into data collectors (gauge talkers).
- Quality software easily constructs $\overline{X}$ and $s$ charts.
- Attribute, as well as variable, studies are facilitated by using the computer.

- Batch processes, especially, may be improved by the use of expert systems. Most batch processes are enhanced by using correlation (e.g., scatter diagram) or CuSum type statistics.
- The quality control practitioner should be familiar with process control strategies such as being able to block-diagram open-loop control and closed-loop control with an expert system.

**Figure 9.17**
Collecting attribute data using QC software. (*Source:* Reprinted by permission from the *DataMyte Handbook—A Practical Guide to Computerized Data Collection for Statistical Process Control*, 4th ed., DataMyte division of Allen-Bradley Company, Inc., 1989.)

- The quality control practitioner must be involved in decisions regarding the automation of processes and the inclusion of automated SPC systems.

## CASE STUDY · CHAPTER 9

You are employed by a major helicopter manufacturer. Your specific job is to manage the printed wire boards and harness assembly area. The circuit elements have a variety of types (up to 100 different types) of nonconformities that must be monitored and controlled. Check sheets have been developed to compile the data, but the number of items inspected per day and the variety of nonconformities to be recorded have made the data collection tedious and error prone. The management also would like weekly reports to show what nonconformances have been found, their location, and other information.

You have been asked to research and recommend computer equipment that can make the task easier. You remember a vendor's handbook, and you can visualize the solution (see Figure 9.17).

### *Solution*

The photograph shows an inspector entering types of nonconformances into a data collector with a bar code reader. The bar code reader picks up the part number and the date of manufacture and automatically enters them into the database. After the circuit element has been identified, a list of nonconformities that the inspector may select for that particular part is presented. The frequency of the various types of nonconformities is recorded in separate databases according to location. Rework tickets, data reports, and attribute charts are generated by the data collector. The data can also be downloaded to a PC.

## *Chapter Review Problems*

1. Using a periodical index, preferably a technical index like the *Applied Science and Technology Index,* research and locate journals advertising QC software. List the major tasks that such software will accomplish.

2. Research and describe commercially available gauges capable of transmitting data electronically.
3. Describe the data collector or gauge talker. List the advantages of such a device.
4. Research one type of data collector by consulting supplier catalogs or by writing to suppliers advertising in engineering journals.
5. *Show by a diagram how the data collector you use in the quality laboratory is connected to measure more than one quality characteristic. List all part numbers including those for the gauges used and the cables connecting the gauges to the data collector.
6. *Prepare a step-by-step procedure for connecting the data collector to the PC.
7. *Using QC software, construct and print out an $\overline{X}$ and $R$ chart and/or an $\overline{X}$ and $s$ chart using data provided by the instructor or from one of the problems in Chapter 4.
8. *Perform a capability analysis of the data in problem 7 with the QC software.
9. List the equations that should be answered before purchasing QC software for your business or industry.
10. List the advantages of using the computer with QC software to perform attribute studies.
11. Explain in one page or less why SPC may be used most effectively with the expert process control system.
12. Research and describe the difference between a "batch process" and a "continuous process."
13. See Example 9.4. Locate and read the article referred to, for further understanding on the use of expert systems in quality control. Prepare a one page report.
14. Using a periodical index, preferably a technical index like the *Applied Science and Technology Index,* research and locate a journal article concerning the use of SPC charts to control a process. Use descriptors in your search such as predictive control, SPC optimization, intelligent alarms, artificial intelligence in process control, and expert systems in process control. Either copy the article for your instructor or report on the contents (instructor's determination).

* Problems 5, 6, 7, and 8 require specialized digital electronic equipment and computer software.

15. What is the role of quality control departments in implementing the automation of a particular plant process? What needs to be considered?
16. What is the role of quality control departments in implementing a CIM system? What needs to be considered?

# LABORATORY · CHAPTER 9

## Automatic Data Collection with Computer Charting and Capability

*Materials needed:* A computer with SQC software and an electronic data collector. The data collector is connected to one of the computer's serial communication ports. The data collector is plugged in, and its power is turned on. DIP switches on the data collector may have to be set. See the operator's manual.

Access the computer software. The computer software will query you as to the type of data collector. Select the data collector you are using. Set parameters—that is, baud rate . . . 1200, transfer data from . . . computer serial port ID, and so on. Enter the sample size desired. Select a file name for the data to be transferred to. Transfer the data to the file as each part is measured. If you are using multiple data collectors, a file must be established for each.

When finished, return to the main menu. You are ready to begin charting or assessing the capability of your measurements.

Chart the data collected by the process. Also, use the computer to draw a histogram of the data or any other data set provided by the instructor. Confirm the computer's calculations of $C_p$ and $C_{pk}$.

Use the software to convert a chart to a CuSum chart. With your instructor, analyze the CuSum chart.

# 10

# *Total Quality Management (TQM)*


**Elements of the TQM System** 10.1
**Recognizing Quality Efforts** 10.2
**Forming a Personal TQM Philosophy** 10.3


## 10.1 *Elements of the TQM System*

The history of the quality revolution, presented in Section 1.1, culminates in the total quality system known today as **total quality management,** or simply as **TQM.** A review of the decade-by-decade history (Table 1.1 and associated discussion) reveals a few of the many people contributing to a complex field that is not only scientific, but philosophical as well. Also, the history shows how rapidly the quality field is changing. *Quality control will continue to experience dynamic change over the foreseeable future, making it a challenging career filled with opportunity.*

The few contributors mentioned in Section 1.1 developed the quality tools we use today—the Shewhart cycle and the seven tools of quality most notable among them. These quality tools have been used in this text to form the basis for quality problem solving. Yet, none of these tools would have much or, indeed, any value without the cooperation of people working together in teams to attain difficult goals—people like you and your fellow workers during this course. **Teamwork** results in high-quality products, and *teamwork is the primary ingredient of TQM.*

## The Organization

To see how TQM promotes teamwork, which in turn dissolves the lines between departments in a company, we will begin by describing an organizational model that could fit most industries. Consider the four major elements shown in Figure 10.1. These major elements are not connected in an ordered configuration but in the three-dimensional form of the atom's nucleus to show the interactive nature of the TQM organization. The customer forms the vital center of the model. Each of the separate elements, "the protons and neutrons," and their interactions with each other and the customer are described in the following pages.

TQM begins with the **chief executive officer (CEO).** Without top management's involvement, the organization will not join together as one system to promote quality as the key to the organization's survival in the future. Top management is not shown in the nuclear model in Figure 10.1, for it should be transparent and all-encompassing. Top management includes the **quality staff function** for the company, which provides information on quality issues to the organization, along with training in quality tools, such as how to construct and understand control charts. The quality staff also conducts **audits,** or statistically based measurements of the quality level at various points in the organization. Top management, along with the quality function, is analogous to the "strong force," the force in the nucleus of an atom that binds the positively charged protons together.

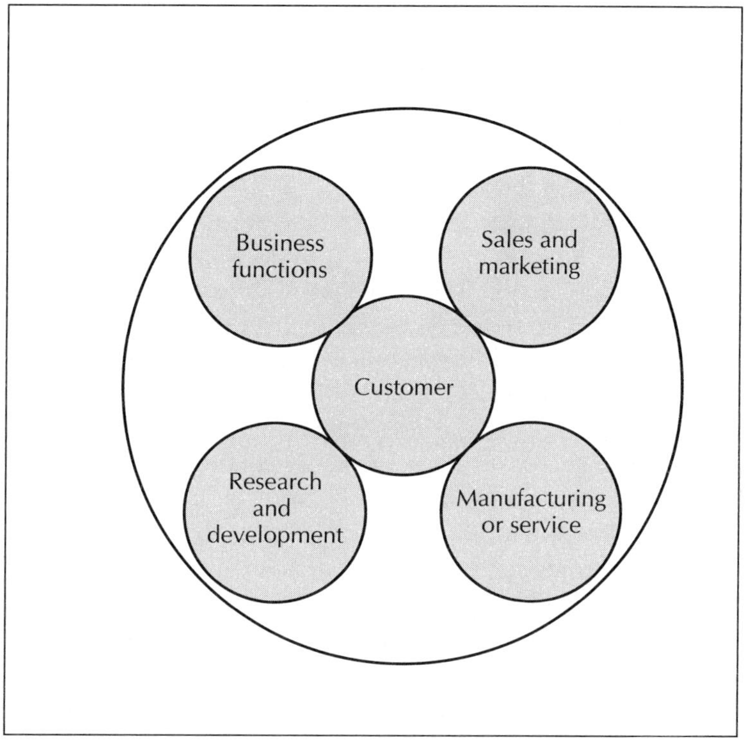


**FIGURE 10.1**
A TQM organization is bound closely together by top management with a customer focus.

## *Sales and Marketing*

We will begin and end with sales and marketing. Sales and marketing is the "glue" interfacing the entire industry with the all-important customer. A successful product must fill the customer's needs. To be successful the product must also be affordable to the customer. Marketing is charged to determine *the totality of features of a product or service that bear on its ability to satisfy stated or implied needs* (Section 1.1).

Benchmarking is a vital component of a quality marketing strategy. **Benchmarking** simply is a system that allows a company or individual to learn from others. Benchmarking may include analyzing a competitor's products and production methods, or analyzing a non-competitor's performance in a certain area of expertise.

Benchmarking was first defined by the Xerox Corporation to identify

- their competitor's level of quality;
- their competitor's cost of manufacturing copying machines.

The company found that foreign competitors were beating them on both price and quality in the 1970s. *After developing measurement processes to collect data on how good the other companies were,* Xerox identified goals and set targets to improve themselves.

Benchmarking may involve looking at companies that are not competitors. For example, Xerox studied the warehousing and distribution practices of L. L. Bean. After measurement processes were again implemented, the study resulted in an improved distribution system for Xerox.

Other companies learned from Xerox's successes. Numerous benchmarking studies by U.S. automobile manufacturers determined that productivity—the number of labor hours to build each car—was not competitive with some foreign manufacturers. For instance, the studies showed that Japanese manufacturers of luxury cars spent one-half to one-third less time on each car than U.S. manufacturers did. In determining the causes for lower productivity, the most significant was that *half of the cars assembled by U.S. companies needed to be repaired before they left the plant.* Could the production people be blamed for a rework rate much greater than that of world-class manufacturers? *Managers of U.S. companies concluded that it was not the worker but the management system that created the major problems!*

Recognizing the need to improve in order to compete in the U.S. automobile industry, the "Group of 99," the original staff (management and union members) of Saturn Corporation in Spring Hill, Tennessee, traveled the world in the 1980s to find the best automobile designs for manufacturing and the most efficient assembly methods. They also interviewed car dealers and owners—all of this, to determine how best to compete in the highly competitive automobile market. Quality standards resulting from their studies formed the basis for product measurements later.

Benchmarking may utilize customer surveys. **Customer surveys** measure how well **external customers,** the purchasers of the product or service, are satisfied with the features of the product or service. Some surveys are implemented immediately after the customer has received the product or service. The restaurant forms you may fill out after dining are an example. The use of 800 numbers placed on product containers will quickly alert an organization of dissatisfied customers. Other surveys may take place months or years after the product has been received, especially if the product is a large investment like an automobile. Whether it is immediate or after some time, one of the most important results of feedback of customer information is to identify causes for customer dissatisfaction. Always remember the following: *It is much easier to keep a present customer than it is to gain a new customer*.

Successful customer surveys are well planned, with input from all departments of the organization regarding the goals of the survey. Each department will wish to identify the quality characteristics under its control that are most important to the customer. Then, when the survey results are in, the department can use statistical methods to best control the vital characteristics identified.

Customer surveys use many of the same statistical methods to analyze the data that you have learned in this statistical quality control course. As a technologist you should be interested in the methodology used in a study, as well as in the conclusions and recommendations of the study.

**Frequency-of-repair records** are more objective measures of customer satisfaction. Appliance manufacturers collect these data from warranty records. Some independent testing laboratories sell journals directly to the consumer that list frequency-of-repair records for appliances or automobiles in different model years (*Consumer Reports*).

EXAMPLE 10.1 A large corporation manufactures truck axles and the connecting shafts for large trucks. The axle assemblies are integrated with transmission and clutch assemblies from other companies. Customer complaints reveal the need for better integration of the axle assemblies with the transmission and clutch assemblies. The company decides to manufacture its own clutches and transmissions, and to design both to better meet customer expectations. How is the company to begin such an ambitious undertaking?

*Solution:* The company decides that a benchmarking strategy is essential to successful implementation of such a financially risky venture. The benchmarking involves the purchase of the soon-to-be competitor's transmissions and clutches. The assemblies are disassembled, and all part tolerances and assembly methods are studied to provide baseline information for future SPC studies. Customers (truck assemblers and truck drivers) are surveyed as to what quality characteristics are most essential to them. The truck drivers, for example, want a clutch that has less spring compression, so less effort needs to be expended during frequent shifting of gears in the city. After compiling the results of the various studies, the company decides to convert one of the older plants, which was scheduled for closing, into a manufacturing facility for the new products.

Later, frequency-of-repair records and customer surveys show that the company has made a correct strategic decision. The investment in new equipment and other resources has paid dividends and increased the company's competitive position.

Benchmarking may also involve an individual selecting the best individual traits to have in his or her career field and determining how to instill these traits or characteristics in his or her life. This type of benchmarking will be discussed in Section 10.3.

## Business Functions

The business functions of any industry must be closely linked to all other functional areas. Accounting must not only keep track of costs, but must also help to categorize costs of manufacturing and service, so that quality costs may be determined. Purchasing must work with manufacturing to buy raw materials and parts for assembly that best meet the needs of the operators. Human resources must select and hire personnel who will best serve quality improvement efforts. All three departments have as their customers the **internal customers** of the organization. In a TQM environment the employees, as the internal customers, are just as important as the buyer—the external customer.

*Accounting* For each functional area the **accounting department** should track the four categories of **quality costs.** The four categories and the costs associated with them are described in the following paragraphs.

First, the losses that occur prior to shipment are known as **internal failure costs.** These include producing scrap or rework, retesting and troubleshooting, added inspection, and disposition of nonconforming material. In Section 4.1 the following was pointed out: Every one dollar of scrap or rework saved is a saving of an estimated six dollars of hidden costs, such as inventory control of scrap and material handling of rework.

Second, losses that occur after shipment are included in the category **external failure costs.** Examples of such costs are customer complaints (negative **word-of-mouth advertising** is the most difficult cost to track), repair costs in the field or on product returned, replacements, and warranty charges.

Costs due to external and internal failure may both be classified as **consequential costs.** Consequential costs are unplanned and represent poor quality efforts. Consequential costs can run as high as 20–25 percent of manufacturing sales and 40 percent of operating expenses in service organizations.

The third category is **appraisal costs,** the costs incurred while inspecting, testing, and otherwise evaluating how well the product conforms to requirements in the normal manufacturing cycle. Acceptance sampling and SPC are examples of appraisal costs.

Fourth, **prevention costs** are tied to those practices intended to prevent nonconformances from occurring. An example of a prevention cost is a statistical quality control (SQC) training program.

Prevention and appraisal costs may be classified as **discretionary costs.** Discretionary costs are planned costs that improve quality efforts. Some researchers have shown that *appraisal costs should not exceed 50 percent of total quality costs and that prevention costs should be approximately 10 percent of quality costs.*

You should be able to list and define the four types of quality costs. Accounting department staff must be available to assist those directly responsible for the product or service, to establish understandable systems that track quality costs. Scrap, rework, warranty, and other failure costs can no longer be buried in the total cost of production. Inspection and test costs must also be separately accounted for. The cost of quality should be reported to all employees of the organization, not just to the CEO and the board of directors.

***Purchasing*** **Purchasing** is responsible for the material quality that is supplied by manufacturers known as **suppliers.** A fundamental rule, best articulated by W. Edwards Deming, is to "never buy on price tag alone." Purchasing cheap typewriter or printer paper may cause so many time-consuming handling problems and so much costly waste in an office that it costs more in the long run than the savings realized from buying the cheaper paper. A motel that cuts corners on cleaning chemicals can lessen the quality of the housekeeping staff's work and cause them to be inefficient.

Many companies are working with their suppliers to help them implement statistical quality control techniques that will improve quality. Some suppliers become certified as **single-source suppliers,** and their customer, the company buying their products, depends upon one supplier alone to provide materials or parts. All incoming products from multiple suppliers should be identified and carefully traced, so as to know the quality level of each supplier. Figure 10.2 shows histograms of the thickness of a metal from two different suppliers. From

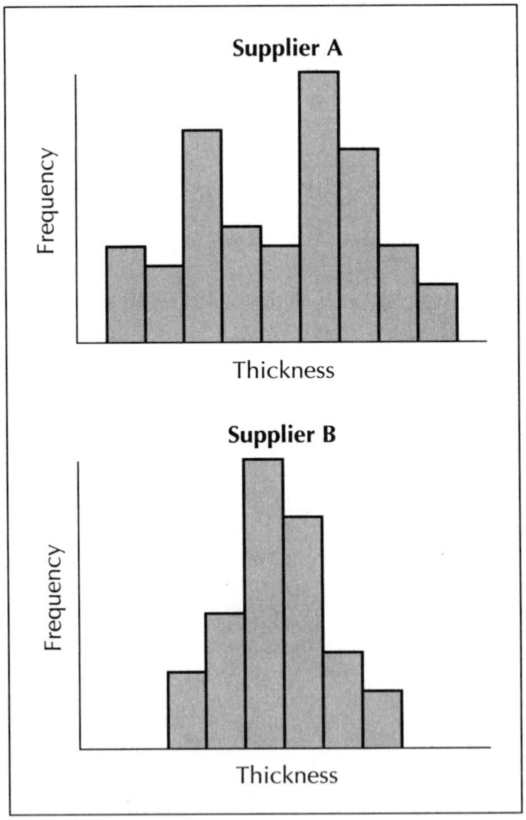


**FIGURE 10.2**
Supplier B's product costs less to use in a manufacturing process.

the histograms, it can be seen that Supplier B's material will cost less to use in a manufacturing process. Also see problem 8.

***Human Resources*** The management of the **human resources** function is key to the maintenance of the quality of the workforce. Are appropriate procedures in place to provide necessary education and training for personnel? Are applicants' résumés or applications checked for accurate experience? Are government regulations in hiring practices being followed? If there is a union, are the human resources staff working to keep the union leadership involved in quality efforts? These are a few of the concerns of a personnel department that can contribute to the overall quality of the product or service. The human resources department should work closely with other departments to best meet their needs for adequately trained and experienced personnel.

**Employee involvement (EI)** programs allow employees to meet regularly with employers to discuss quality of product or service, new technology, and management practices. In some companies EI programs have replaced unions. This may not be the best approach, especially for large manufacturers with branch plants. A 1992 Carnegie-Mellon University study of 1,015 such companies found that large metalworking and machinery industries that are unionized are more productive than nonunion manufacturers of the same products, even if the nonunion manufacturers have EI programs.

In companies with collective bargaining, the human resources staff and the union leadership must be alert to and limit adversarial relationships between management and the workforce that could damage quality efforts. Union leadership must be apprised of all evidence confirming the need for a more competitive stance. To initiate change in bargaining agreements that will enhance competitiveness, union leadership must be convinced by sound research that such policies as strict job classifications and intricate grievance procedures protecting marginal workers can destroy a company's profitability.

Recognizing the need for cooperation between management and labor, General Motors and the United Auto Workers formed the UAW-GM Quality Network in 1987. The network has joint union and management quality councils at all levels of the organization—from corporate to plant level. Cadillac Motor Car, the winner of the 1990 Malcolm Baldrige National Quality Award (see Section 10.2), recognizes the Quality Network as the beginning of its successful program known as the Cadillac People Strategy. The People Strategy is designed to meet the needs of the people of Cadillac while achieving, at the same time, the goals of the business. The People Strategy provides appropriate placement, frequent training and development, and clear communication to all personnel.

The ratio of **indirect labor** to direct labor should also be a concern of the human resources department. Indirect labor does not directly add value to a product and is considered waste. Hence, the move to better train production workers, floor managers, and technicians, who contribute directly to the product or service. Provisions should be made for updating personnel files with education and training experiences. Technologists working in manufacturing should make sure they are directly adding value to the product. In some cases, they should be spending more time out on the factory floor and not as much in their offices. Time spent listening to the production workers and helping them to solve problems is usually more valuable than time spent working and planning in the absence of others.

## *Research and Development (R&D)*

Technologists in **research and development** design products for manufacturing or study how to improve services. In manufacturing, if their design tolerances fit the capabilities of the manufacturing processes, then the product can be manufactured successfully. To understand quality of design we must distinguish between design quality and conformance quality.

**Design quality** refers to the level of quality that is designed into the product. The concept of design quality is illustrated by the difference between a Cadillac and a Yugo. Obviously, the Cadillac motor car is designed to closer tolerances and will require better materials than the foreign-made economy car. The Cadillac is said to have a higher level of design quality than the Yugo.

**Conformance quality** means conformance-to-the-design quality. Both the Cadillac and the Yugo have to conform to the differing design tolerances of the designers. A Yugo could have a higher conformance quality than a Cadillac.

EXAMPLE 10.2 From Example 10.1: How does the company manufacturing the new transmissions and clutches proceed to design the new subassemblies?

*Solution:* Using sound mechanical design practices, the design team first determines how to design the clutches and transmissions to better work with the axle systems (system integration improvement). Using customer feedback, the features desired by the truck drivers and truck assemblers are designed into the new subassemblies.

Concurrently, the design team considers how the product can be best manufactured. This **design-for-manufacturing** process includes meetings with the manufacturing team—consisting of managers, technologists, and operators—to ascertain machine tool capabilities. Necessary modifications are made to improve manufacturability.

***Reliability*** The R&D department will design for the reliability of a product, but reliability is also the responsibility of all departments of the company. **Reliability** is most objectively defined as "the probability of a product performing without failure a specified function under given conditions for a given period of time." The definition states that for reliability to be measured the product must be tested under given conditions and that statistical probability is to be determined. See Appendix B for the **exponential probability distribution** that may be used to predict reliability.

One of the best models used to illustrate the reliability of a product over time is the **bathtub curve** shown in Figure 10.3. The **infant mortal-**

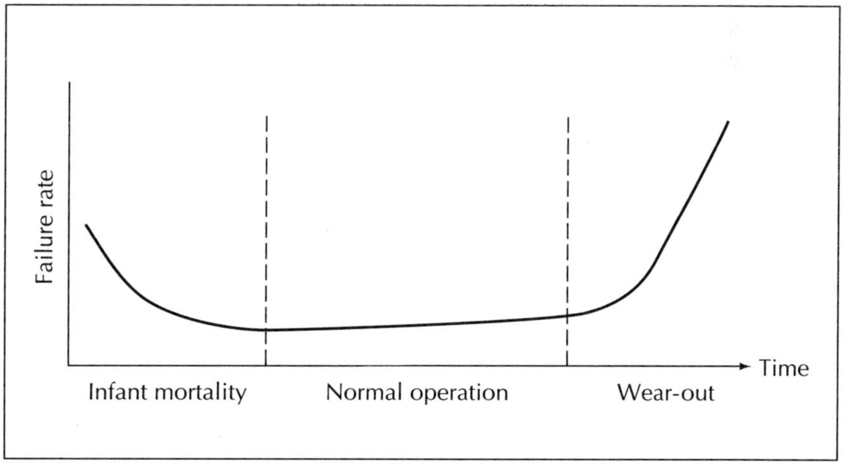


**FIGURE 10.3**
The "bathtub curve" is representative of the life history of many products.

**ity** region represents the failure of new products. Failure of new products may be higher during this period, and appliance manufacturers, for instance, may eliminate products that fail early by testing them before they reach the customer. The **normal operating** region shows the expected failure rate of products that have operated beyond the infant mortality region, but prior to the wear-out period. The **wear-out** period shows the increasing failure rate of the product as it operates beyond its design lifetime.

What is the chance that a product will fail if the independent reliabilities are known? The probability calculation is similar to the independent probability calculations discussed in Chapter 6 (Examples 6.7 and 6.8). Using $R$ instead of $P$ for the probability,

$$R(A, B, C\ldots) = R(A) \times R(B) \times R(C)\ldots$$

EXAMPLE 10.3 The policies issued to customers of an insurance company pass through three distinct processes—from sales representative to office staff to computer keypunch. The company has calculated the probability of a policy successfully passing each step. It is known that the reliability of each of three steps in writing a correct policy is 0.87, 0.96, and 0.98, respectively. What is the probability of obtaining a correct policy from this system?

*Solution:* $R(A, B, C) = 0.87 \times 0.96 \times 0.98 = 0.818$ or 81.8%

## Manufacturing or Service

The heart of any industry is the people that make the product reality. In a manufacturing industry it is the operators and maintenance personnel in the plant. For a service industry it is the office workers and computer personnel. One prominent service industry, Federal Express, refers to the people who interface with the external customers as the front-line employees. The key to TQM is to view the organization as an inverted triangle as shown in Figure 10.4. In the triangle, external customers are served by the front-line employees, who in turn are served by the technologists, middle managers, and top management. To make this model work, front-line employees need training, clear communication channels, and empowerment.

**Training** must be instituted when adopting TQM. Deming (see Section 1.1, "1950—Total Quality Control") has often referred to the need for "profound knowledge" in the workforce. For example,

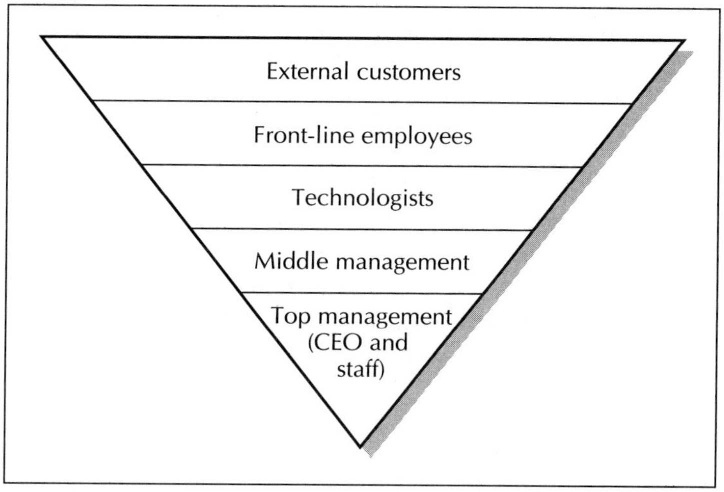


**FIGURE 10.4**
The front-line employees are served by the technologists and management.

machine operators in a manufacturing operation need training in SPC, which may include the following elements:

- process flowcharting;
- judging and measurement of quality characteristics;
- variable and attribute charting;
- determining process capability and why it is important.

**Clear communication** to management depends on knowledgeable front-line employees. It is the workforce that knows the processes and how to improve them. To communicate necessary changes, they must be trained in the language of the industry they serve.

**Empowerment** means giving all employees of the company the power to contribute in the decision-making process. *Empowerment will not be realized until every employee becomes knowledgeable concerning the goals of the organization and how they may best be achieved.* Statistical thinking, central to effective quality control efforts, cannot be achieved without statistical knowledge being shared by all people in the organization. If profit sharing is implemented, the employees should be able to understand a balance sheet—the "bottom line."

The formation of effective quality teams and other TQM efforts will produce positive results only if individual team members know what it takes for the company to compete. Take, for instance, a suggestion from a quality team in manufacturing that the purchase of a new overhead crane can speed up the processing of large metal doors. Engineering has calculations showing that the structure of the building cannot support such a crane without extensive renovation, and the financial personnel know that the cost would be prohibitive. Top management must be sure to communicate what went into the decision not to purchase the overhead crane, so that the team understands why their idea was rejected. The appropriate and timely communication to quality teams takes time away from the business-as-usual tasks that middle managers face every day. This additional time commitment must be realized and accepted as necessary. The technologist or manager should be aware of the responsibilities that go along with the conversion of the workforce to teamwork, a major component of TQM.

## *Sales and Marketing*

The marketing function has already been discussed. But what of the sales function? How can it contribute to the TQM effort? Sales personnel best know what the customer wants. They deal on a day-to-day basis with customers, and their career depends on pleasing customers and gaining repeat sales. Salespersons also hear the complaints concerning a particular product. Their feedback to manufacturing and design is invaluable.

Put yourself in the place of a new car buyer. After purchasing an automobile from a salesperson who has assured you of "years of trouble-free service," you find that the engine is not running well. After making a time-consuming trip back to the dealer, you find that the salesperson seems concerned about your problem, but then simply refers you to the service manager. The service manager explains how busy they are and tells you to return later and be prepared to leave the car all day. After three such appointments (the first appointment was used to determine that the correct part was not in stock, the second involved the repair not being inspected and a part not properly connected), the automobile finally seems to "run okay." Sound familiar?

Obviously, if you were a busy person, this scenario would sour you on any further new car buys from this dealer. The scenario that would bring you back is as follows: The new car needs repair and you return to the dealer. The salesperson who sold you the automobile personally takes you back to the service department, and your car is given a brief inspection to see what needs to be done. During the inspection you learn that the salesperson has been trained at the manufacturing plant to recognize and report technical problems. The salesperson gives you the keys to a rental car and tells you that it is yours free of charge until your new car is fixed. The car is repaired in short order, and you return to pick it up. It performs well, so you feel the repair was inspected and found to be okay. A week later, you receive a letter from the salesperson apologizing for any inconvenience. The letter has an attachment, forwarded to the manufacturer, explaining the problem in detail.

Under TQM, the latter scenario would bring you back as a repeat customer. The salesperson demonstrated commitment to you as a customer and reduced your anxiety over the less-than-perfect product. Just as importantly, the salesperson's attachment to the letter of apology demonstrated that the manufacturer was linked to sales and that your concern was heard by the part of the organization that created the problem. The salesperson could call on you in another three years and have a good chance of selling you another new car.

EXAMPLE 10.4 From Examples 10.1 and 10.2: How does the company manufacturing the new transmissions and clutches proceed to market the new subassemblies?

*Solution:* After manufacturing begins, the sales team spends a week at the manufacturer to learn the design rationale, manufacturing processes, and technical details concerning the product. The result of building a knowledgeable sales force is fewer customer complaints and lower warranty costs, and the capturing of 25 percent of the market for all clutches and transmissions.

The key to all elements of TQM is **teamwork.** Whether in a manufacturing or service industry, teamwork, good communication, and cooperation between departments are essential to successful implementation of total quality management. Top management and the quality staff represent the "strong forces" that bind the organization together. With their demonstrated commitment, the technologists and managers throughout the organization can implement TQM.

## Section 10.1 Review Problems

1. Benchmarking may involve
   a. examining a competitor's product or service
   b. examining a noncompetitor's product or service
   c. customer surveys
   d. all of the above
2. Of the four classifications of quality costs, which may be classified as a discretionary cost?
   a. appraisal costs
   b. internal failure costs
   c. prevention costs
   d. both a and c are discretionary
3. SPC control charting is accounted for by which quality cost area?
   a. appraisal costs
   b. internal failure costs
   c. prevention costs
   d. external failure costs
4. What is the probability of a toaster working if its four major parts have a reliability of 0.991, 0.986, 0.990, and 0.999, respectively?
5. Calculate the probability of a space shuttle's launch if 1,000 subsystems are checked by the launch computers and any one subsystem failure would scrub the launch. Estimate the reliability of each subsystem as 0.999. Would you want a ride in this shuttle?

# 10.2 Recognizing Quality Efforts

**Standards** are developed by governments and professional societies to provide guidelines for specific ways of doing business. Offering a baseline of minimum efforts required to compete in today's regional and global markets, standards are being constantly revised to satisfy the ever-increasing demands of the customer. Standards may be mandatory or voluntary. Food, drug, and safety standards are examples of mandatory standards. Examples of voluntary standards are product sizes (the volume and height of bottles), fluid viscosities (lubricating oils and paints), and engineering drawing symbols (process and instrumentation drawings). Voluntary standards are often mandatory, in an economic sense—for if products are not standardized, they will not satisfy the customer.

## ISO 9000 (ANSI/ASQC Q90)

The International Organization for Standardization (ISO) headquartered in Europe has developed a series of quality standards. Known as the **ISO 9000 Series**, they are voluntary standards that many companies worldwide are complying with in order to do business with the European Community and other member countries. The ISO 9000 standards have been endorsed and adopted by 53 countries. The standards apply to a broad range of businesses, including manufacturing, processing, distribution, and service industries. The United States version is known as the American National Standards Institute / American Society for Quality Control (ANSI / ASQC) **Q90 Series.**

The 20 major elements found in ISO 9000 / Q90 are presented in Table 10.1. The table also shows typical percentages for the deficiencies registered during past audits. For instance, document control is cited most often during independent audits, 18 percent of the time, as not adequate for ISO 9000 standards.

The 20 major elements are not self-explanatory. For instance, management responsibility includes a written policy statement including quality ojectives, an organizational chart with responsibilities defined, and a plan to evaluate the system. An extremely detailed manual known as a **quality manual** will flow from this work. The quality manual will document how your company assures quality control in every facet of the organization—from design of the product through the handling of customer complaints. To obtain further information, the ANSI / ASQC Q90 Series, *identical to the ISO 9000 Series,* is available from the ASQC, Standards Development, 611 East Wisconsin Avenue, P.O. Box 3005, Milwaukee, WI 53201–3005; or call (414) 272–8575.

Managers and technologists employed by companies seeking ISO 9000 registration must view the standard as an important and effective tool for the company's survival. Indeed, many companies implementing the standard have made concurrent gains in job satisfaction, productivity, and quality. During implementation, training will focus on cross-functional teams. People will be asked to learn, and be sensitive to, how their specific responsibilities relate to those of others. Managers and technologists must also be aware that the standard may be modified

TABLE 10.1
**Elements of the ISO 9000/Q90 Quality Standards with Percentage of System Deficiencies Registered**

| *Element* | *Deficiencies (%)* |
|---|---|
| Management responsibility | 4 |
| Quality system documentation | 8 |
| Contract review | 5 |
| Design control | 12 |
| Document control | 18 |
| Purchasing | 9 |
| Purchaser-supplied product | 1 |
| Product identification and traceability | 2 |
| Process control | 7 |
| Equipment maintenance and calibration | 7 |
| Inspection and testing | 9 |
| Inspection and test status | 3 |
| Control of nonconforming product | 3 |
| Corrective action | 5 |
| Handling, storage, packaging, and delivery | 4 |
| Quality records | 4 |
| Internal quality audits | 4 |
| Training | 4 |
| Servicing | 1 |
| Statistical techniques | 1 |

for the particular type of business seeking registration. For instance, audit points may be far fewer for an appliance manufacturer than for an aerospace company.

# *Quality Awards*

A company that is implementing ISO 9000 standards may aspire to winning one of two prestigious quality awards—the Malcolm Baldrige National Quality Award, of the United States; or Japan's Deming Prize. The ISO 9000 criteria provide an excellent foundation on which a company may prepare to win quality awards.

The **Malcolm Baldrige National Quality Award,** initiated in 1987, was named for a late secretary of commerce of the United States. Its purposes are to

- make the public aware of the importance of quality in the society;
- motivate U.S. companies to produce high-quality products and services.

The National Institute of Standards and Technology (NIST) administers the program and authorizes awards. Applicants for the award submit a report describing such items as leadership practices, training programs, quality control systems (including process control), personnel policies, and customer satisfaction measures. Companies scoring 750 points or more, out of a possible 1,000, become finalists and are visited by volunteer examiners from other industries. The examiners make recommendations to judges, also volunteers from industry.

Each year the judges may select two winners in three award categories: large manufacturers, large service companies, and businesses with not more than 500 full-time employees. However, only large manufacturers captured two awards every year from 1988 to 1991. The large service category has only received three awards and the small business category only four awards during this period. Cadillac Motor Car Company (large manufacturing) and Federal Express (large service) were winners in 1990. Federal Express cites the following as reasons for their success:

- Customer satisfaction is of highest priority.
- Statistical and other mathematical measures are used to track nonconformances and to promote continuous quality improvement.
- Employees are well trained and empowered to make decisions.

Sounds familiar by now, doesn't it?

Japan's **Deming Prize** is much more difficult to win. Whereas the Baldrige Award aspirants only have to submit a 75-page application, the Deming application can run up to 1,000 pages filled with data. Companies must also work with representatives from the **Union of Japanese Scientists and Engineers (JUSE)** for up to several years before making application. Only one company other than a Japanese company has won the Deming. That company, which won in 1989, was already mentioned in Section 1.3. It is the U.S. company, Florida Power and Light. Customer complaints have dropped markedly in the past decade.

Both of the awards have been subject to criticism. Florida Power and Light, for instance, lost its chairman after winning the Deming. In his book *Winning with Quality,* John Hudiburg describes the trials and tribulations of preparing for the award. Employees spent massive amounts of overtime and were extensively trained to answer possible questions put forth by JUSE inspectors in three minutes or less. The CEO of the parent FPL group discovered widespread employee dissatisfaction over the focus on process and not on people. Baldrige Award winners have complained similarly of the focus on process. In this case, the complaint was that judges for the Baldrige judged the process and not the product or service itself. Key elements of business, such as creativity, profitability, and long-term planning, are not a primary focus. Managers and technologists should be aware of the stress generated in the company that aspires to an award.

What managers and technologists must realize, however, is that working toward an award can be of great benefit to a company. In the case of the Baldrige, a recent survey (fall 1991 to winter 1992) by the ASQC found, in part, that "the award fosters quality awareness, promotes the understanding of the requirements for quality excellence, promotes sharing of information on successful quality strategies, and recognizes U.S. companies that excel in quality achievement and quality management" (*Quality Progress,* May 1993, page 49; see problem 16).

Other awards come from manufacturers to their suppliers. General Motors has the **Supplier Performance Evaluation and Reporting (SPEAR) Award** and Ford has the **Q101 Award.** Both awards recognize suppliers for superiority in all facets of their operations and for providing an atmosphere in which high-quality processes are confirmed by mathematical methods.

Working to achieve quality awards usually results in positive growth for the organization. Major benefits that can be realized are lower production costs, fewer quality costs such as scrap and rework, lower inventories and floor space, decreasing costs from suppliers, and less time for new-product development.

## *Section 10.2 Review Problems*

1. ______________________________ ______________________________ standards have been adopted by more than 50 countries. The United States version is known as the ANSI/ASQC, Q90 Series.

2. The purposes of the ______________________________

______________________ ______________________

Quality Award are to

- make the public aware of the importance of quality in the society;
- motivate U.S. companies to produce high-quality products and services.

3. To receive the ______________________ Prize, companies must work with representatives from the JUSE before making application.

## 10.3 *Forming a Personal TQM Philosophy*

Your quality course work has given you many models and tools that may be useful not only in your career, but in your personal life as well. In fact, the greatest achievers are those who feel very little separation between work time and leisure time. Both may be enjoyed, and both can be constantly improved.

One of the tools presented in Chapter 1 (Figure 1.3) was the Shewhart cycle. The cycle is a simple model that shows you how to recognize problems, select the most important to work on, work on the problem, collect data as you attempt solutions, evaluate your data, and decide whether to continue to work on the present problem or go on to a more pressing one. The Shewhart cycle can be used for personal goals from correcting health problems to improving your ability to participate in your favorite sport.

EXAMPLE 10.5 As we age, many of us will have fluctuating blood pressure, known as labile hypertension, that may cause heart problems if not corrected. Once the problem has arisen, it must be recognized as serious, and appropriate action should follow. Blood pressure medicine may not be the best solution, even though some medical practitioners will recommend it after the first few higher than normal measurements. The medicine is costly, may have side effects, and once begun will probably have to be used for the rest of your life.

*Solution:* A better alternative is to recognize that blood pressure is variable and may be improved by changes in lifestyle. Being variable, it should be studied by using a variable chart such as an $\overline{X}$ and $R$ chart. As you learn to measure your blood pressue, with an inexpensive sphygmomanometer that can be purchased at your local pharmacist, you could also use some of the seven tools of quality presented in Chapter 1 (Figure 1.4), in order to decide on what changes to make in diet, exercise, or even a change in career that will reduce stress. In fact, the major bones of the "fish-bone diagram" (cause-and-effect diagram) may be labeled as exercise, diet, stress, and so on.

As effective changes are implemented, such as a loss of weight, these changes will be reflected on the chart as out-of-control to the low side. An evaluation of the chart may lead you to a more healthful lifestyle.

Of course, the scatter diagram may be used as well. It will show correlations, or no correlations, between blood pressure and specific changes in lifestyle (Section 2.2).

Benchmarking, discussed in Section 10.1, may be used to improve your personal life. A prominent person, a manager in your company, or a coworker—anyone whom you respect and admire—may be someone that you could emulate. We all have role models and subconsciously attempt to follow their lead. Why not consciously list the attributes of him or her and try to make appropriate changes in your behavior? If you look up to a manager who never misuses his or her power, but instead listens to employees and acts on their input, why not write a note every evening as to how you behaved that day—either how you satisfied that attribute or how you did not.

No matter how you approach it, to lead a life filled with quality experiences you must work hard and use your time wisely. You must be prepared to take risks in order to reap rewards. Hopefully, some of the information gained from your quality course work may be used to "live the good life," serving yourself, your family, and your community to the best of your ability.

## Summary

- The TQM system encompasses and binds together all departments of the organization, whether it be public or private, service or manufacturing. A strong emphasis on people creates a constant focus on satisfying both internal and external customers.

- Marketing uses benchmarking, or compares the organization to other businesses, to assure that the organization is competitive.
- Everyone in the organization should be aware of and able to track the four quality costs—internal failure costs, external failure costs, appraisal costs, and prevention costs.
- Purchasing should never buy on price tag alone and should work with suppliers and internal customers (the people in their organization) to assure high-quality materials.
- The human resources department should work closely with employees through EI programs or with the union leadership to assure a satisfied workforce.
- R&D departments must design for manufacturing as well as designing for the needs of the external customer.
- Reliability is the probability that a product or service will perform without failure under given conditions for a given period of time.
- Front-line employees must be empowered to best satisfy the external customer. Continuous training to assure a strong knowledge base is necessary in order to empower internal customers and to have effective teamwork.
- Sales personnel must be familiar with the product they sell. They are best positioned to offer realist customer feedback.
- Satisfying ISO 9000 (ANSI / ASQC Q90) standards builds a solid foundation for attaining quality goals and satisfying many other quality award criteria.
- Meeting the criteria of the Malcolm Baldrige National Quality Award may increase productivity while decreasing costs.
- A personal TQM philosophy will help you in pursuing the goals you need to assure the "good life."

## CASE STUDY · CHAPTER 10

You have recently been promoted to quality technician at a manufacturing company of less than 500 employees. The promotion followed successful completion of a quality course at your local community college. Your first assignment is to research the value of achieving ISO 9000 registration for your company and to recommend possible implementation steps if registration seems to be appropriate. List the steps you will use to fulfill your assignment.

### *Solution Steps*

- Seminars on ISO 9000 are offered frequently in many convenient locations. You apply for funding to attend a workshop of your choosing. You also contact the ASQC for information and a list of qualified independent registrar agencies.
- After returning you report to the CEO how the organization may or may not benefit from applying for registration. The CEO is also informed that the registration process can take 9 to 18 months and cost well into the tens of thousands of dollars.
- If the CEO decides to seek registration, you next recommend members to serve on cross-functional teams or one ISO implementation team to work on the criteria. You may decide at this time to ask a consultant in to assist you in implementation or in training personnel.
- The team members and most other employees will be offered training on what ISO 9000 means to your organization, what their responsibilities are in gaining registration, and a description of the process to gain registration.
- The ISO teams perform audits of existing operating procedures and quality standards in light of the ISO 9000 criteria.
- A quality manual is developed that satisfies ISO 9000 standards.
- The teams assess existing procedures and quality standards against the new manual. Any discrepancies are rectified.
- A registrar (an accredited independent third party) is hired to assess compliance with the ISO 9000 standards. Upon successful completion of this final audit, your company will receive registration.

## *Chapter Review Problems*

1. List and define five of the terms or abbreviations (your choice) that you have learned from this chapter.
2. Interview an employee of a local service or manufacturing industry and draw an organizational diagram of the organization. Recommend improvements to the organizational structure that may improve chances for implementing a TQM environment.
3. Report on one article pertaining to TQM from the journal *Quality Progress.* The report may consist of two to three pages.
4. Henry Ford's description of a suitable accounting system goes something like, "Its a big bag in which you throw the receipts and then reach in and take out what is needed to pay the bills; what's left is profit." Discuss the merits of this statement. Recommend an

alternative statement based on your knowlege of quality costing gained from Section 10.1.

5. It was stated in Section 10.1, under "Accounting," that costs due to external or internal failure "can run as high as 20–25 percent of manufacturing sales and 40 percent of operating expenses in service organizations." What is significant about these figures? Is there an opportunity revealed by the figures that could increase a company's competitive position?

6. The quality costs for a manufacturing operation are as follows:

   - Internal failure = $67,000
   - External failure = $35,000
   - Appraisal = $25,000
   - Prevention = $3,000

   The sales for the company are $1 million for the same period. Analyze the quality costs by calculating the percent of total quality costs for each category. Also, calculate the ratio of the total quality costs to the sales figure. Make recommendations based on your calculations.

7. List three products that, if you bought them on least cost alone, could cause your organization increased costs by using them. Explain how the increased costs could be generated. See Figure 10.2.

8. Refer to Section 3.5, especially Example 3.13. What advantages may Supplier B in Figure 10.2 offer to its customers by working to target?

9. Compare and contrast two of the same type of products by discussing their differences in terms of design quality and conformance quality. See Section 10.1, "R&D."

10. In one paragraph define the term "empowerment."

11. The registration process at a community college consists of seven steps. According to college records the reliability of each step—the chance of a student not being delayed by any one of the steps—is 95 percent. What is the overall reliability of the registration process? Would you want your college to have an equally reliable process, or would you appreciate the administration working to improve the process?

12. The reliability of the components that form a CIM system (Section 9.3) have been measured by one study as follows:

    - Software reliability = 0.90
    - Hardware reliability = 0.97
    - Maintenance reliability = 0.99

    Calculate the overall reliability of the CIM system based on these data.
13. List typical occupational titles for front-line employees of a restaurant, a motel, and a discount store.
14. Construct a Pareto diagram (see Section 2.5) of Table 10.1. If your company was seeking ISO 9000 registration, what are the items you would be sure to check against your procedures?
15. Compare the elements of the Malcolm Baldrige National Quality Award with the ISO 9000 / Q90 requirements by listing the elements that are similar. (In some cases you will have to make assumptions.)
16. Write a one-page report on the May 1993 *Quality Progress* article, "What Does the U.S. Business Community Really Think about the Baldrige Award?"
17. Report on one of the self-improvement articles by Harry Forsha (*Quality Progress*, 1992: January, p. 57; February, p. 61; March, p. 98; April, p. 111; May, p. 67). The report may consist of two to three pages (instructor's determination).
18. Describe one aspect of your personal life that might be improved by using the Shewhart cycle and a few of the seven quality tools—it might even be your golf game. List the tools you would most likely use, along with a brief explanation of how they would be used. See Example 10.5.
19. Construct a cause-and-effect diagram for a problem that you recognize as important to solve for your personal self-improvement. See Example 10.5.
20. In today's technological world, we tend to think that technology can solve any problem. This chapter seems different from the others because it deals mostly with nontechnical ideas. Discuss your feelings regarding the need for the nontechnical ideas of this chapter. Do you feel they are mostly faddish and short-term, or are they useful concepts that must be implemented in order for a company

or business to be competitive? Explain your answer in a one-page report.

# LABORATORY · CHAPTER 10
## Project Report to Industry

*Materials needed:* A word processor and transparencies.

Assuming that the class completed a project for a local business or industry, a report will be prepared. The report may be mailed to the company or organization that supplied the project, or the project report may be presented to representatives of the organization.

The report should include the steps of the Shewhart cycle or a similar model of sound project management. The steps of the Shewhart cycle are included in brackets in the suggested outline which follows:

*Introduction:* List the members of the quality team (i.e., the class and the professor). Acknowledge the representatives from the company that provided information and tours of the facility. Describe the product or service studied, the need for the study [project selection], and offer other pertinent background information, such as how critical is the quality characteristic studied (Section 8.1).

*Objectives:* List a few specific objectives for the study. Typical objectives might be:

- Establish gauge accuracy and precision.
- Establish a procedure for measuring parts.
- Establish process capability.
- Complete a report and/or presentation for the project.

*Procedure* [*Project Implementation*]*:* The procedure should follow the objectives of the study and explain step-by-step how the project was completed. Supporting charts, graphs, and other quality tools, such as cause-and-effect diagrams, may be referenced and attached.

*Conclusion* [*Project Evaluation*]*:* Conclusions should be presented as a series of concise statements, each statement supported by data collected and included in the attachments.

*Recommendations* [*Continuing Improvement*]: Any weaknesses or incomplete findings involving the project should be reported in this section. Specific recommendations should be made concerning a continuation of the present project or a recommendation to continue the Shewhart cycle by selecting a next project in the quest for never-ending improvement.

*The Presentation:* If a presentation is requested by the organization, transparencies should be prepared of all critical information, especially charts and graphs. Determine as accurately as possible the audience: for instance, will your presentation be to business managers that have never seen a control chart or to engineers that are quite familiar with SPC? The members of the quality team should decide who is to present what material and on what topics (if possible, all team members should have some part in the presentation). A "dry run" is always vital to a successful presentation.

*APPENDIX*

# A

# The Normal Distribution

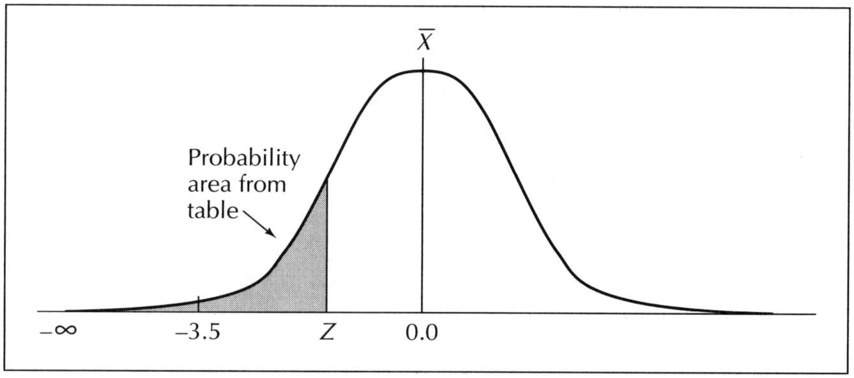


After solving for $Z$, find the area from the table's left side ($-\infty$) to the $Z$ of interest. (Table is following this page.) The area represents the probability of occurrence of the variable $X$, used to calculate $Z$.

$$Z = \frac{X - \overline{X}}{s}$$

**Probability Areas Under the Standardized Normal Curve**

| *Z* | *0.09* | *0.08* | *0.07* | *0.06* | *0.05* | *0.04* | *0.03* | *0.02* | *0.01* | *0.00* |
|---|---|---|---|---|---|---|---|---|---|---|
| −3.5 | 0.00017 | 0.00017 | 0.00018 | 0.00019 | 0.00019 | 0.00020 | 0.00021 | 0.00022 | 0.00022 | 0.00023 |
| −3.4 | 0.00024 | 0.00025 | 0.00026 | 0.00027 | 0.00028 | 0.00029 | 0.00030 | 0.00031 | 0.00033 | 0.00034 |
| −3.3 | 0.00035 | 0.00036 | 0.00038 | 0.00039 | 0.00040 | 0.00042 | 0.00043 | 0.00045 | 0.00047 | 0.00048 |
| −3.2 | 0.00050 | 0.00052 | 0.00054 | 0.00056 | 0.00058 | 0.00060 | 0.00062 | 0.00064 | 0.00066 | 0.00069 |
| −3.1 | 0.00071 | 0.00074 | 0.00076 | 0.00079 | 0.00082 | 0.00085 | 0.00087 | 0.00090 | 0.00094 | 0.00097 |
| −3.0 | 0.00100 | 0.00104 | 0.00107 | 0.00111 | 0.00114 | 0.00118 | 0.00122 | 0.00126 | 0.00131 | 0.00135 |
| −2.9 | 0.0014 | 0.0014 | 0.0015 | 0.0015 | 0.0016 | 0.0016 | 0.0017 | 0.0017 | 0.0018 | 0.0019 |
| −2.8 | 0.0019 | 0.0020 | 0.0021 | 0.0021 | 0.0022 | 0.0023 | 0.0023 | 0.0024 | 0.0025 | 0.0026 |
| −2.7 | 0.0026 | 0.0027 | 0.0028 | 0.0029 | 0.0030 | 0.0031 | 0.0032 | 0.0033 | 0.0034 | 0.0035 |
| −2.6 | 0.0036 | 0.0037 | 0.0038 | 0.0039 | 0.0040 | 0.0041 | 0.0043 | 0.0044 | 0.0045 | 0.0047 |
| −2.5 | 0.0048 | 0.0049 | 0.0051 | 0.0052 | 0.0054 | 0.0055 | 0.0057 | 0.0059 | 0.0060 | 0.0062 |
| −2.4 | 0.0064 | 0.0066 | 0.0068 | 0.0069 | 0.0071 | 0.0073 | 0.0075 | 0.0078 | 0.0080 | 0.0082 |
| −2.3 | 0.0084 | 0.0087 | 0.0089 | 0.0091 | 0.0094 | 0.0096 | 0.0099 | 0.0102 | 0.0104 | 0.0107 |
| −2.2 | 0.0110 | 0.0113 | 0.0116 | 0.0119 | 0.0122 | 0.0125 | 0.0129 | 0.0132 | 0.0136 | 0.0139 |
| −2.1 | 0.0143 | 0.0146 | 0.0150 | 0.0154 | 0.0158 | 0.0162 | 0.0166 | 0.0170 | 0.0174 | 0.0179 |

| | | | | | | | | | | |
|---|---|---|---|---|---|---|---|---|---|---|
| −2.0 | 0.0183 | 0.0188 | 0.0192 | 0.0197 | 0.0202 | 0.0207 | 0.0212 | 0.0217 | 0.0222 | 0.0228 |
| −1.9 | 0.0233 | 0.0239 | 0.0244 | 0.0250 | 0.0256 | 0.0262 | 0.0268 | 0.0274 | 0.0281 | 0.0287 |
| −1.8 | 0.0294 | 0.0301 | 0.0307 | 0.0314 | 0.0322 | 0.0329 | 0.0336 | 0.0344 | 0.0351 | 0.0359 |
| −1.7 | 0.0367 | 0.0375 | 0.0384 | 0.0392 | 0.0401 | 0.0498 | 0.0418 | 0.0427 | 0.0436 | 0.0446 |
| −1.6 | 0.0455 | 0.0465 | 0.0475 | 0.0485 | 0.0495 | 0.0505 | 0.0516 | 0.0526 | 0.0537 | 0.0548 |
| −1.5 | 0.0559 | 0.0571 | 0.0582 | 0.0594 | 0.0606 | 0.0618 | 0.0630 | 0.0643 | 0.0655 | 0.0668 |
| −1.4 | 0.0681 | 0.0694 | 0.0708 | 0.0721 | 0.0735 | 0.0749 | 0.0764 | 0.0778 | 0.0793 | 0.0808 |
| −1.3 | 0.0823 | 0.0838 | 0.0853 | 0.0869 | 0.0885 | 0.0901 | 0.0918 | 0.0934 | 0.0951 | 0.0968 |
| −1.2 | 0.0895 | 0.1003 | 0.1020 | 0.1038 | 0.1057 | 0.1075 | 0.1093 | 0.1112 | 0.1131 | 0.1151 |
| −1.1 | 0.1170 | 0.1190 | 0.1210 | 0.1230 | 0.1251 | 0.1271 | 0.1292 | 0.1314 | 0.1335 | 0.1357 |
| −1.0 | 0.1379 | 0.1401 | 0.1423 | 0.1446 | 0.1469 | 0.1492 | 0.1515 | 0.1539 | 0.1562 | 0.1587 |
| −0.9 | 0.1611 | 0.1635 | 0.1660 | 0.1685 | 0.1711 | 0.1736 | 0.1762 | 0.1788 | 0.1814 | 0.1841 |
| −0.8 | 0.1867 | 0.1894 | 0.1922 | 0.1949 | 0.1977 | 0.2005 | 0.2033 | 0.2061 | 0.2090 | 0.2119 |
| −0.7 | 0.2148 | 0.2177 | 0.2207 | 0.2236 | 0.2266 | 0.2297 | 0.2327 | 0.2358 | 0.2389 | 0.2420 |
| −0.6 | 0.2451 | 0.2483 | 0.2514 | 0.2546 | 0.2578 | 0.2611 | 0.2643 | 0.2676 | 0.2709 | 0.2743 |
| −0.5 | 0.2776 | 0.2810 | 0.2843 | 0.2877 | 0.2912 | 0.2946 | 0.2981 | 0.3015 | 0.3050 | 0.3085 |
| −0.4 | 0.3121 | 0.3156 | 0.3192 | 0.3228 | 0.3264 | 0.3300 | 0.3336 | 0.3372 | 0.3409 | 0.3446 |
| −0.3 | 0.3483 | 0.3520 | 0.3557 | 0.3594 | 0.3632 | 0.3669 | 0.3707 | 0.3745 | 0.3783 | 0.3821 |
| −0.2 | 0.3859 | 0.3897 | 0.3936 | 0.3974 | 0.4013 | 0.4052 | 0.4090 | 0.4129 | 0.4168 | 0.4207 |
| −0.1 | 0.4247 | 0.4286 | 0.4325 | 0.4364 | 0.4404 | 0.4443 | 0.4483 | 0.4522 | 0.4562 | 0.4602 |
| −0.0 | 0.4641 | 0.4681 | 0.4721 | 0.4761 | 0.4801 | 0.4840 | 0.4880 | 0.4920 | 0.4960 | 0.5000 |

**Probability Areas Under the Standardized Normal Curve** (*continued*)

| Z | *0.00* | *0.01* | *0.02* | *0.03* | *0.04* | *0.05* | *0.06* | *0.07* | *0.08* | *0.09* |
|---|---|---|---|---|---|---|---|---|---|---|
| +0.0 | 0.5000 | 0.5040 | 0.5080 | 0.5120 | 0.5160 | 0.5199 | 0.5239 | 0.5279 | 0.5319 | 0.5359 |
| +0.1 | 0.5398 | 0.5438 | 0.5478 | 0.5517 | 0.5557 | 0.5596 | 0.5636 | 0.5675 | 0.5714 | 0.5753 |
| +0.2 | 0.5793 | 0.5832 | 0.5871 | 0.5910 | 0.5948 | 0.5987 | 0.6026 | 0.6064 | 0.6103 | 0.6141 |
| +0.3 | 0.6179 | 0.6217 | 0.6255 | 0.6293 | 0.6331 | 0.6368 | 0.6406 | 0.6443 | 0.6480 | 0.6517 |
| +0.4 | 0.6554 | 0.6591 | 0.6628 | 0.6664 | 0.6700 | 0.6736 | 0.6772 | 0.6808 | 0.6844 | 0.6879 |
| +0.5 | 0.6915 | 0.6950 | 0.6985 | 0.7019 | 0.7054 | 0.7088 | 0.7123 | 0.7157 | 0.7190 | 0.7224 |
| +0.6 | 0.7257 | 0.7291 | 0.7324 | 0.7357 | 0.7389 | 0.7422 | 0.7454 | 0.7486 | 0.7517 | 0.7549 |
| +0.7 | 0.7580 | 0.7611 | 0.7642 | 0.7673 | 0.7704 | 0.7734 | 0.7764 | 0.7794 | 0.7823 | 0.7852 |
| +0.8 | 0.7881 | 0.7910 | 0.7939 | 0.7967 | 0.7995 | 0.8023 | 0.8051 | 0.8079 | 0.8106 | 0.8133 |
| +0.9 | 0.8159 | 0.8186 | 0.8212 | 0.8238 | 0.8264 | 0.8289 | 0.8315 | 0.8340 | 0.8365 | 0.8389 |
| +1.0 | 0.8413 | 0.8438 | 0.8461 | 0.8485 | 0.8508 | 0.8531 | 0.8554 | 0.8577 | 0.8599 | 0.8621 |
| +1.1 | 0.8643 | 0.8665 | 0.8686 | 0.8708 | 0.8729 | 0.8749 | 0.8770 | 0.8790 | 0.8810 | 0.8830 |
| +1.2 | 0.8849 | 0.8869 | 0.8888 | 0.8907 | 0.8925 | 0.8944 | 0.8962 | 0.8980 | 0.8997 | 0.9015 |
| +1.3 | 0.9032 | 0.9049 | 0.9066 | 0.9082 | 0.9099 | 0.9115 | 0.9131 | 0.9147 | 0.9162 | 0.9177 |
| +1.4 | 0.9192 | 0.9207 | 0.9222 | 0.9236 | 0.9251 | 0.9265 | 0.9279 | 0.9292 | 0.9306 | 0.9319 |
| +1.5 | 0.9332 | 0.9345 | 0.9357 | 0.9370 | 0.9382 | 0.9394 | 0.9406 | 0.9418 | 0.9429 | 0.9441 |
| +1.6 | 0.9452 | 0.9463 | 0.9474 | 0.9484 | 0.9495 | 0.9505 | 0.9515 | 0.9525 | 0.9535 | 0.9545 |
| +1.7 | 0.9554 | 0.9564 | 0.9573 | 0.9582 | 0.9591 | 0.9599 | 0.9608 | 0.9616 | 0.9625 | 0.9633 |
| +1.8 | 0.9641 | 0.9649 | 0.9656 | 0.9664 | 0.9671 | 0.9678 | 0.9686 | 0.9693 | 0.9699 | 0.9706 |
| +1.9 | 0.9713 | 0.9719 | 0.9726 | 0.9732 | 0.9738 | 0.9744 | 0.9750 | 0.9756 | 0.9761 | 0.9767 |
| +2.0 | 0.9773 | 0.9778 | 0.9783 | 0.9788 | 0.9793 | 0.9798 | 0.9803 | 0.9808 | 0.9812 | 0.9817 |

| | | | | | | | | | | |
|---|---|---|---|---|---|---|---|---|---|---|
| +2.1 | 0.9821 | 0.9826 | 0.9830 | 0.9834 | 0.9838 | 0.9842 | 0.9846 | 0.9850 | 0.9854 | 0.9857 |
| +2.2 | 0.9861 | 0.9864 | 0.9868 | 0.9871 | 0.9875 | 0.9878 | 0.9881 | 0.9884 | 0.9887 | 0.9890 |
| +2.3 | 0.9893 | 0.9896 | 0.9898 | 0.9901 | 0.9904 | 0.9906 | 0.9909 | 0.9911 | 0.9913 | 0.9916 |
| +2.4 | 0.9918 | 0.9920 | 0.9922 | 0.9925 | 0.9927 | 0.9929 | 0.9931 | 0.9932 | 0.9934 | 0.9936 |
| +2.5 | 0.9938 | 0.9940 | 0.9941 | 0.9943 | 0.9945 | 0.9946 | 0.9948 | 0.9949 | 0.9951 | 0.9952 |
| +2.6 | 0.9953 | 0.9955 | 0.9956 | 0.9957 | 0.9959 | 0.9960 | 0.9961 | 0.9962 | 0.9963 | 0.9964 |
| +2.7 | 0.9965 | 0.9966 | 0.9967 | 0.9968 | 0.9969 | 0.9970 | 0.9971 | 0.9972 | 0.9973 | 0.9974 |
| +2.8 | 0.9974 | 0.9975 | 0.9976 | 0.9977 | 0.9977 | 0.9978 | 0.9979 | 0.9979 | 0.9980 | 0.9981 |
| +2.9 | 0.9981 | 0.9982 | 0.9983 | 0.9983 | 0.9984 | 0.9984 | 0.9985 | 0.9985 | 0.9986 | 0.9986 |
| +3.0 | 0.99865 | 0.99869 | 0.99874 | 0.99878 | 0.99882 | 0.99886 | 0.99889 | 0.99893 | 0.99896 | 0.99900 |
| +3.1 | 0.99903 | 0.99906 | 0.99910 | 0.99913 | 0.99915 | 0.99918 | 0.99921 | 0.99924 | 0.99926 | 0.99929 |
| +3.2 | 0.99931 | 0.99934 | 0.99936 | 0.99938 | 0.99940 | 0.99942 | 0.99944 | 0.99946 | 0.99948 | 0.99950 |
| +3.3 | 0.99952 | 0.99953 | 0.99955 | 0.99957 | 0.99958 | 0.99960 | 0.99961 | 0.99962 | 0.99964 | 0.99965 |
| +3.4 | 0.99966 | 0.99967 | 0.99969 | 0.99970 | 0.99971 | 0.99972 | 0.99973 | 0.99974 | 0.99975 | 0.99976 |
| +3.5 | 0.99977 | 0.99978 | 0.99978 | 0.99979 | 0.99980 | 0.99981 | 0.99981 | 0.99982 | 0.99983 | 0.99983 |

APPENDIX

# B

# Frequently Occurring Probability Distributions

Groups of measurements from a process will form patterns known as frequency distributions or probability distributions. The graphs describing the patterns of the most frequently occurring distributions are shown below.

## Variable Data Distributions

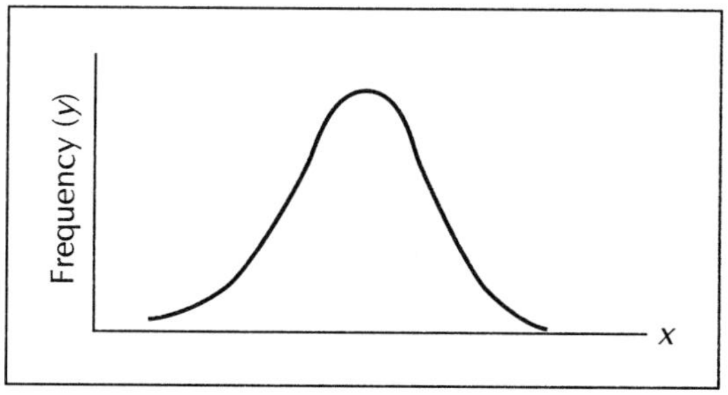


The normal distribution is used to describe data that are mostly due to chance causes (Appendix A).

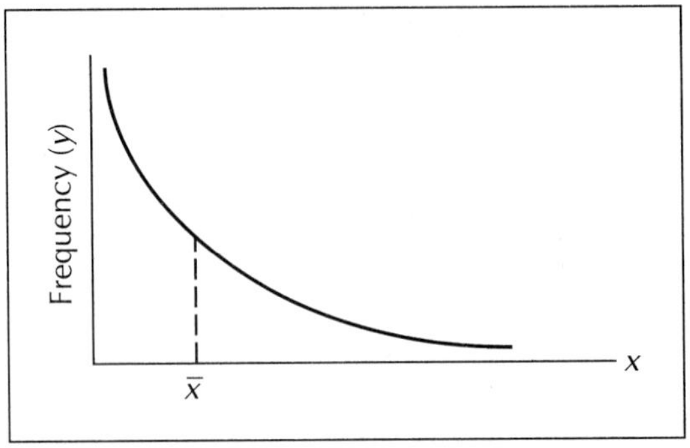


The exponential distribution is used to describe data that occur below the average more than above it—i.e., reliability ($p$ = probability of occurrence).

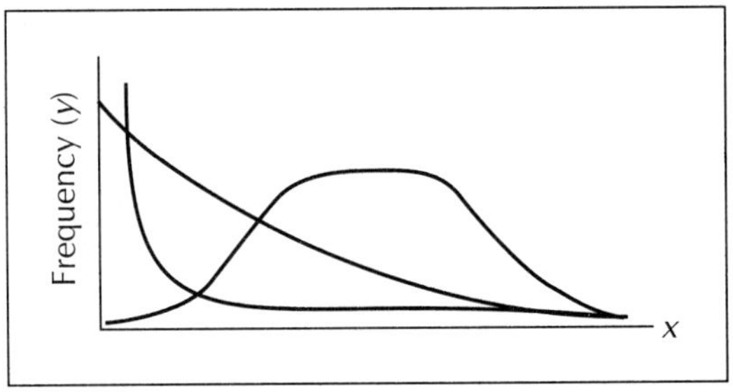


The Weibull distribution is used to describe a variety of patterns of data. Curves vary as to scale, shape, and location.

# *Attribute Data Distributions*

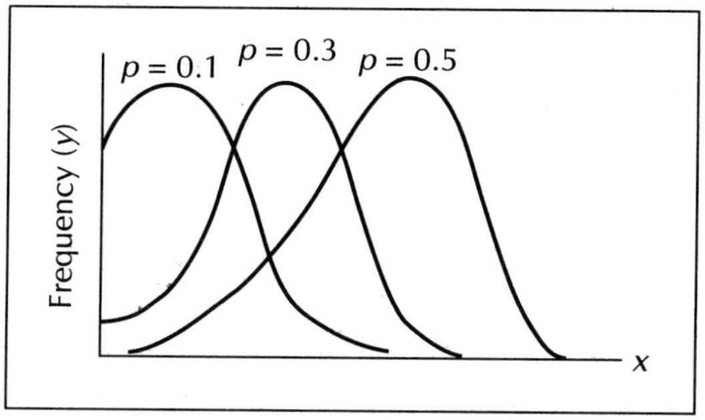


The binomial distribution is used to predict the number of occurrences of an event, given the number of trials, where each independent trial has a constant probability of occurrence.

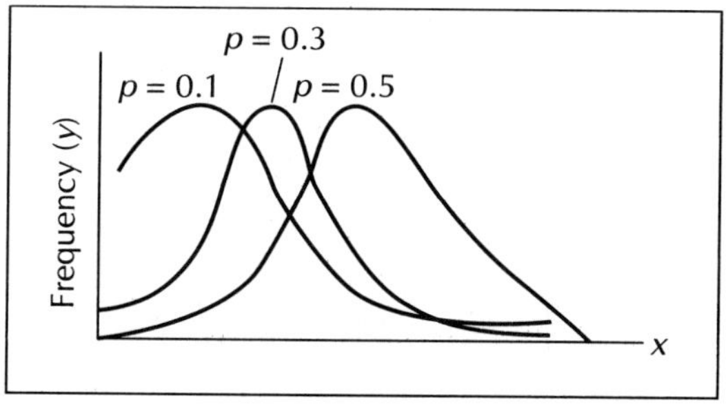


The Poisson distribution is similar to the binomial, but used mostly when the number of trials is large and the probability of the event is low (Appendix D).

*APPENDIX*

# C

# *Control Chart Factors*

| | Factors for $\overline{X}$ Control Limits | | Factors for R Control Limits | | Capability Factor |
|---|---|---|---|---|---|
| *Subsample Size, n* | $A_2$ | $A_3$ | $D_3$ | $D_4$ | $d_2$ |
| 2 | 1.880 | 2.659 | 0 | 3.267 | 1.128 |
| 3 | 1.023 | 1.954 | 0 | 2.574 | 1.693 |
| 4 | 0.729 | 1.628 | 0 | 2.282 | 2.059 |
| 5 | 0.577 | 1.427 | 0 | 2.114 | 2.326 |
| 6 | 0.483 | 1.287 | 0 | 2.004 | 2.534 |
| 7 | 0.419 | 1.182 | 0.076 | 1.924 | 2.704 |
| 8 | 0.373 | 1.099 | 0.136 | 1.864 | 2.847 |
| 9 | 0.337 | 1.032 | 0.184 | 1.816 | 2.970 |
| 10 | 0.308 | 0.975 | 0.223 | 1.777 | 3.078 |
| 11 | 0.285 | 0.927 | 0.256 | 1.744 | 3.173 |
| 12 | 0.266 | 0.886 | 0.283 | 1.717 | 3.258 |
| 13 | 0.249 | 0.850 | 0.307 | 1.693 | 3.336 |
| 14 | 0.235 | 0.817 | 0.328 | 1.672 | 3.407 |
| 15 | 0.223 | 0.789 | 0.347 | 1.653 | 3.472 |

| | Standard Deviation Factors for Control Limits | | | | Capability Factor |
|---|---|---|---|---|---|
| *Subsample Size, n* | $B_3$ | $B_4$ | $B_5$ | $B_6$ | $c_4$ |
| 2 | 0 | 3.267 | 0 | 2.606 | 0.7979 |
| 3 | 0 | 2.568 | 0 | 2.276 | 0.8862 |
| 4 | 0 | 2.266 | 0 | 2.088 | 0.9213 |
| 5 | 0 | 2.089 | 0 | 1.964 | 0.9400 |
| 6 | 0.030 | 1.970 | 0.029 | 1.874 | 0.9515 |
| 7 | 0.118 | 1.882 | 0.113 | 1.806 | 0.9594 |
| 8 | 0.185 | 1.815 | 0.179 | 1.751 | 0.9650 |
| 9 | 0.239 | 1.761 | 0.232 | 1.707 | 0.9693 |
| 10 | 0.284 | 1.716 | 0.276 | 1.669 | 0.9727 |
| 11 | 0.321 | 1.679 | 0.313 | 1.637 | 0.9754 |
| 12 | 0.354 | 1.646 | 0.346 | 1.610 | 0.9776 |
| 13 | 0.382 | 1.618 | 0.374 | 1.585 | 0.9794 |
| 14 | 0.406 | 1.594 | 0.399 | 1.563 | 0.9810 |
| 15 | 0.428 | 1.572 | 0.421 | 1.544 | 0.9823 |

APPENDIX

# D

# The Poisson Distribution

$$P(c) = \frac{(n\bar{p})^c}{c!} e^{-n\bar{p}}$$

The tables follow this page. Cumulative values are in parentheses.

| $c$ \ $n\bar{p}$ | | 0.02 | 0.04 | 0.06 | 0.08 |
|---|---|---|---|---|---|
| 0 | | 0.980 (0.980) | 0.961 (0.961) | 0.942 (0.942) | 0.923 (0.923) |
| 1 | | 0.002 (1.000) | 0.038 (0.999) | 0.056 (0.998) | 0.074 (0.997) |
| 2 | | | 0.001 (1.000) | 0.002 (1.000) | 0.003 (1.000) |

| $c$ \ $n\bar{p}$ | 0.1 | 0.2 | 0.3 | 0.4 | 0.5 |
|---|---|---|---|---|---|
| 0 | 0.905 (0.905) | 0.819 (0.819) | 0.741 (0.741) | 0.670 (0.670) | 0.607 (0.607) |
| 1 | 0.091 (0.996) | 0.164 (0.983) | 0.222 (0.963) | 0.268 (0.938) | 0.303 (0.910) |
| 2 | 0.004 (1.000) | 0.016 (0.999) | 0.033 (0.996) | 0.054 (0.992) | 0.076 (0.986) |
| 3 | | 0.010 (1.000) | 0.004 (1.000) | 0.007 (0.999) | 0.013 (0.999) |
| 4 | | | | 0.001 (1.000) | 0.001 (1.000) |

| $c$ \ $n\bar{p}$ | 0.6 | 0.7 | 0.8 | 0.9 | 1.0 |
|---|---|---|---|---|---|
| 0 | 0.549 (0.549) | 0.497 (0.497) | 0.449 (0.449) | 0.406 (0.406) | 0.368 (0.368) |
| 1 | 0.329 (0.878) | 0.349 (0.845) | 0.359 (0.808) | 0.366 (0.772) | 0.368 (0.736) |
| 2 | 0.099 (0.977) | 0.122 (0.967) | 0.144 (0.952) | 0.166 (0.938) | 0.184 (0.920) |
| 3 | 0.020 (0.997) | 0.028 (0.995) | 0.039 (0.991) | 0.049 (0.987) | 0.061 (0.981) |
| 4 | 0.003 (1.000) | 0.005 (1.000) | 0.008 (0.999) | 0.011 (0.998) | 0.016 (0.997) |
| 5 | | | 0.001 (1.000) | 0.002 (1.000) | 0.003 (1.000) |

| $c$ \ $n\bar{p}$ | *1.1* | *1.2* | *1.3* | *1.4* | *1.5* |
|---|---|---|---|---|---|
| 0 | 0.333 (0.333) | 0.301 (0.301) | 0.273 (0.273) | 0.247 (0.247) | 0.223 (0.223) |
| 1 | 0.366 (0.699) | 0.361 (0.662) | 0.354 (0.627) | 0.345 (0.592) | 0.335 (0.558) |
| 2 | 0.201 (0.900) | 0.217 (0.879) | 0.230 (0.857) | 0.242 (0.834) | 0.251 (0.809) |
| 3 | 0.074 (0.974) | 0.087 (0.966) | 0.100 (0.957) | 0.113 (0.947) | 0.126 (0.935) |
| 4 | 0.021 (0.995) | 0.026 (0.992) | 0.032 (0.989) | 0.039 (0.986) | 0.047 (0.982) |
| 5 | 0.004 (0.999) | 0.007 (0.999) | 0.009 (0.998) | 0.011 (0.997) | 0.014 (0.996) |
| 6 | 0.001 (1.000) | 0.001 (1.000) | 0.002 (1.000) | 0.003 (1.000) | 0.004 (1.000) |

| $c$ \ $n\bar{p}$ | *1.6* | *1.7* | *1.8* | *1.9* | *2.0* |
|---|---|---|---|---|---|
| 0 | 0.202 (0.202) | 0.183 (0.183) | 0.165 (0.165) | 0.150 (0.150) | 0.135 (0.135) |
| 1 | 0.323 (0.525) | 0.311 (0.494) | 0.298 (0.463) | 0.284 (0.434) | 0.271 (0.406) |
| 2 | 0.258 (0.783) | 0.264 (0.758) | 0.268 (0.731) | 0.270 (0.704) | 0.271 (0.677) |
| 3 | 0.138 (0.921) | 0.149 (0.907) | 0.161 (0.892) | 0.171 (0.875) | 0.180 (0.857) |
| 4 | 0.055 (0.976) | 0.064 (0.971) | 0.072 (0.964) | 0.081 (0.956) | 0.090 (0.947) |
| 5 | 0.018 (0.994) | 0.022 (0.993) | 0.026 (0.990) | 0.031 (0.987) | 0.036 (0.983) |
| 6 | 0.005 (0.999) | 0.006 (0.999) | 0.008 (0.998) | 0.010 (0.997) | 0.012 (0.995) |
| 7 | 0.001 (1.000) | 0.001 (1.000) | 0.002 (1.000) | 0.003 (1.000) | 0.004 (0.999) |
| 8 | | | | | 0.001 (1.000) |

| $c$ \ $n\bar{p}$ | 2.1 | 2.2 | 2.3 | 2.4 | 2.5 |
|---|---|---|---|---|---|
| 0 | 0.123 (0.123) | 0.111 (0.111) | 0.100 (0.100) | 0.091 (0.091) | 0.082 (0.082) |
| 1 | 0.257 (0.380) | 0.244 (0.355) | 0.231 (0.331) | 0.218 (0.309) | 0.205 (0.287) |
| 2 | 0.270 (0.650) | 0.268 (0.623) | 0.265 (0.596) | 0.261 (0.570) | 0.256 (0.543) |
| 3 | 0.189 (0.839) | 0.197 (0.820) | 0.203 (0.799) | 0.209 (0.779) | 0.214 (0.757) |
| 4 | 0.099 (0.938) | 0.108 (0.928) | 0.117 (0.916) | 0.125 (0.904) | 0.134 (0.891) |
| 5 | 0.042 (0.980) | 0.048 (0.976) | 0.054 (0.970) | 0.060 (0.964) | 0.067 (0.958) |
| 6 | 0.015 (0.995) | 0.017 (0.993) | 0.021 (0.991) | 0.024 (0.988) | 0.028 (0.986) |
| 7 | 0.004 (0.999) | 0.005 (0.998) | 0.007 (0.998) | 0.008 (0.996) | 0.010 (0.996) |
| 8 | 0.001 (1.000) | 0.002 (1.000) | 0.002 (1.000) | 0.003 (0.999) | 0.003 (0.999) |
| 9 | | | | 0.001 (1.000) | 0.001 (1.000) |

| $c$ \ $n\bar{p}$ | 2.6 | 2.7 | 2.8 | 2.9 | 3.0 |
|---|---|---|---|---|---|
| 0 | 0.074 (0.074) | 0.067 (0.067) | 0.061 (0.061) | 0.055 (0.055) | 0.050 (0.050) |
| 1 | 0.193 (0.267) | 0.182 (0.249) | 0.170 (0.231) | 0.160 (0.215) | 0.149 (0.199) |
| 2 | 0.251 (0.518) | 0.245 (0.494) | 0.238 (0.469) | 0.231 (0.446) | 0.224 (0.423) |
| 3 | 0.218 (0.736) | 0.221 (0.715) | 0.223 (0.692) | 0.224 (0.670) | 0.224 (0.647) |
| 4 | 0.141 (0.877) | 0.149 (0.864) | 0.156 (0.848) | 0.162 (0.832) | 0.168 (0.815) |
| 5 | 0.074 (0.951) | 0.080 (0.944) | 0.087 (0.935) | 0.094 (0.926) | 0.101 (0.916) |
| 6 | 0.032 (0.983) | 0.036 (0.980) | 0.041 (0.976) | 0.045 (0.971) | 0.050 (0.966) |
| 7 | 0.012 (0.995) | 0.014 (0.994) | 0.016 (0.992) | 0.019 (0.990) | 0.022 (0.988) |
| 8 | 0.004 (0.999) | 0.005 (0.999) | 0.006 (0.998) | 0.007 (0.997) | 0.008 (0.996) |
| 9 | 0.001 (1.000) | 0.001 (1.000) | 0.002 (1.000) | 0.002 (0.999) | 0.003 (0.999) |
| 10 | | | | 0.001 (1.000) | 0.001 (1.000) |

| $c$ \ $n\bar{p}$ | 3.1 | 3.2 | 3.3 | 3.4 | 3.5 |
|---|---|---|---|---|---|
| 0 | 0.045 (0.045) | 0.041 (0.041) | 0.037 (0.037) | 0.033 (0.033) | 0.030 (0.030) |
| 1 | 0.140 (0.185) | 0.130 (0.171) | 0.122 (0.159) | 0.113 (0.146) | 0.106 (0.136) |
| 2 | 0.216 (0.401) | 0.209 (0.380) | 0.201 (0.360) | 0.193 (0.339) | 0.185 (0.321) |
| 3 | 0.224 (0.625) | 0.223 (0.603) | 0.222 (0.582) | 0.219 (0.558) | 0.216 (0.537) |
| 4 | 0.173 (0.798) | 0.178 (0.781) | 0.182 (0.764) | 0.186 (0.744) | 0.189 (0.726) |
| 5 | 0.107 (0.905) | 0.114 (0.895) | 0.120 (0.884) | 0.126 (0.870) | 0.132 (0.858) |
| 6 | 0.056 (0.961) | 0.061 (0.956) | 0.066 (0.950) | 0.071 (0.941) | 0.077 (0.935) |
| 7 | 0.025 (0.986) | 0.028 (0.984) | 0.031 (0.981) | 0.035 (0.976) | 0.038 (0.973) |
| 8 | 0.010 (0.996) | 0.011 (0.995) | 0.012 (0.993) | 0.015 (0.991) | 0.017 (0.990) |
| 9 | 0.003 (0.999) | 0.004 (0.999) | 0.005 (0.998) | 0.006 (0.997) | 0.007 (0.997) |
| 10 | 0.001 (1.000) | 0.001 (1.000) | 0.002 (1.000) | 0.002 (0.999) | 0.002 (0.999) |
| 11 | | | | 0.001 (1.000) | 0.001 (1.000) |

| $c$ \ $n\bar{p}$ | 3.6 | 3.7 | 3.8 | 3.9 | 4.0 |
|---|---|---|---|---|---|
| 0 | 0.027 (0.027) | 0.025 (0.025) | 0.022 (0.022) | 0.020 (0.020) | 0.018 (0.018) |
| 1 | 0.098 (0.125) | 0.091 (0.116) | 0.085 (0.107) | 0.079 (0.099) | 0.073 (0.091) |
| 2 | 0.177 (0.302) | 0.169 (0.285) | 0.161 (0.268) | 0.154 (0.253) | 0.147 (0.238) |
| 3 | 0.213 (0.515) | 0.209 (0.494) | 0.205 (0.473) | 0.200 (0.453) | 0.195 (0.433) |
| 4 | 0.191 (0.706) | 0.193 (0.687) | 0.194 (0.667) | 0.195 (0.648) | 0.195 (0.628) |
| 5 | 0.138 (0.844) | 0.143 (0.830) | 0.148 (0.815) | 0.152 (0.800) | 0.157 (0.785) |
| 6 | 0.083 (0.927) | 0.088 (0.918) | 0.094 (0.909) | 0.099 (0.899) | 0.104 (0.889) |
| 7 | 0.042 (0.969) | 0.047 (0.965) | 0.051 (0.960) | 0.055 (0.954) | 0.060 (0.949) |
| 8 | 0.019 (0.988) | 0.022 (0.987) | 0.024 (0.984) | 0.027 (0.981) | 0.030 (0.979) |
| 9 | 0.008 (0.996) | 0.009 (0.996) | 0.010 (0.994) | 0.012 (0.993) | 0.013 (0.992) |
| 10 | 0.003 (0.999) | 0.003 (0.999) | 0.004 (0.998) | 0.004 (0.997) | 0.005 (0.997) |
| 11 | 0.001 (1.000) | 0.001 (1.000) | 0.001 (0.999) | 0.002 (0.999) | 0.002 (0.999) |
| 12 | | | 0.001 (1.000) | 0.001 (1.000) | 0.001 (1.000) |

| $c$ \ $n\bar{p}$ | 4.1 | 4.2 | 4.3 | 4.4 | 4.5 |
|---|---|---|---|---|---|
| 0 | 0.017 (0.017) | 0.015 (0.015) | 0.014 (0.014) | 0.012 (0.012) | 0.011 (0.011) |
| 1 | 0.068 (0.085) | 0.063 (0.078) | 0.058 (0.072) | 0.054 (0.066) | 0.050 (0.061) |
| 2 | 0.139 (0.224) | 0.132 (0.210) | 0.126 (0.198) | 0.119 (0.185) | 0.113 (0.174) |
| 3 | 0.190 (0.414) | 0.185 (0.395) | 0.180 (0.378) | 0.174 (0.359) | 0.169 (0.343) |
| 4 | 0.195 (0.609) | 0.195 (0.590) | 0.193 (0.571) | 0.192 (0.551) | 0.190 (0.533) |
| 5 | 0.160 (0.769) | 0.163 (0.753) | 0.166 (0.737) | 0.169 (0.720) | 0.171 (0.704) |
| 6 | 0.110 (0.879) | 0.114 (0.867) | 0.119 (0.856) | 0.124 (0.844) | 0.128 (0.832) |
| 7 | 0.064 (0.943) | 0.069 (0.936) | 0.073 (0.929) | 0.078 (0.922) | 0.082 (0.914) |
| 8 | 0.033 (0.976) | 0.036 (0.972) | 0.040 (0.969) | 0.043 (0.965) | 0.046 (0.960) |
| 9 | 0.015 (0.991) | 0.017 (0.989) | 0.019 (0.988) | 0.021 (0.986) | 0.023 (0.983) |
| 10 | 0.006 (0.997) | 0.007 (0.996) | 0.008 (0.996) | 0.009 (0.995) | 0.011 (0.994) |
| 11 | 0.002 (0.999) | 0.003 (0.999) | 0.003 (0.999) | 0.004 (0.999) | 0.004 (0.998) |
| 12 | 0.001 (1.000) | 0.001 (1.000) | 0.001 (1.000) | 0.001 (1.000) | 0.001 (0.999) |
| 13 | | | | | 0.001 (1.000) |

| $c$ \ $n\bar{p}$ | 4.6 | 4.7 | 4.8 | 4.9 | 5.0 |
|---|---|---|---|---|---|
| 0 | 0.010 (0.010) | 0.009 (0.009) | 0.008 (0.008) | 0.008 (0.008) | 0.007 (0.007) |
| 1 | 0.046 (0.056) | 0.043 (0.052) | 0.039 (0.047) | 0.037 (0.045) | 0.034 (0.041) |
| 2 | 0.106 (0.162) | 0.101 (0.153) | 0.095 (0.142) | 0.090 (0.135) | 0.084 (0.125) |
| 3 | 0.163 (0.325) | 0.157 (0.310) | 0.152 (0.294) | 0.146 (0.281) | 0.140 (0.265) |
| 4 | 0.188 (0.513) | 0.185 (0.495) | 0.182 (0.476) | 0.179 (0.460) | 0.176 (0.441) |
| 5 | 0.172 (0.685) | 0.174 (0.669) | 0.175 (0.651) | 0.175 (0.635) | 0.176 (0.617) |
| 6 | 0.132 (0.817) | 0.136 (0.805) | 0.140 (0.791) | 0.143 (0.778) | 0.146 (0.763) |
| 7 | 0.087 (0.904) | 0.091 (0.896) | 0.096 (0.887) | 0.100 (0.878) | 0.105 (0.868) |
| 8 | 0.050 (0.954) | 0.054 (0.950) | 0.058 (0.945) | 0.061 (0.939) | 0.065 (0.933) |
| 9 | 0.026 (0.980) | 0.028 (0.978) | 0.031 (0.976) | 0.034 (0.973) | 0.036 (0.969) |
| 10 | 0.012 (0.992) | 0.013 (0.991) | 0.015 (0.991) | 0.016 (0.989) | 0.018 (0.987) |
| 11 | 0.005 (0.997) | 0.006 (0.997) | 0.006 (0.997) | 0.007 (0.996) | 0.008 (0.995) |
| 12 | 0.002 (0.999) | 0.002 (0.999) | 0.002 (0.999) | 0.003 (0.999) | 0.003 (0.998) |
| 13 | 0.001 (1.000) | 0.001 (1.000) | 0.001 (1.000) | 0.001 (1.000) | 0.001 (0.999) |
| 14 | | | | | 0.001 (1.000) |

| $c$ \ $n\bar{p}$ | 6.0 | 7.0 | 8.0 | 9.0 | 10.0 |
|---|---|---|---|---|---|
| 0 | 0.002 (0.002) | 0.001 (0.001) | 0.000 (0.000) | 0.000 (0.000) | 0.000 (0.000) |
| 1 | 0.015 (0.017) | 0.006 (0.007) | 0.003 (0.003) | 0.001 (0.001) | 0.000 (0.000) |
| 2 | 0.045 (0.062) | 0.022 (0.029) | 0.011 (0.014) | 0.005 (0.006) | 0.002 (0.002) |
| 3 | 0.089 (0.151) | 0.052 (0.081) | 0.029 (0.043) | 0.015 (0.021) | 0.007 (0.009) |
| 4 | 0.134 (0.285) | 0.091 (0.172) | 0.057 (0.100) | 0.034 (0.055) | 0.019 (0.028) |
| 5 | 0.161 (0.446) | 0.128 (0.300) | 0.092 (0.192) | 0.061 (0.116) | 0.038 (0.066) |
| 6 | 0.161 (0.607) | 0.149 (0.449) | 0.122 (0.314) | 0.091 (0.206) | 0.063 (0.129) |
| 7 | 0.138 (0.745) | 0.149 (0.598) | 0.140 (0.454) | 0.117 (0.324) | 0.090 (0.219) |
| 8 | 0.103 (0.848) | 0.131 (0.729) | 0.140 (0.594) | 0.132 (0.456) | 0.113 (0.332) |
| 9 | 0.069 (0.917) | 0.102 (0.831) | 0.124 (0.718) | 0.132 (0.588) | 0.125 (0.457) |
| 10 | 0.041 (0.958) | 0.071 (0.902) | 0.099 (0.817) | 0.119 (0.707) | 0.125 (0.582) |
| 11 | 0.023 (0.981) | 0.045 (0.947) | 0.072 (0.889) | 0.097 (0.804) | 0.114 (0.696) |
| 12 | 0.011 (0.992) | 0.026 (0.973) | 0.048 (0.937) | 0.073 (0.877) | 0.095 (0.791) |
| 13 | 0.005 (0.997) | 0.014 (0.987) | 0.030 (0.967) | 0.050 (0.927) | 0.073 (0.864) |
| 14 | 0.002 (0.999) | 0.007 (0.994) | 0.017 (0.984) | 0.032 (0.959) | 0.052 (0.916) |
| 15 | 0.001 (1.000) | 0.003 (0.997) | 0.009 (0.993) | 0.019 (0.978) | 0.035 (0.951) |
| 16 | | 0.002 (0.999) | 0.004 (0.997) | 0.011 (0.989) | 0.022 (0.973) |
| 17 | | 0.001 (1.000) | 0.002 (0.999) | 0.006 (0.995) | 0.013 (0.986) |
| 18 | | | 0.001 (1.000) | 0.003 (0.998) | 0.007 (0.993) |
| 19 | | | | 0.001 (0.999) | 0.004 (0.997) |
| 20 | | | | 0.001 (1.000) | 0.002 (0.999) |
| 21 | | | | | 0.001 (1.000) |

| $c$ \ $n\bar{p}$ | 11.0 | 12.0 | 13.0 | 14.0 | 15.0 |
|---|---|---|---|---|---|
| 0 | 0.000 (0.000) | 0.000 (0.000) | 0.000 (0.000) | 0.000 (0.000) | 0.000 (0.000) |
| 1 | 0.000 (0.000) | 0.000 (0.000) | 0.000 (0.000) | 0.000 (0.000) | 0.000 (0.000) |
| 2 | 0.001 (0.001) | 0.000 (0.000) | 0.000 (0.000) | 0.000 (0.000) | 0.000 (0.000) |
| 3 | 0.004 (0.005) | 0.002 (0.002) | 0.001 (0.001) | 0.000 (0.000) | 0.000 (0.000) |
| 4 | 0.010 (0.015) | 0.005 (0.007) | 0.003 (0.004) | 0.001 (0.001) | 0.001 (0.001) |
| 5 | 0.022 (0.037) | 0.013 (0.020) | 0.007 (0.011) | 0.004 (0.005) | 0.002 (0.003) |
| 6 | 0.041 (0.078) | 0.025 (0.045) | 0.015 (0.026) | 0.009 (0.014) | 0.005 (0.008) |
| 7 | 0.065 (0.143) | 0.044 (0.089) | 0.028 (0.054) | 0.017 (0.031) | 0.010 (0.018) |
| 8 | 0.089 (0.232) | 0.066 (0.155) | 0.046 (0.100) | 0.031 (0.062) | 0.019 (0.037) |
| 9 | 0.109 (0.341) | 0.087 (0.242) | 0.066 (0.166) | 0.047 (0.109) | 0.032 (0.069) |
| 10 | 0.119 (0.460) | 0.105 (0.347) | 0.086 (0.252) | 0.066 (0.175) | 0.049 (0.118) |
| 11 | 0.119 (0.579) | 0.114 (0.461) | 0.101 (0.353) | 0.084 (0.259) | 0.066 (0.184) |
| 12 | 0.109 (0.688) | 0.114 (0.575) | 0.110 (0.463) | 0.099 (0.358) | 0.083 (0.267) |
| 13 | 0.093 (0.781) | 0.106 (0.681) | 0.110 (0.573) | 0.106 (0.464) | 0.096 (0.363) |
| 14 | 0.073 (0.854) | 0.091 (0.772) | 0.102 (0.675) | 0.106 (0.570) | 0.102 (0.465) |
| 15 | 0.053 (0.907) | 0.072 (0.844) | 0.088 (0.763) | 0.099 (0.669) | 0.102 (0.567) |
| 16 | 0.037 (0.944) | 0.054 (0.898) | 0.072 (0.835) | 0.087 (0.756) | 0.096 (0.663) |
| 17 | 0.024 (0.968) | 0.038 (0.936) | 0.055 (0.890) | 0.071 (0.827) | 0.085 (0.748) |
| 18 | 0.015 (0.983) | 0.026 (0.962) | 0.040 (0.930) | 0.056 (0.883) | 0.071 (0.819) |
| 19 | 0.008 (0.991) | 0.016 (0.978) | 0.027 (0.957) | 0.041 (0.924) | 0.056 (0.875) |
| 20 | 0.005 (0.996) | 0.010 (0.988) | 0.018 (0.975) | 0.029 (0.953) | 0.042 (0.917) |
| 21 | 0.002 (0.998) | 0.006 (0.994) | 0.011 (0.986) | 0.019 (0.972) | 0.030 (0.947) |
| 22 | 0.001 (0.999) | 0.003 (0.997) | 0.006 (0.992) | 0.012 (0.984) | 0.020 (0.967) |
| 23 | 0.001 (1.000) | 0.002 (0.999) | 0.004 (0.996) | 0.007 (0.991) | 0.013 (0.980) |
| 24 | | 0.001 (1.000) | 0.002 (0.998) | 0.004 (0.995) | 0.008 (0.988) |
| 25 | | | 0.001 (0.999) | 0.003 (0.998) | 0.005 (0.993) |
| 26 | | | 0.001 (1.000) | 0.001 (0.999) | 0.003 (0.996) |
| 27 | | | | 0.001 (1.000) | 0.002 (0.998) |
| 28 | | | | | 0.001 (0.999) |
| 29 | | | | | 0.001 (1.000) |

# Answers to Section Review Problems

## Chapter 1

Section 1.1

1. a
2. b
3. d
4. d
5. c

Section 1.2

1. quality Improvement
2. graphs, cause-and-effect diagrams, histograms, check sheets, Pareto diagrams, control charts, and scatter diagrams
3. materials, methods, machinery, measurement, environment, people

4. b
5a. 60
5b. 0.0007
5c. 3,400

Section 1.3

1. d
2. b
3. American Society for Quality Control
4. Certified Quality Technician

# *Chapter 2*

Section 2.1

1. c
2. *Y*
3. d

Section 2.2

1. b
2. d
3. a
4. d

Section 2.3

1. b
2. range
3. interval
4. midpoint (class mark)

Section 2.4

1. d
2. random, small
3. nonrandom, large
4. a
5. Both assignable and chance causes are present.

Section 2.5

1. d
2. 80, 20
3. by scrap amount or dollar amount

# Chapter 3

Section 3.1

1. Me, Mo, $\overline{X}$
2. Me is not affected by extreme values, but is difficult to calculate. Mo is the most frequently occurring value, but some distributions have two or more modes. $\overline{X}$ is a well-known measure of central tendency, but is affected by extreme values.
3. 71.2 in.
4. 5′ 11″ or 71 in.
5. Answer depends on students in a particular class.

Section 3.2

1. $s = 0.355$
2. $s^2$ = variance = 0.126
3. 0.83
4. $\sigma = 0.318$, the sample standard deviation uses $n - 1$ in the denominator, and the population standard deviation uses $n$.

Section 3.3

1. negatively
2. upscale
3. normal
4. b

Section 3.4

1. d
2. 0.9761, 0.5871, 0.0022, 0.00135, 0.4129
3. −0.84, −1.55 (not interpolated), 3.06, 0.200, 0.00
4. 64%
5. No, less; the distribution is not normal and more data are clustered around the mean using the normal table.

Section 3.5

1. chance
2. assignable
3. Sorting costs the company more and causes a lot of product to be near the specification limits.

# Chapter 4

Section 4.1

1. costs, control charts
2. c
3. assignable
4. identify quality characteristic, allocate resources, prepare personnel

Section 4.2

1. b
2. a
3. $UCL_{\overline{X}} = 73.7$, $LCL_{\overline{X}} = 56.3$, $UCL_R = 27.4$, $LCL_R = 0$
4. $UCL_{\overline{X}} = 0.101$, $LCL_{\overline{X}} = -0.0153$, $UCL_R = 0.183$, $LCL_R = 0$
5. $n$ is number in each subsample, and $k$ is number of samples

Section 4.3

1. sevens
2. thirds
3. b
4. increases
5. The spread of the individual values of the data will always be more extreme than the spread of the averages.

Section 4.4

1. $C_p = 1.79$, $C_{pk} = 0.571$
2. c
3. a
4. b

# Chapter 5

Section 5.1

1. short
2. b
3. destructive
4. best for short runs, quick feedback on assignable causes

Section 5.2

1. temporary
2. middle
3. b
4. $X$-tilde-bar
5. odd

Section 5.3

1. c
2. approximately 0.06, slightly less than the slope for all of the data
3. The control limits may be used to set up the machine at the beginning of a run. In the case of a die or tooling change, the control limit can indicate when to end a run.
4. Processes with obvious trends must be controlled by the trend chart in order to identify assignable causes, and it can be used for setup and to predict tool or die changes.

Section 5.4

1. c
2. attribute
3. short runs, less training for operators, fewer parts inspected

# Chapter 6

Section 6.1

1. classical = events counted without experimentation; experimental = events determined from experiment
2. 0.111

3. the seven
4. 0.077
5. b

Section 6.2

1. 50%
2. 50%
3. 30.8%
4. 72%
5. 55%

Section 6.3

1. b
2. d
3. A combination lock is opened by a *sequenced set of numbers*; this is a permutation.

Section 6.4

1. Yes, it is to be used when sampling without replacement.
2. d
3. c
4. a
5. a

# *Chapter 7*

Section 7.1

1. Use text criteria.
2. Advantage—no need for expensive measurement equipment; disadvantage—need for more inspected product.
3. See Figure 7.1 and memorize.

Section 7.2

1. b
2. c
3. a

4. d
5. 10.539

Section 7.3

1. nonconformities, varying
2. nonconformities, constant
3. c

# Chapter 8

Section 8.1

1. d
2. b
3. true

Section 8.2

1. c
2. a
3. a

Section 8.3

1. d
2. d
3. radiographic

# Chapter 9

Section 9.1

1. c
2. b
3. d
4. b
5. a

Section 9.2

1. c
2. $s$ or $\sigma_{n-1}$ key

3. b
4. Points out: 1, 2, 8, 14, 17, and 18 on $\overline{X}$. Rule of thirds: first part of *R* chart. Rule of sevens (below mean): last part of *R* chart.
5. The region from point 13 to point 21 (point 15 is on the mean) is below the mean. Joe may use this region of the chart to study what has changed in the process in order to improve it.

Section 9.3

1. b
2. d
3. d

# Chapter 10

Section 10.1

1. d
2. d
3. a
4. 0.966
5. Use $0.999^{1,000}$ to find an astonishing low probability of 0.368.

Section 10.2

1. ISO 9000
2. Malcolm Baldrige National
3. Deming

Section 10.3

No problems.

# Answers to Odd-Numbered Chapter Review Problems

## Chapter 1

Problems 1 through 7 require personal responses.

9. See Section 1.1.
11. Anyone paying taxes is a customer of the IRS. Employers are customers of the Federal Tax Deposit System.
13. Requires personal design.
15. Requires personal research.

## Chapter 2

1. The independent variable, RPM, is placed on the *X* axis. Axes should be labeled and a smooth curve drawn through the plotted points.
3. The coefficient of correlation, *r*, may vary slightly from the 0.9 found in Example 2.1.

5. See Section 2.3.
7. The histogram should have approximately 7 cells, be centered close to 84 mm of Hg, and the midpoint values of the lowest and highest cells will range from 74 to 98 mm of Hg.
9. Both vendors have similar levels of fat content, but Vendor A is sorting (i.e., a skewed left distribution with one sample missed with a fat content of 1.0). Vendor B is working to target and is most likely to improve the fat removal process.
11. Gauge is the top priority based on scrap reports. Bubble breaks is the top priority based on cost analysis, and this problem should be addressed first.
13. The Pareto diagram should reflect that the management-by-objective strategy, that is not integrated with the company's overall objectives, is most often cited as an impediment to SPC implementation.

# *Chapter 3*

1. $\overline{X} = 69.6$ in., median $= 72$ in.
3. $\overline{X} = 1.79$ m, $s = 0.11$ m.
5. $\overline{X} = 54.2$, $s = 0.64$.
7. The distribution, near normal, indicates they are working to target.
9. $g_1 = 0.317$ (moderately skewed to right).
   $g_2 = 2.19$ (moderately flat).
11. 5.82 percent of product is below the LSL.
13. 93.84 percent of product is within specification.
15. 0.135 percent of product is below the LSL.
17. 99.73 percent of product is within specification.
19. 19.5 percent, 4.60 percent. Result was less scrap and more rework. The scrap rate was affected most.
21. Your new grade point average would be 3.36.
23. Your paragraph should explain that gauge error is influenced by two factors (the R & R factors), and that it is referenced to the specification limits. You should also explain how the gauge error affects the actual distribution of the measured part.

# Chapter 4

1. $\overline{\overline{X}} = 5.2546$; $LCL_{\bar{x}} = 5.2513$; $UCL_{\bar{x}} = 5.2579$; $\overline{R} = 0.0019$; $UCL_R = 0.0060$.
3. Between points 7 and 16, both charts reflect out of control conditions due to the rule of thirds. The $\overline{X}$ chart shows out of control due to rule of sevens in the same region.
5. Points 10 and 11 on the $R$ chart are both greater than 1 standard deviation, and could be considered to disobey the rule of thirds. The $\overline{X}$ chart seems to follow this signal with points 12 through 16 showing out of control by rule of thirds. Also, excluding point number 16, there are many points in a series that are below the $\overline{\overline{X}}$ (rule of sevens).
7. a. $C_p = 1.37$, $C_{pk} = 1.10$.
   b. $C_p = 2.12$, $C_{pk} = 1.41$.
   c. $C_p = 2.16$, $C_{pk} = 1.44$.
   d. $C_p = 0.267$, $C_{pk} = 0.267$.
9. $\overline{\overline{X}} = 2.0095$, $LCL_{\overline{X}} = 2.0050$, $UCL_{\overline{X}} = 2.0140$, $\overline{R} = 0.0041$, $UCL_R = 0.0113$.
11. The machine produces product with too much variation. The bimodal distribution indicates that assignable causes are operating.
13. a. c
    b. f
    c. e
    d. d
    e. b
    f. a
15. Control limits are calculated by using averages of data, and can never be wider than the limits for the individuals (see Figure 4.12).
17. a. Assignable causes could be present.
    b. Chance causes only.
    c. Assignable causes would be present on most of the machinery.
    d. Chance causes only.
    e. Assignable causes could be present.
    f. A definite assignable cause.
    g. All *duffers* recognize this as an assignable cause, but they usually use other language to describe it.
    h. Chance causes operating and par for the course.
19. Requires individual research.

# Chapter 5

1. $\overline{X} = 1.66$, $LCL_X = 1.61$, $UCL_X = 1.72$, $\overline{R} = 0.021$, $UCL_R = 0.068$.
3. $C_{pk} = 0.733$.
5. $\overline{X} = 2.010$, $LCL_X = 2.000$, $UCL_X = 2.020$, $\overline{R} = 0.0037$, $UCL_R = 0.012$. The control limits for machine A are much wider than the chart for the three machines in problem 9, Chapter 4.
7. $\bar{\tilde{X}} = 6.24$, $LCL_{\tilde{x}} = 6.20$, $UCL_{\tilde{x}} = 6.29$, $\overline{R} = 0.043$, $UCL_R = 0.0113$.
9. 0.007 inches/hr.
11. a. Continue to run the process.
    b. Only 1 piece needs to be checked; continue to run the process.
    c. Only 1 piece needs to be checked; stop the process.
    d. Stop the process.
    e. Stop the process.
13. Individual response required.

# Chapter 6

1. Individual student response required.
3. $P$(no births) = 0.
5. $P$(win lottery) = $3.87 \times 10^{-8}$, odds of winning the lottery = 25,827,165.
7. $P$(passing 4 inspections) = $(1 - 0.05)^4 = 0.815$ or 81.5 percent.
9. $P$(*n.c.* unit from machine C) = 0.188 or 18.8 percent.
11. $P$(finding a trend of seven points down) = 0.0078 or 7.8 chances in 1000—Yes, the process has likely changed for the better.
13. The hypergeometric is only used in sampling without replacement. $P$(*n.c.*) = 0.01 or 1 percent as was calculated in Example 6.7.
15. $P$(two fours) = 0.116.
17. $P(0) = 0.941$, $P(1) = 0.0571$, $P(2) = 0.0014$.
19. $P$(exactly nine *n.c.*) = 0.032, $P$(fewer than nine *n.c.*) = 0.037, $P$(nine or more *n.c.*) = 1 − 0.037 = 0.963.

# Chapter 7

1. a. $u$
   b. $c$

c. $\overline{X}$ and $R$
d. $p$
e. $p$
f. $np$

3. $\overline{p} = 3.7$ percent, with $LCL_p = 1.4$ percent and $UCL_p = 6.0$ percent (for $\overline{n}$ of 597).
5. See Figure 7.7.
7. One of the points out of control to the high side, point 5, follows a spool change. The feed mechanism and the spool holding fixture should be checked by maintenance after every spool change.
9. Points 12 to 18 are out of control due to the rule of sevens. All points are below the centerline, indicating improved quality for the period.
11. A trend between points 8 and 14 indicates that quality is deteriorating. The chart also shows too much variation overall (rule of thirds). The system (MMMMEP) should be checked for causes of variation.
13. $LCL_u = 0.59$, $UCL_u = 2.2$, see Figure 7.11.
15. $\overline{c} = 10.6$ per each 100 $yd^2$ of carpet, $LCL_c = 0.833$, $UCL_c = 20.4$, points 9 and 17 are out of control to the high side.
17. The process has shifted and nonconformities per automobile has declined. Management should continue SPC implementation and training.

## Chapter 8

1. attribute, variable
3. Individual response required.
5. The welds on the reactor at a nuclear power plant require 100 percent inspection.
7. Individual response required.
9. $P(A) = 90.5$ percent (from Poisson Table). The ratio of inspections passed (no ace of spades) to total inspections (at least 150) should approximate this figure.
11. $n = 80$, $Ac = 0$, $Re = 1$.
13. $n = 20$, $Ac = 0$, $Re = 1$.
15. $\overline{p}$ at AQL $= 0.03$ percent, $\overline{p}$ at LQL $= 1.5$ percent.
17. Individual response required.

## Chapter 9

1. List major tasks, such as types of charts or other SPC tools constructed, design of experiment (DOE), gauge calibration, and real-time data collection.
3. See Section 9.2.
5. Requires a description of a particular computer system.
7. Requires individual response.
9. List questions, such as those found in Section 9.2, "Purchasing Computer Equipment."
11. Requires individual response, and overcontrol should be mentioned.
13. Requires individual research.
15. See Section 9.3, "When to Automate."

## Chapter 10

1. Individual response required.
3. Personal research required.
5. Consequential costs may be decreased. A twenty to forty percent potential for cost reduction represents a great opportunity to increase profits.
7. For instance, inexpensive solder for an automatic soldering process could cause enormous scrap and rework.
9. Individual response required.
11. $R$(registration process) $= 0.95^7 = 0.698$ or 69.8 percent. College administrators should attempt to improved reliability of each step to 99 percent, making overall reliability 93 percent. Then, they should shoot for 100 percent reliability.
13. Some examples are as follows: waiter; maintenance person, housekeeper; salesperson.
15. Both include training.
17. Personal research required.
19. Individual response required.

# Index

Acceptable quality level (AQL), 278
Acceptance sampling, 8, 266, 268
Accounting department, 63, 336
Accuracy, 116, 303
Air Force Logistic Command, 13
American National Standards Institute (ANSI), 4, 347
American Society for Quality Control (ASQC), 9, 20, 119, 347
Analog to digital (A/D) conversion, 305
Appraisal costs, 337
Associate degree, 21
Attribute data
  computerized data collection, 327
  control charts for, 228, 233, 249
  defined, 17
  sampling plans for, 270, 274
Audit, quality, 332
Automated process control, 316
Average. *See* Mean
Average outgoing quality limit (AOQL), 279

Batch process, 322
Bathtub curve
  infant mortality, 341
  normal operating, 342
  wear-out period, 342
Benchmarking, 333, 352
Binomial probability distribution, 211, 276
Blood pressure, 69, 351
Brainstorming, 234
Business functions, 336

Cadillac
  motor car company, 349
  people strategy, 339

Capability
of gauge, 108, 116
index, 142
of process, 131, 139, 237, 245, 251, 311
Case study
acceptance sampling, 294
automobile manifold, 65
automobile weatherstripping, 150
bomb fuse, 188
camera, 222
car wash, 257
collecting attribute data using QC software, 327
dairy producer fill weight, 23
hotel, time to register, 102
ISO 9000, 353
Cause-and-effect diagram, 16, 27, 301
Causes
assignable, 30, 56, 58, 99, 114, 129, 136, 168, 246
chance, 30, 56, 96, 115, 319
*c* chart, 229, 253, 314
Cell
interval, 49
midpoint, 53
Central tendency, 66
Certified quality engineer (CQE), 21
Certified quality technician (CQT), 21
Check sheet, 16, 224, 230
Chief executive officer, 332
Coefficient
of correlation, 45, 69
of determination, 45
Collective bargaining, 339
Combinations, 207
Communication, 343
Companywide quality control (CWQC), 10. *See also* TQM
Compliance reports, 24, 25
Computer-integrated manufacturing (CIM), 323
Consequential costs, 337
*Consumer Reports*, 335
Continuous improvement, 10, 33, 143
Control
artificial intelligence, 320
closed loop, 317
by expert system, 320
open loop, 316
Control chart
attribute, 227
variable, 8, 16, 113, 166
Control limits, 122, 139
Controlled experiment, 36
Correlation, 40, 42
Cost(s), 10, 19, 188, 248, 257, 324, 336
Cumulative frequency distribution, 63
Cumulative sum chart, 323
Customer(s)
external, 334
importance of, 11, 333
internal, 11, 336
survey, 334
tracking number of, 192

Data
acquisition, 305
attribute, 17, 185, 208, 228, 327
coded, 48, 119
collection of, 17, 230, 309
collectors, electronic, 305
editing the, 135
grouped, 50, 52
historical, 25, 234
individual, 136
significant figures, 17, 309
variable, 17, 208
Deming Prize, 10, 348
Deming, W. Edwards, 10, 342
Dependent variable, 39
Design for manufacturing, 341
Destructive testing, 169, 188
Die wear, 176
Discrete probability distributions, 208
binomial, 211
hypergeometric, 209
Poisson, 215, 223
Discretionary costs, 337

Eddy current testing, 291
Electronic data collectors, 305
Employee involvement, 339
Empowerment, 10, 343
Errors, type 1 and type 2, 278

Experimentation
controlled, 36
scatter diagram, 41
used to determine probability, 195
Exponential probability distribution, 341
External failure costs, 337

Feedback loop, 269, 317
Feigenbaum, Armand, 10
Florida Power and Light, 19, 349
Flowchart, 25, 221, 318
Food and Drug Administration, 23
*Fourteen points for management*, 10
Frequency distribution, 49
Frequency of repair, 335
Frerichs, Donald K., 320

Gauge
air, 307
capability of, 108, 116, 158, 302
dial snap, 230
narrow limit, 186
snap, 230
talkers, 306
Grade point, 107
Graph
bathtub curve, 341
control chart, 117
line, 36
line-of-best-fit, 43, 178
operating characteristic curve, 274
scatter diagram, 16, 41
slope, 178
Grinding operation, 315
Grouped data, 50, 52

Haze, plastic film, 166
Histogram, 16, 31, 47, 137, 143
cells or classes, 48
Hotel management, 102
Human resources, 339
Hypergeometric probability distribution, 209
Hysteresis errors, 41

Incoming materials, 232, 268
Independent event, 203
Independent variable, 39
Individuals chart, 166
Inspection
early history of, 5
lack of, 271
normal, reduced, tightened, 283
one hundred percent, 270
tollgate, 6
*Intech*, 320
Intelligent alarms, 320
Internal failure costs, 336
Ishikawa, Kaoru, 10
ISO 9000, 289, 347

Japan, 9, 349
Juran, Joseph, 6
Just-in-time manufacturing, 268

Kurtosis, 56, 84

Labor
collective bargaining, 339
direct, 340
indirect, 340
training, 342
Laboratory
acceptance sampling, 297
attribute chart, 263
automatic data collection, 329
constructing histogram, 73
data collection, 35
gauge capability, 108, 158
gauge repeatability and reproducibility (R&R), 158
individuals chart, 191
project report, 357
QC software, 226
Limiting quality level (LQL), 278
Limits for individuals, 140
Linear regression, 43, 178
Liquid penetrant dyes, 290
Lot, 8, 267
size, 277
Lot tolerance percent defective (LTPD), 278
Lottery, 203

Magnetic particle tester, 290
Malcolm Baldridge National Quality Award, 348
Marketing, 333, 344

Material review board, 295
Maximum allowable range card, 174
Mean, 76
Median, 77
Median chart, 170, 303
MIL-STD-105 sampling tables, 283
MMMMEP, 27, 53, 268
Mode, 56, 77, 79
Moving-average (or moving-range) chart, 189
Multiplexer, 308

National Institute of Standards and Technology (NIST), 349
Nonconformance, 228
 critical, 267
 major, 267
 minor, 267
Nonconforming unit, 204, 228
Nonconformity, 228, 249
Nondestructive testing (NDT), 290
Normal probability distribution, 54, 84, 88, 148, 218, 359
*np* chart, 229, 243

Operating characteristic (OC) curve, 274, 284
Optical NDT, 291
Other probability distributions, 365
Out-of-control, 129, 130
Overcontrol, 27, 113, 320

Paper-making process, 321
Pareto
 diagram, 16, 24, 61, 300
 principle, 60
*p* chart, 229, 235
Permutation, 206
Plan-do-check-act cycle, 8
Plastic film, 61
Poisson probability distribution, 215, 223, 251, 276, 371
Population statistics, 76, 82, 141, 195
Precontrol
 chart, 182
 form, 181
 rules, 181
Prevention costs, 337
Probability, 90, 131, 139, 184, 193
 of acceptance, 274, 296
 of dependent events, 201, 202
 flowchart, 221
 of independent events, 202, 205
 of mutually exclusive events, 198
 of non-mutually exclusive events, 200
 and reliability, 342
Process
 capability, 31, 139
 controller, 153, 317
 quality characteristic, 3
Product quality characteristic, 2
Profilometer, 45
Profit sharing, 343
Profound knowledge, 342
Project
 evaluation, 30, 31, 67
 implementation, 25, 67
 management, 14, 67
 selection, 24, 25, 67
 team, 23, 67
Purchasing
 computer software, 315
 department, 337

Q90 Series, 347
Q101 Award, 350
Quality
 assurance, 10, 67
 audit, 332
 *Bill of Rights*, 13
 characteristic, 2, 267
 of conformance, 340
 control, 6
 definition of, 4
 of design, 340
 history of, 5, 331
 manual, 347
 staff function, 332
 tools of, 10, 15, 332, 352
*Quality Control Handbook*, 6
Quality control software, 299, 309
*Quality Progress*, 21, 309

Radiography, 291
Random sample, 136, 196, 223, 272
Range, 48, 81

Reduced workload, 243
Rejectable quality level (RQL), 278
Reliability, 341
Repeatability, 108, 116, 302
Reproducibility, 108, 302
Research and development (R&D), 340
Rework. *See* Scrap and rework
Rule(s)
of precontrol, 181
of sevens, 130, 132, 246
of thirds, 130, 134, 246

Sampling
attribute and variable plans, 283
constant percentage, 271
one hundred percent, 270
perfect plan, 275
random, 136, 196, 223, 272
space, 196
systematic, 273
Sample statistics, 76, 195
Scatter diagram, 16, 41
Scrap and rework, 10, 289, 295
Seriousness classification, nonconformance, 267
Service industries, 19, 102, 192, 342, 344
Shewhart cycle, 7, 15, 67, 247, 351
Shewhart, Walter, 7, 8, 14
Short runs, 169, 186
Significant figures, 17
Skewness, 57, 58, 84
Software
communications, 308
quality control, 299, 309
Sorting, 57
SPC optimization, 320
Specification limits, 139
Standard deviation, 81, 123
of the averages, 138
Standards, 4, 346, 349
Statistical process control (SPC), 7, 72, 149, 266, 320
Statistical quality control (SQC), 7, 266
Subsample, 122, 231
Supplier
chain, 7
lots, 232
single sourcing, 267
multiple sourcing, 338
Supplier Performance Evaluation and Reporting (SPEAR) Award, 350
Surface finish, 43
Systematic sample, 273

Tamarack Farms Dairy, Newark, Ohio, 26, 121
Tare, 30, 127
Teamwork, 12, 67, 116, 332, 344
Technologists, 20, 172, 340, 347
Temporary workers, 170
Time
to register, 102
of service, 192
Tool life, 38
Total quality control (TQC), 10
Total quality management (TQM), 12, 295, 331, 342
Traceability, 108
Training, 342, 347
Trend, 133
Trend chart, 176

*u* chart, 229, 249
Ultrasonic testing, 291
Unacceptable quality level (UQL), 278
Union of Japanese Scientists and Engineers (JUSE), 349
Unions, 339

Variable data
defined, 17
for gauge, 304
sampling plans for, 266, 283
Variance, 81

Warranty, 188
Word-of-mouth advertising, 337
Working to target, 57, 99, 319

$\overline{X}$ and $R$ chart, 117, 128
$\overline{X}$ and $s$ chart, 310
Xerox Corporation, 334

# Glossary of Symbols

$A_2$, $A_3$ control chart factors
AOQ average outgoing quality
AOQL average outgoing quality limit
AQL acceptable quality level
ASQC American Society for Quality Control

$B_3$, $B_4$, $B_5$, $B_6$ control chart factors

$c$ number of nonconformities in a sample (the acceptance number in sampling plans)
$\bar{c}$ (pronounced "c-bar") average number of nonconformities in several samples ($c$ chart)
$c_4$ capability factor
${}_nC_r$ number of combinations of $n$ things taken $r$ at a time
$C_p$ capability of a process
$C_{pk}$ capability of a process (and considering the location of the center of the process)
CQE certified quality engineer
CQT certified quality technician

$D_3$, $D_4$ control chart factors
$d_2$ capability factor

$f$ frequency

$g_1$ measure of skewness
$g_2$ measure of kurtosis

$k$ number of samples

LCL lower control limit
LSL lower specification limit
LTPD lot tolerance percent defective

MMMMEP machinery, materials, measurement, methods, environment, and people

$N$ population or lot size
$n$ sample size or number in subsample

OOC out of control
OOS out of specification

$np$ number of nonconforming units in a sample
$n\bar{p}$ average number nonconforming in several samples

$p$ fraction nonconforming
$\bar{p}$ (pronounced "p-bar") average fraction nonconforming in several subsamples and an estimate of the population's fraction nonconforming